Springer Proceedings in Mathematics & Statistics

Volume 442

This book series features volumes composed of selected contributions from workshops and conferences in all areas of current research in mathematics and statistics, including data science, operations research and optimization. In addition to an overall evaluation of the interest, scientific quality, and timeliness of each proposal at the hands of the publisher, individual contributions are all refereed to the high quality standards of leading journals in the field. Thus, this series provides the research community with well-edited, authoritative reports on developments in the most exciting areas of mathematical and statistical research today.

Ashokkumar Patel • Nishtha Kesswani •
Bosubabu Sambana
Editors

Advances in Machine Learning and Big Data Analytics II

ICMLBDA 2023, NIT Arunachal Pradesh, India, May 29-30

Springer

Editors
Ashokkumar Patel
Computer & Information Science
University of Massachusetts Dartmouth
Dartmouth, MA, USA

Nishtha Kesswani
Department of Computer Science
Central University of Rajasthan
Ajmer, Rajasthan, India

Bosubabu Sambana
Department of Computer Science and
Engineering (DataScience)
School of Computing, Mohan Babu
University
Tirupathi, Andhra Pradesh, India

International Conference on Machine Learning and Big Data Analytics

ISSN 2194-1009 ISSN 2194-1017 (electronic)
Springer Proceedings in Mathematics & Statistics
ISBN 978-3-031-51341-1 ISBN 978-3-031-51342-8 (eBook)
https://doi.org/10.1007/978-3-031-51342-8

Mathematics Subject Classification: 68T09, 94A16, 62R07

This Springer imprint is published by the registered company Springer Nature Switzerland AG
The registered company address is: Gewerbestrasse 11, 6330 Cham, Switzerland

Preface

In today's data-driven world, machine learning (ML) and big data analytics have become pivotal forces driving innovation and decision-making across various industries. "Key Trends Shaping the Future of ML and Big Data Analytics" offers a comprehensive exploration of the transformative shifts and emerging trends in these domains. From the evolution of deep learning algorithms to the growing importance of ethical AI, this book provides a roadmap for understanding how ML and big data analytics are shaping our future. It brings together leading experts, offering valuable insights and strategies to navigate the dynamic landscape of data science, ensuring that readers stay at the forefront of this ever-evolving field.

Dartmouth, MA, USA	Ashokkumar Patel
Ajmer, India	Nishtha Kesswani
Vizianagaram, Andhra Pradesh, India	Bosubabu Sambana

Organization

Program Committee Chairs

2023, ICCM	IAASSE, LA, India
Mishra, Madhusudhan	NERIST, Electronics and Communication Engineering, Nirjuli, India
Sambana, Bosubabu	Department of Computer Science and Engineering (DataScience), School of Computing, Mohan Babu University, Tirupathi, Andhra Pradesh, India

Program Committee Members

2023, ICCM	IAASSE, LA, India
Ashiq, Abdul	Sri Eshwar College of Engineering, Information Technology, Coimbatore, India
Bokka, Yugandhar	
Bada, Rajasekhar Reddy	New Horizon College of Engineering, Computer Science and Engineering, Bangalore, India
Bandi, Vamsi	Madanapalle Institute of Technology and Science, Artificial Intelligence and Data Science, Madanapalle, India
Bano, Shahana	Robert Gordon University, School of Computing, Aberdeen, UK
Bendi, Dr. Nageswara Rao	Lendi Institute of Engineering and Technology, Department of Mathematics, Viziangaram, India
Bhattacharya, Sovan	DR. B.C. Roy Engineering College, Computer Science and Engineering, Durgapur, India
Bodepudi, Nageswararao	
Borugadda, Dr. Somasekhar	
Bouabdallaoui, Doha	Hassania School of Publics Works (EHTP), Laboratory of Systems Engineering, Casablanca, Morocco

(continued)

Camilleri, Rebecca	University of Malta, Computer Information Systems, Msida, Malta
Chemboli, Vaibhav	GITAM University, Department of Computer Science and Engineering, Visakhapatnam, India
Chowdary, Bhavya	
D, Dakshayani Himabindu	VNRVJIET, Information technology, Hyderabad, India
Dubey, Vineet Kumar	HBTU, COMPUTER SCIENCE, KANPUR, INDIA
Duddupudi, Janardhan	
Santosh Kumar	
Dabbeeru Priyanka	Aditya Institute of Technology and Management, Computer Science and Engineering, Tekkali, India
Dhotre, Sudhirkumar	
G, Sathya Priyanka	Periyar University, Statistics, Salem - 11, India
Gudla, Sateesh	
Gajjala, Nirmala joycee	St Anns college for women, Computer Science, Hyderabad, India
Goyal, Jayanti	
Gubbala, Dr. Kumari	CMR Engineering College, CSE, Hyderabad, India
Hossain, Dr. Syed Akhter	
Inala, Dr. Krishna Pavan	BVRIT Narsapur, Electronics and Communication Engineering Department, Narsapur, India
Ingale, Hemant	
Iyer, Sailesh	
Javvadi, N V R Swarup Kumar	
Jena, Sanjay Kumar	C. V. Raman Global University, Computer Science Engineering, Bhubaneswar, India
K, Vigneswara Reddy	Research Scholar, Annamalai University, CSE, Chidambaram, India
Kale, Kiran	
Krishnareddy, K	Mother Theresa Institute of Engineering & Technology, EEE, Palamaner, India
Karri, Praveen Kumar	Lendi Institute of Engineering and Technology(A), Vizianagaram, Computer Science and Engineering, Vizianagaram, India
Karyemsetty, Nagarjuna	
Khemchandani, Maahi	Dr. Babasaheb Ambedkar Technological University Vidyavihar, Lonere, Maharashtra 402103, Information Technology, Lonere, India
Kumar, Krishan	NIT Kurukshetra, Computer Engineering, Kurukshetra, Indian
Kumar, Achal	Anand Engineering College Agra, Computer Science & Engineering, Agra, India
Lakshmikanthareddy,M	

(continued)

Laxmikanth, Pydipala	Vignan Institute of Technology and Science, Department of Computer Science and Engineering, Yadadri Bhuvanagiri District, India
Lokesh, K	
M, N P Patnaik	GITAM Deemed to be University, Visakhapatnam, Department of Computer Science Engineering, Visakhapatnam, India
Madala, Chandu JaganSekhar	
Mahanty, Dr. Mohan	Vignan's Institute of Information Technology(A), Computer Science and Engineering, Visakhapatnam, India
Majumder, Swanirbhar	Tripura University, Information Technology, Agartala, India
Mishra, Madhusudhan	NERIST, Electronics and Communication Engineering, Nirjuli, India
Mohan, Dr. T Murali	Swarnandhra Institute of Engineering and Technology, Seethrampuram, Computer Science and Science, East Godavari, India
Muttipati, Appala Srinuvasu	
N M, Jyothi	Koneru LAkshmaiah Education Foundation, Computer Science Engineering, Vaddeswaram, India
Nandwalkar, Dr. Jayant	
Netto, Ann Nita	LBSITW, Poojappura, Trivandrum, ECE, Trivandrum, India
PYLA, J V G Prakasa Rao	Lendi Institute of Engineering and Technology, CSE, Vizianagaram, India
Pal, Saurabh	
Pandit, Dipti	
Patil, Mithun	N K Orchid College of Engg & Technology, Solapur MH India, Department of Computer Science and Engg, Solapur, India
Patil, Shashikant	ViMEET, Computer Science & Engineering (AI & ML), Raigad, India
R, Ramyadevi	SRM Institute of Science and Technology, Department of Computer Science and Applications, Chennai, India
Rajesh, P K	
Ramesh, R	Arignar Anna Government Arts College Musiri, Mathematics, Musiri Trichy, India
Ramakrishna, Chithari	CMR institute of Technology (Autonomous), Hyderabad, CSE(DS), Hyderabad, India
SABBI, Simhadri	
Dakshayani Kanaka	
Mahalakshmi	
Sabharwal, Seema	Government Post Graduate College for Women, Department of Computer Science, Panchkula, India
Sagar, Sanjeela	
Sambana, Bosubabu	Department of Computer Science and Engineering (DataScience), School of Computing, Mohan Babu University, Tirupathi, Andhra Pradesh, India

(continued)

Shaik, Mahaboob Hussain	Vishnu Institute of Technology, Computer Science and Engineering, Bhimavaram, India
Sharma, Amita	
Sharma, Anshuman	Kalinga Institute of industrial Technology, CSE, Bhubaneswar, India
Simhadri, Kumaraswamy	Aditya Institute of Technology And Management, EEE, Tekkali, India
Sinha, Anurag	IGNOU, Computer Science, Ranchi, India
Sridhar, B	Lendi Institute of Engineering and Technology, ECE, Vizianagaram, India
Srinivas, Thiruchinapalli	T Srinivas, Associate Professor in Mathematics, Nellore, India
Srinivasa Rao, M	Vignan's Institute of Information Technology, Computer Science and Engineering, Visakhapatnam, India
Sundara Siva Rao, Ivaturi	
Tuduku, Vinod Kumar	Lendi Institute of Engineering and Technology, Management Studies, Viziangaram, India
Talha, Iftakhar Mohammad	
Thirupatirao, Tirlingi	
Tirumala, Pentapati	
Uppalapati, Padma Jyothi	
Usha Bala, Dr. Varanasi	Anil Neerukonda Institute of Technology and Sciences, Computer Science and Engineering, Visakhapatnam, India
Vardhan, Dr. D. Harsha	
Vamsi, T M N	
Vamsi B,	
Dr Vasamsetti, Samuel	
Venunath, M	Pondicherry University, Computer Science, Puducherry, India
Vijaychandra, Joddumahanthi	LIET, Vizizanagaram, India
Yegireddi, Dr. Ramesh	
Kantharaju, Kanaparti	VITAP University, Computer Science And Engineering, Andhra Pradesh, India
Rao, Srinivas	Lendi Institute of Engineering and Technology(A), Vizianagaram, Computer Science and Engineering, Vizianagaram, India
Samatha, Badugu	

Reviewers

2023, ICCM	IAASSE, LA, India
Ashiq, Abdul	Sri Eshwar College of Engineering, Information Technology, Coimbatore, India
Bokka, Yugandhar	
Bada, Rajasekhar Reddy	New Horizon College of Engineering, Computer Science and Engineering, Bangalore, India

(continued)

Bandi, Vamsi	Madanapalle Institute of Technology and Science, Artificial Intelligence and Data Science, Madanapalle, India
Bano, Shahana	Robert Gordon University, School of Computing, Aberdeen, UK
Bendi, Dr. Nageswara Rao	Lendi Institute of Engineering and Technology, Department of Mathematics, Viziangaram, India
Bodepudi, Nageswararao	
Borugadda, Dr.	
Somasekhar	
Bouabdallaoui, Doha	Hassania School of Publics Works (EHTP), Laboratory of Systems Engineering, Casablanca, Morocco
Camilleri, Rebecca	University of Malta, Computer Information Systems, Msida, Malta
Chemboli, Vaibhav	GITAM University, Department of Computer Science and Engineering, Visakhapatnam, India
Chowdary, Bhavya	
Dubey, Vineet Kumar	HBTU, COMPUTER SCIENCE, KANPUR, INDIA
Duddupudi, Janardhan	
Santosh Kumar	
Dabbeeru Priyanka,	Aditya Institute of Technology and Management, Computer Science and Engineering, Tekkali, India
Dhotre, Sudhirkumar	
G, Sathya Priyanka	Periyar University, Statistics, Salem - 11, India
Gudla, Sateesh	
Gajjala, Nirmala Joycee	St Anns college for women, Computer Science, Hyderbad, India
Goyal, Jayanti	
Gubbala, Dr Kumari	CMR Engineering College, CSE, Hyderabad, India
Hossain, Dr. Syed Akhter	
Inala, Dr. Krishna Pavan	BVRIT Narsapur, Electronics and Communication Engineering Department, Narsapur, India
Ingale, Hemant	
Iyer, Sailesh	
Javvadi, N V R Swarup	
Kumar	
Jena, Sanjay Kumar	C. V. Raman Global University, Computer Science Engineering, Bhubaneswar, India
K, Vigneswara Reddy	Research Scholar, Annamalai University, CSE, Chidambaram, India
Kale, Kiran	
KRISHNAREDDY, K	Mother Theresa Institute of Engineering & Technology, EEE, Palamaner, India
Karri, Praveen Kumar	Lendi Institute of Engineering and Technology(A), Vizianagaram, Computer Science and Engineering, Vizianagaram, India
Karyemsetty, Nagarjuna	
Khemchandani, Maahi	Dr. Babasaheb Ambedkar Technological University Vidyavihar, Lonere, Maharashtra 402103, Information Technology, Lonere, India

(continued)

Kumar, Krishan	NIT Kurukshetra, Computer Engineering, Kurukshetra, Indian
Kumar, Achal	Anand Engineering College Agra, Computer Science & Engineering, Agra, India
Lakshmikanthareddy, M	
Laxmikanth, Pydipala	Vignan Institute of Technology and Science, Department of Computer Science and Engineering, Yadadri Bhuvanagiri District., India
Lokesh, K	
M, N P Patnaik	GITAM Deemed to be University, Visakhapatnam, Department of Computer Science Engineering, Visakhapatnam, India
Madala, Chandu Jagan	
Sekhar	
Mahanty, Dr. Mohan	Vignan's institute of information technology(A), Computer science and Engineering, Visakhapatnam, India
Majumder, Swanirbhar	Tripura University, Information Technology, Agartala, India
Mishra, Madhusudhan	NERIST, Electronics and Communication Engineering, Nirjuli, India
Mohan, Dr. T Murali	Swarnandhra Institute of Engineering and Technology, Seethrampuram, Computer Science and Science, East Godavari, India
Muttipati, Appala	
Srinuvasu	
N M, Jyothi	Koneru LAkshmaiah Education Foundation, Computer Science Engineering, Vaddeswaram, India
Nandwalkar, Dr. Jayant	
Pyla, J V G Prakasa Rao	Lendi Institute of Engineering and Technology, CSE, Vizianagaram, India
Pal, Saurabh	
Pandit, Dipti	
Patil, Mithun	N K Orchid College of Engg & Technology, Solapur MH India, Department of Computer Science and Engg, Solapur, India
Patil, Shashikant	ViMEET, Computer Science & Engineering (AI & ML), Raigad, India
R, Ramyadevi	SRM Institute of Science and Technology, Department of Computer Science and Applications, Chennai, India
Rajesh, P K	
Ramesh, R	Arignar Anna Government Arts College Musiri, Mathematics, Musiri Trichy, India
Ramakrishna, Chithari	CMR institute of Technology (Autonomous), Hyderabad, CSE(DS), Hyderabad, India
SABBI, Simhadri	
Dakshayani Kanaka	
Mahalakshmi	
Sabharwal, Seema	Government Post Graduate College for Women, Department of Computer Science, Panchkula, India
Sagar, Sanjeela	
Sambana, Bosubabu	Department of Computer Science and Engineering (DataScience), School of Computing, Mohan Babu University, Tirupathi, Andhra Pradesh, India

(continued)

Shaik, Mahaboob Hussain	Vishnu Institute of Technology, Computer Science and Engineering, Bhimavaram, India
Sharma, Amita	
Sharma, Anshuman	Kalinga Institute of industrial Technology, CSE, Bhubaneswar, India
Simhadri, Kumaraswamy	Aditya Institute of Technology And Management, EEE, Tekkali, India
Sinha, Anurag	IGNOU, COMPUTER SCIENCE, RANCHI, INDIA
Sridhar, B	Lendi Institute of Engineering and Technology, ECE, Vizianagaram, India
Srinivas, Thiruchinapalli	T Srinivas, Associate Professor in Mathematics, Nellore, India
Srinivasa Rao, M	Vignan's Institute of Information Technology, Computer Science and Engineering, Visakhapatnam, India
Sundara Siva Rao, Ivaturi	
Tuduku, Vinod Kumar	Lendi Institute of Engineering and Technology, Management Studies, Viziangaram, India
Talha, Iftakhar Mohammad	
Thirupatirao, Tirlingi	
Tirumala, Pentapati	
Uppalapati, Padma Jyothi	
Usha Bala, Dr. Varanasi	Anil Neerukonda Institute of Technology and Sciences, Computer Science and Engineering, Visakhapatnam, India
Vardhan, DR. D. Harsha	
Vamsi, T M N	
Vamsi B, Dr.	
Vasamsetti, Samuel	
Venunath, M	Pondicherry University, Computer Science, Puducherry, India
Vijaychandra,	
Joddumahanthi	LIET, Vizizanagaram, India
Yegireddi, Dr Ramesh	
Kantharaju, Kanaparti	VITAP University, Computer Science and Engineering, Andhra Pradesh, India
Rao, Srinivas	Lendi Institute of Engineering and Technology(A), Vizianagaram, Computer Science and Engineering, Vizianagaram, India
Samatha, Badugu	

Contents

Index [illegible]

Optimized Neural Networks Archetype for Prediction of Socio-economic Class of Women in India

N. M. Jyothi, B. K. Rajya Lakshmi, Pavani Gonnuri, and A. Pavani

Abstract Women play a vital role in the national economy, and their Socio-Economic Class (SEC) is essential for their empowerment and the nation's progress. This research aims to build an Artificial Intelligence (AI) model that classifies and compares the SEC of women across different regions of India. Numerous social and economic factors impact women's SEC, including education, employment, and community support. We employed an Artificial Neural Network (ANN) model, which is enhanced by a new optimization technique involving Segmenting, Screening, Integrating, Enhancing, and Categorizing. This innovative method makes more precise feature selection possible and improves computational efficiency. The research outcomes reveal that this innovative technique significantly boosts the ANN's performance, achieving an impressive classification accuracy of 99.34%. Compared to previous work, this model performs exceptionally well.

Keywords ANN · Socio-economic parameters · SEC classification · Machine learning

1 Introduction

Research underscores the substantial impact of SEC on a nation's economic well-being, influencing both mental and physical health, additionally the overall Gross Domestic Product (GDP). Bhaskar V. et al. emphasize the importance of identifying and addressing socioeconomic disparities through extensive research utilizing AI

N. M. Jyothi (✉) · A. Pavani
Koneru Lakshmaiah Education Foundation, Guntur, India

B. K. R. Lakshmi
Hyderabad Institute of Technology and Management, Hyderabad, India

P. Gonnuri
Velagapudi Ramakrishna Siddhartha Engineering College, Vijayawada, India

A. Patel et al. (eds.), *Advances in Machine Learning and Big Data Analytics II*,
Springer Proceedings in Mathematics & Statistics 442,
https://doi.org/10.1007/978-3-031-51342-8_1

and models from Machine Learning (ML). This approach aims to eliminate socioeconomic inequalities and foster equitable growth for all citizens [5]. Nasejje et al. highlight that SEC reflects an individual's purchasing power for essential needs, including education. Their study demonstrated that accurately predicting SEC using relevant data can significantly affect a country's global economic standing. By analyzing data related to occupation and app usage through an ML model, they achieved an 80% accuracy rate in predicting SEC [7]. The aim is to forecast the SEC of women using a novel optimized ANN. This optimization identifies the most impactful features affecting SEC and their relationships with other features. A new model is designed with the process of Segmenting, Screening, Integrating, Enhancing, and Categorizing for this purpose. These sequences of processes enhance prediction speed by removing insignificant features and preserving only those needed for classifying the SEC into low, medium, and high categories.

2 Literature Survey

Firoozeh Karimi et al. proposed a binary SVM model with an RBF kernel, utilizing a dataset of 19 attributes, to predict dynamic urban expansion in Guilford County, USA. This study aids city corporations and government agencies in developing policies and plans to mitigate the adverse effects of rapid urbanization. The model achieved an accuracy of 85% on the test data [1]. Machicao et al. used socio-economic indicators from images acquired by Google Street View through remote observations. They trained a Convolutional Neural Network (CNN) with this data, achieving an 80% accuracy rate in predicting higher income classes [2]. Christos and colleagues used Gradient Boosting as an ML method to forecast the physical well-being of veterans. They analyzed data from the US Daily Poll and the Census Bureau, resulting in an 80.4% accuracy rate for predicting SEC [3]. Winston et al. explored the link between SEC and COVID-19 prevalence using a multi-country dataset. They applied Random Forest, linear regression, and adaptive boosting, achieving a maximum prediction accuracy of 76%. Their analysis included attribute distribution, correlations, and outlier detection, using data from SEC indicators and Human Development Metrics [4]. Bhaskar V. and team used an array of ML algorithms combined with statistical analysis methods to investigate SEC levels in certain rural regions, achieving a high overall accuracy of 96.67% [5]. Aritbol et al. identified rapid urbanization as a challenge that widens the SEC gap despite economic growth. Using remote sensing and satellite images in a CNN, they effectively predicted SEC disparities in French cities, achieving promising results [6]. Nasejje et al. studied key factors influencing the survival index in sub-Saharan Africa, focusing on education, child sex, and wealth. Deploying Deep Survival Network and Random Survival Forest, they achieved 80% accuracy in linking survival outcomes to these SEC factors [7]. Individual income was classified using a dataset of mobile phone buyers, incorporating features like phone usage, top-up patterns, handset model, social network activity, and portability. Deep learning techniques were applied, achieving an accuracy of 72% [8].

The authors used French Twitter users' data, including tweet content, social networks, habitat, and occupation, to study SEC with machine learning. They found SEC is influenced by personal traits, environment, and social networks, and their model effectively analyzed social inequalities and satisfaction [9]. The Light GBM technique yielded an F1-Score of 81%. Household expenditure and income survey data were analyzed by employing multiple types of predictive modeling techniques to classify poverty in Jordan [10, 11]. Using satellite images and CNN, Shetty et al. classified the SEC of areas based on available infrastructure facilities [12]. The study used SEC indicators to predict asthma worsening in children. By applying ML and comparing the outcome to real data, it turned out that kids from lower SEC backgrounds were more prone to asthma than those from higher SEC backgrounds, highlighting the SEC's influence on children's health [13, 14, 15].

3 Experimentation

3.1 Data Preparation

The dataset comprises information from 2000 subjects and incorporates 30 different features. Table 1 shows the attribute list which has both direct and indirect effects on the SEC. The Kaggle dataset is considered.

3.2 Proposed Methodology

Following data preprocessing, the dataset is ready for training. A novel architecture has been developed, named Segment, Screen, Integrate, Enhance, and Categorize technique, to identify the most effective features for classification. Figure 1 illustrates the structure of this technique.

3.3 Experimentation

Segmentation: Divide the features obtained after data preprocessing into two halves. Feed the first set of features into one ANN and the second set into another. Run the first ANN fifteen times, shuffling one feature each time and recording the error values.

Table 1 Features of dataset

Feature no.	Feature name	Feature no.	Feature name	Feature no.	Feature name
1	Age	11	Mother (housewife /working)	21	Age at the time of marriage
2	Place of birth	12	Number of family members	22	Number of children
3	Gender	13	Family type (joint /nuclear)	23	Savings/investment
4	Religion	14	Community (BC/OBC/GM)	24	Education
5	Residence in the rural area	15	Involved in family decision	25	Employed/entrepreneur/self-employment
6	Province (city /metropolitan)	16	Major health issues in family members, if any	26	Not employed
7	Family member's addiction to alcohol	17	Self-property owned	27	Personal income
8	Family income	18	Family background (agriculture/non-agriculture)	28	Father's income
9	Father's occupation	19	Ancestral property owned	29	Dependent members
10	Mother's education	20	Marital status	30	Spouse income

Screening: Sort these errors to identify crucial features. Repeat with the second ANN. Retain the top features from both models for final classification and discard the less significant ones. This filters unwanted features.

Integrate: Merging the selected features from both sets and feeding them into the final classification model.

Enhancing: Train the final model using the combined features while monitoring performance metrics. Adjust and retrain until significant error improvements are achieved and record the final accuracy and error rates. Use the Optimization technique until error and accuracy stabilize. Compare the model's efficiency with other ML models.

Segmentation: Classification of SEC is obtained as the final output.

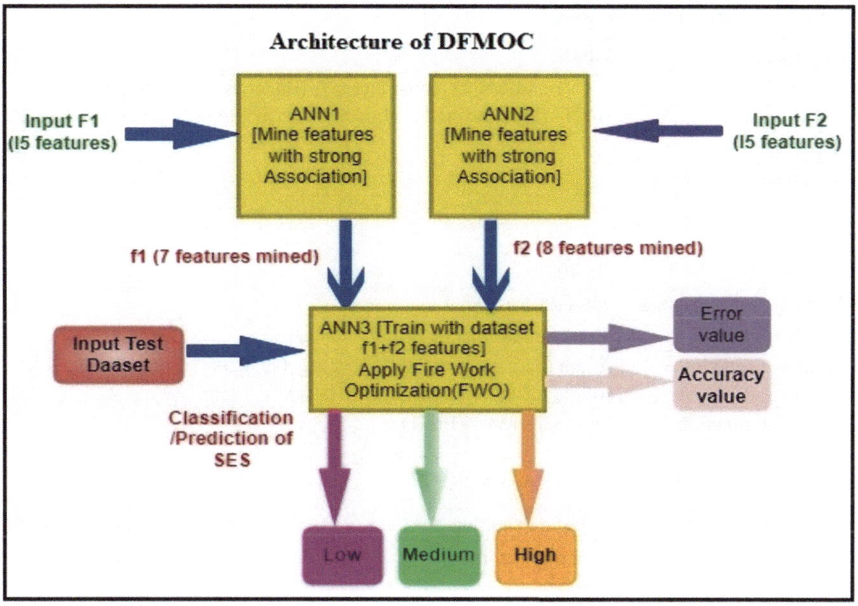

Fig. 1 Model architecture

3.4 *Setting Hyperparameters of ANN*

Hyperparameters are key settings in ANN that are defined before the learning procedure starts. They encompass variables such as the number of batches, training rate, layer configurations (e.g., number of layers, neurons per layer), and activation functions. These choices directly influence how quickly the model learns, its accuracy, and how well it generalizes to new data. Therefore, carefully adjusting these hyperparameters is essential for optimizing the network's performance and ensuring it effectively learns from the dataset provided.

3.5 *Implementation*

The first ANN utilizes the initial portion of the features (1–15), while the second ANN uses the second half (16–30), detailed in Table 1. Hyperparameter configurations are specified in Table 2. MSE values obtained after shuffling each feature are recorded and sorted for the first and second ANN in Tables 3 and 4, respectively. Features that significantly increase MSE are retained. Initial MSE values are 0.2152 for the first ANN and 0.1531 for the next ANN, serving as benchmarks for comparing subsequent MSE values in successive runs.

Table 2 Hyperparameters settings

Hyperparameters	Values
Number of layers	ANN1 and ANN2 (Input layer: 7; Hidden layer: 3; Output layer: 3) ANN3 (Input layer: 15; Hidden layer: 5; Output layer: 3)
Activation	0, 7
Optimizer	0,5
Learning rate	0.02, 1
Batch size	250, 1500
Epochs	50, 70
Training data, Test data	75%, 25%
Loss functions	Mean Square Error (MSE)
Optimizer function	Adam (params, lr = 0.001, betas = (0.9, 0.999), eps = 1e-08, weight_decay = 0, amsgrad = false)
Activation function	Sigmoid (from −1 to 1)

Table 3 Feature selection from the first ANN

Iteration no.	Feature no.	Feature shuffled	MSE	Feature retained
1	8	Family income	0.2875	Yes
2	1	Age	0.2812	Yes
3	12	Number of family members	0.2715	Yes
4	15	Involved in family decision	0.2676	Yes
5	6	Province (city /metropolitan)	0.2652	Yes
6	14	Community (BC/OBC/GM)	0.2591	Yes
7	9	Father's occupation	0.2321	Yes
8	5	Residence in the rural area	0.2158	No
9	4	Religion	0.2154	No
10	7	Family member's addiction to alcohol	0.2151	No
11	13	Family type (joint /nuclear)	0.2149	No
12	11	Mother (housewife/working)	0.2147	No
13	10	Mother's education	0.2145	No
14	2	Place of birth	0.2139	No
15	3	Gender (entire dataset female)	0.2133	No

The initial ANN converged at 0.15, the subsequent ANN at 0.17, and the enhanced ANN at 0.014 after incorporating the selected features from both preceding ANNs for predicting SEC. Figure 2 shows the enhanced architecture of the final ANN.

Table 4 Feature selection from the second ANN

Iteration no.	Feature no.	Feature shuffled	MSE	Feature retained
1	27	Personal income/salary/from other sources	0.2869	Yes
2	17	Self-property owned	0.2751	Yes
3	19	Ancestral property owned	0.2662	Yes
4	23	Savings/investment	0.2660	Yes
5	24	Education (UG/PG/professional)/10th/dropout)	0.2599	Yes
6	25	Employed/entrepreneur/self-employment	0.2572	Yes
7	28	Father's income	0.1952	Yes
8	30	Spouse income	0.1800	Yes
9	21	Age at the time of marriage	0.1571	No
10	29	Dependent members	0.1564	No
11	20	Marital status	0.1564	No
12	18	Family background (agriculture/ non-agriculture)	0.1550	No
13	22	Number of children	0.1543	No
14	26	Not employed	0.1543	No
15	16	Major health issues in family members, if any	0.1541	No

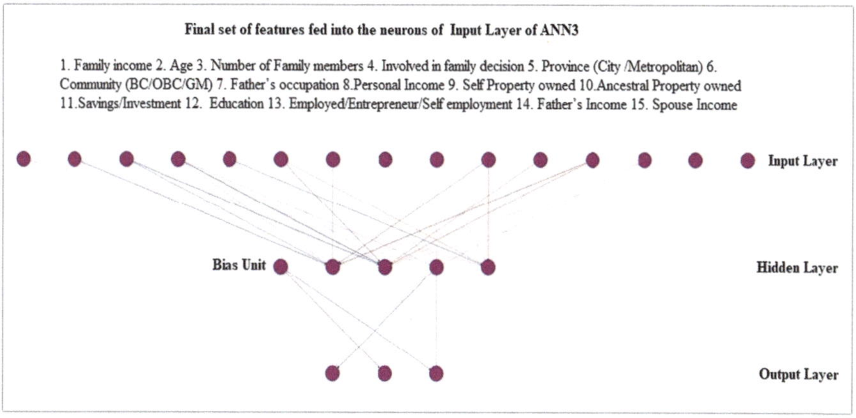

Fig. 2 Enhanced ANN

4 Results

Classification using a single ANN utilized all features from Table 1, achieving an accuracy of 84.5% within 3.76 s, as seen in Fig. 3. After applying the new framework, the accuracy improved significantly to 99.34%, illustrated in Fig. 4.

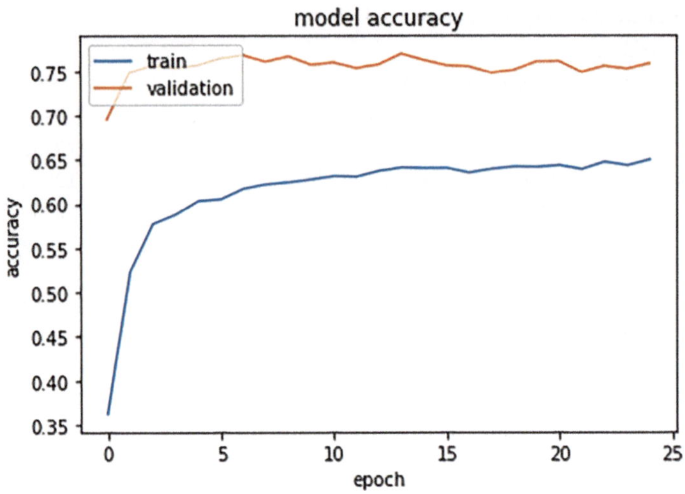

Fig. 3 Performance on single ANN

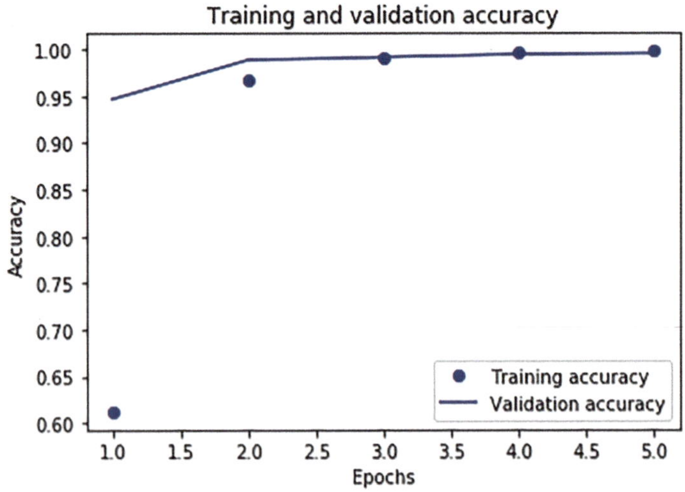

Fig. 4 ANN performance on the proposed model

4.1 Discussions

The model achieved a classification accuracy of 94% using one benchmark dataset and 0.93% using another focusing on women-related socio-economic data. Cross-evaluation with rural socio-economic and caste census data yielded 95% accuracy. The model's accuracy in classifying SEC levels for women in the Guntur district

Table 5 Comparison with other ML Models

Deployed models	MSE	Accuracy (%)
Single ANN with all features	0.245	98.45
ANN with the proposed technique	0.120	99.34
Logistic regression	0.297	91.83
Support vector	0.292	94.14
Random forest	0.294	93.34
Naive bayes	0.299	92.51
Decision tree	0.211	93.95
Xtreme gradient boosting	0. 212	92.45

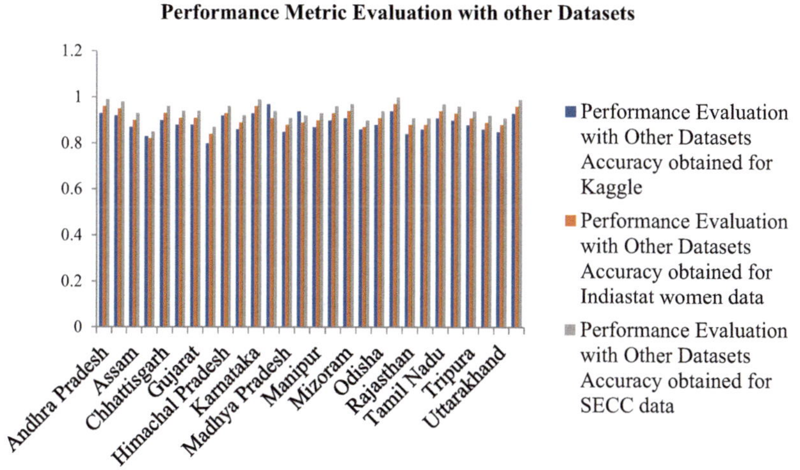

Fig. 5 Performance evaluation with other datasets

aligned closely with official government statistics. Figure 5 displays the model's performance metrics across states using these datasets.

Table 5 presents accuracy and MSE details on ML models for evaluation and comparison, while Fig. 5 illustrates a comparative analysis of their performance. The suggested model combined with ANN achieved the highest accuracy of 99.34% and the lowest MSE of 0.120 among the models evaluated.

Table 6 presents the SEC classification measures on the test dataset, including True Positives (TP) and True Negatives (TN). ANN, with the proposed updated model achieved high rates of correctly identifying positive and negative instances with 99.63% confidence and 99.45% support. In contrast, ANN without the updated model showed slightly lower performance, achieving 94.23% confidence in correctly identifying positive instances and 92.34% support in correctly identifying negative instances. Figure 6 provides a visual synopsis of the classification findings for all models (Fig. 7).

Table 6 Confusion matrix showing classification summary

ANN with the proposed model	Class	High	Medium	Low	ANN without the proposed model	High	Medium	Low
	High	533	12	9		533	12	9
	Medium	8	123	5		8	123	5
	Low	11	9	90		11	9	90

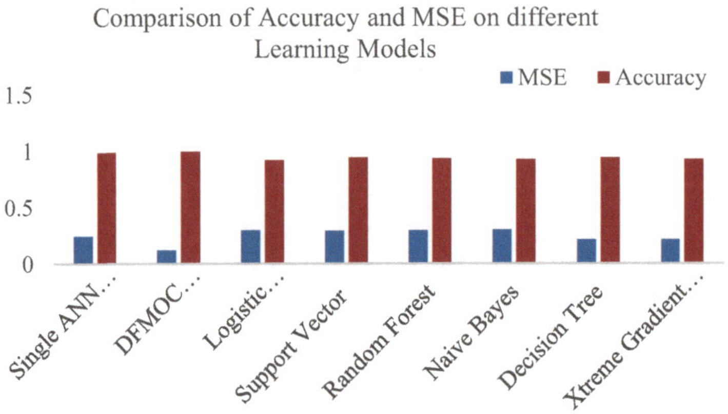

Fig. 6 Performance evaluation of accuracy and MSE on different ML models

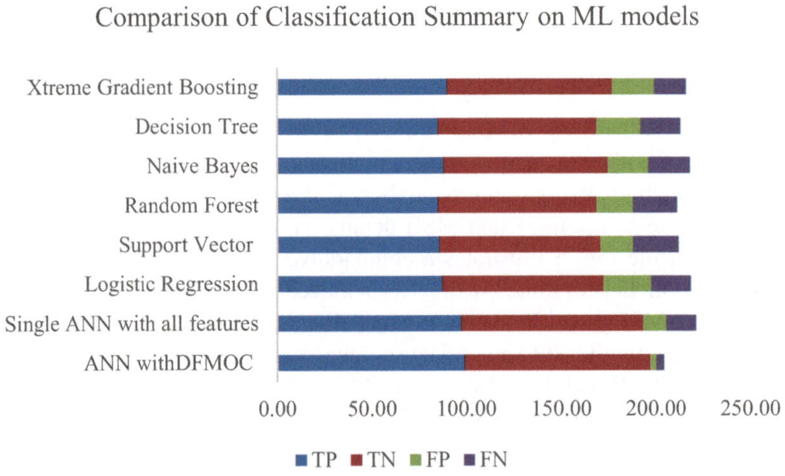

Fig. 7 Classification metric on ML models

5 Conclusion

The Artificial Neural Network, based on current methods, excels in classifying a dataset of socio-economic statuses of women into high, medium, and low categories with an impressive accuracy of 99.34%. This model efficiently identifies and utilizes aspects with a strong correlation with the prediction task, thereby enhancing both prediction speed and accuracy. Cross-evaluation with other datasets further validates its capability to deliver accurate real-time predictions. The current approach proves particularly effective in scenarios involving large feature sets, significantly improving prediction efficiency.

References

1. Firoozeh Karimi, Selima Sultana, Ali Shirzadi Babakan, Shan Suthaharan, An enhanced support vector machine model for urban expansion prediction, Computers, Environment and Urban Systems, Volume 75, 2019, Pages 61–75, ISSN 0198-9715, https://doi.org/10.1016/j.compenvurbsys.2019.01.001
2. Jeaneth Machicao, Alison Specht, Danton Vellenich, Leandro Meneguzzi, Romain David, et al. A Deep-Learning Method for the Prediction of Socio-Economic Indicators from Street-View Imagery Using a Case Study from Brazil. CODATA Data Science Journal, Committee on Data for Science and Technology (CODATA), 2022, 21, ⟨https://doi.org/10.5334/dsj-2022-006⟩. ⟨hal-03663799⟩, https://hal.archives-ouvertes.fr/hal-03663799
3. Christos A. Makridis, David Y. Zhao, Cosmin A. Bejan, Gil Alterovitz, Leveraging machine learning to characterize the role of socio-economic determinants on physical health and well-being among veterans, Computers in Biology and Medicine, Volume 133, 2021, 104354, ISSN 0010-4825, https://doi.org/10.1016/j.compbiomed.2021.104354
4. Winston L, McCann M, Onofrei G. Exploring Socioeconomic Status as a Global Determinant of COVID-19 Prevalence, Using Exploratory Data Analytic and Supervised Machine Learning Techniques: Algorithm Development and Validation Study. JMIR Form Res. 2022 Sep 27;6(9): e35114. Doi: https://doi.org/10.2196/35114. PMID: 36001798; PMCID: PMC9518652, https://pubmed.ncbi.nlm.nih.gov/36001798
5. Balasankar V, Penumatsa SV, Terlapu PRV. (2020) Intelligent socio-economic status prediction system using machine learning models on Rajahmundry A.P., SES dataset. *Indian* Journal of Science and Technology. 13(37):3820–3842. https://doi.org/10.17485/IJST/v13i37.1435
6. Abitbol, J.L., Karsai, M. Interpretable socioeconomic status inference from aerial imagery through urban patterns. Nat Mach Intell 2, 684–692 (2020). https://doi.org/10.1038/s42256-020-00243-5
7. Nasejje JB, Mbuvha R, Mwambi H Use of a deep learning and random forest approach to track changes in the predictive nature of socioeconomic drivers of under-5 mortality rates in sub-Saharan Africa BMJ Open 2022;12: e049786. doi: https://doi.org/10.1136/bmjopen-2021-049786
8. Pal Sundsøy, Johannes Bjelland, Bjørn-Atle Reme, Asif M. Iqbal, Eaman Jahani, 2016/01, Deep Learning Applied to Mobile Phone Data for Individual Income Classification, Proceedings of the 2016 International Conference on Artificial Intelligence: Technologies and Applications, Atlantis Press, 96–99, SN - 1951-6851, https://doi.org/10.2991/icaita-16.2016.24
9. J. Levy Abitbol, M. Karsai and E. Fleury, Location, Occupation, and Semantics Based Socioeconomic Status Inference on Twitter 2018 IEEE International Conference on Data Mining Workshops (ICDMW), 2018, 1192–1199, https://doi.org/10.1109/ICDMW.2018.00171

10. Alsharkawi, Adham, Mohammad Al-Fetyani, Maha Dawas, Heba Saadeh, and Musa Alyaman. 2021. "Poverty Classification Using Machine Learning: The Case of Jordan" Sustainability 13, no. 3: 1412. https://doi.org/10.3390/su13031412
11. Dehtiarova YV, Yevdokimov YURI. Data Mining Methods and Models for Social and Economic Processes Forecasting. 2018. Available from: https://doi.org/10.21272/mer.2018.80.03
12. A. Shetty, A. Thorat, R. Singru, M. Shigawan and V. Gaikwad, "Predict Socio-Economic Status of an Area from Satellite Image Using Deep Learning," 2020 International Conference on Electronics and Sustainable Communication Systems (ICESC), 2020, pp. 177–182, doi: https://doi.org/10.1109/ICESC48915.2020.9155696
13. Juhn YJ, Ryu E, Wi CI, King KS, Malik M, Romero-Brufau S, Weng C, Sohn S, Sharp RR, Halamka JD. Assessing socioeconomic bias in machine learning algorithms in health care: a case study of the HOUSES index. J Am Med Inform Assoc. 2022 Jun 14;29(7):1142–1151. doi: https://doi.org/10.1093/jamia/ocac052. PMID: 35396996; PMCID: PMC9196683
14. Adel Daoud, Rockli Kim, S.V. Subramanian, Predicting women's height from their socioeconomic status: A machine learning approach, Social Science & Medicine, Volume 238,2019, 112486, ISSN 0277-9536, https://doi.org/10.1016/j.socscimed.2019.112486
15. Islam, Md & Mustaquim, Md. (2014). Socio-Economic Status of Rural Population: An Income Level Analysis. Asian Academic Research Journal of Multidisciplinary. 1. 98–106.

Improvement of UPFC Performance in Power Systems Using Artificial Intelligence

Mamatha Deenakonda, Padma Jyothi Uppalapati, Sridevi Bonthu, and Phanikumar Chittala

Abstract To address various challenges in power generation and distribution such as power quality deterioration, total harmonic distortion (THD), stability issues, and transient response, FACTS (Flexible Alternating Current Transmission Systems) devices are commonly employed. The Unified Power Flow Controller (UPFC) plays a significant role among these devices. Artificial Intelligence techniques, particularly Artificial Neural Networks (ANNs), are utilized to enhance their performance. In this paper, we propose a novel approach combining an ANN controller with UPFC. This combination aims to significantly improve upon previous methodologies by reducing THD values and enhancing power system stability. In conjunction with a shunt filter and transformer, the ANN controller effectively manages power transient conditions, thereby demonstrating superior performance compared to conventional methods.

Keywords Unified power flow controller · Artificial neural network · Total harmonic distortion · Flexible AC transmission device · Multi-layer perceptron

1 Introduction

The need for electrical power has been steadily growing in recent years. Nevertheless, the construction of new power transmission lines is being slowed down by natural, financially feasible, and time-sensitive issues. The current transmission lines should be used to the very end of their operational range in order to overcome these obstacles. The current electricity transmission cables' main benefit is diminished by their obstinateness and heated limit. Equal activity on both lines

M. Deenakonda (✉) · P. Chittala
EEE Department, Vishnu Institute of Technology, Bhimavaram, AP, India
e-mail: mamatha.d@vishnu.edu.in

P. J. Uppalapati · S. Bonthu
CSE Department, Vishnu Institute of Technology, Bhimavaram, AP, India

A. Patel et al. (eds.), *Advances in Machine Learning and Big Data Analytics II*,
Springer Proceedings in Mathematics & Statistics 442,
https://doi.org/10.1007/978-3-031-51342-8_2

is necessary for financially viable activity, and both lines must be pushed to the absolute maximum. To leverage the controllability and adaptability of the intensity framework, regulators must be used. Quick-acting regulators are needed to alter the power framework components for managing the solidity. FACTS devices play a crucial role in enhancing the operational limits of power systems, leveraging high-capacity static switches. They are integrated into power networks to bolster transmission line capabilities, thereby improving voltage stability, transient stability, voltage regulation, system reliability, and thermal limits. Initially developed FACTS technologies like SVC, TSSC, TCSC, and thyristor-controlled phase shifters utilize thyristor-based switches for control mechanisms. The devices of the second generation, such as STATCOM, SSSC, UPFC, and IPFC, are converter-based and utilize switching static converters as quick-controllable sources of voltage and current. The UPFC [1–6], which has comprehensive capabilities for voltage regulation, series compensation, and phase shifting, is the most adaptable FACTS controller ever created.

The main goal of this study is to utilize UPFC both as a shunt active power filter and as a reactive power compensation device to ensure a near-unity power factor. Flowchart depicting a system with interconnected components. The process starts with a "Source" leading to a "Shunt Transformer." From there, it connects to "UPF," which is also linked to "ANN" below it. The "UPF" connects to a "Series Transformer," which then leads to a "Non-linear" component, completing the loop back to the "Source." Arrows indicate the flow direction between components. This technology aims to enhance power quality within transmission lines, thereby contributing to grid stability and reliability. It is possible to achieve unbalanced voltage correction more successfully. The project's main goals are to keep the power factor close to unity and reduce the THD value of current flowing over feeders. The other goals are to enhance power quality and offer better current tracking capabilities with quicker dynamic response. Both UPFC [1–6], which functions as a shunt active power filter to effectively provide compensation, and ANN are used to accomplish these goals. One of the key elements in systems for generating renewable energy is the active power filter. They are employed to improve power quality, rectify power factors, and compensate for reactive power. For active power filter applications, using UPFC as a shunt active power filter will result in superior solutions. A recent innovation that will soon be applied to several applications is the use of UPFC for shunt active power filtering.

2 Problem Identification and Why Selection of Artificial Neural Network

1. Detecting higher Total Harmonic Distortion (THD) is a significant concern addressed in previous research on FACT devices. While these devices aim

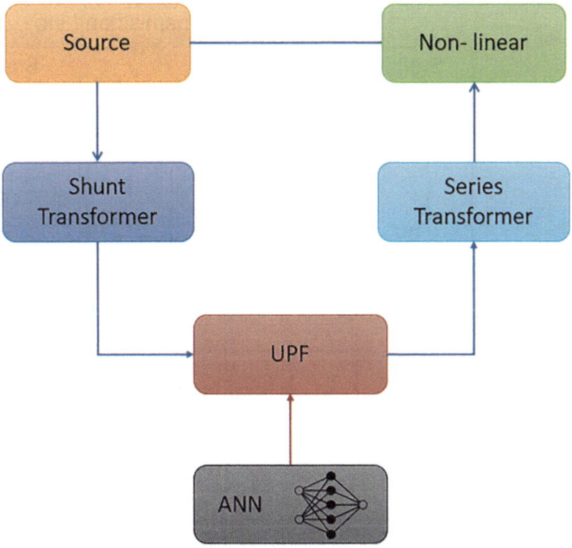

Fig. 1 Block diagram of the proposed system

to mitigate THD in power systems by improving power flow characteristics, increased transmission losses have exacerbated THD levels.
2. Fluctuation in the voltage and steadiness of the current.
3. Lower steady state time and longer transient time.
4. Problem with voltage harmonic SAG and swell.

3 Research Impact and Originality

The significance of this proposed work lies in its utilization of a UPFC system integrated with an ANN [10] controller to significantly reduce Total Harmonic Distortion (THD) parameters compared to earlier methodologies. Previous studies employing conventional FACT devices such as SRF (Synchronous Reference Frame) and STATCOM typically achieve THD levels ranging from 1% to 2%. In contrast, the incorporation of an ANN-based controller in this research enhances reliability and intelligence, offering a more effective solution for THD reduction in power systems.

4 Proposed ANN Controller

With its extensive capabilities for voltage regulation, series compensation, and phase shifting, the UPFC is the most adaptable FACTS controller yet created. In

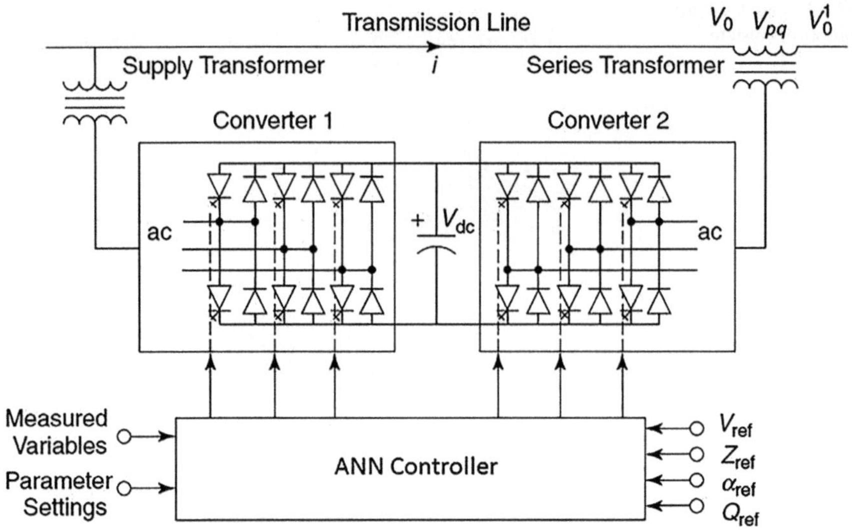

Fig. 2 Circuit diagram of UPFC

a transmission line, it can independently and very quickly regulate both real- and reactive power flows. Two AC/DC converters which are shown in Fig. 2 make up the uninterruptible power supply (UPFC) which may generate and consume reactive as well as actual power. The transmission line is connected to one of the two converters in series and the other in parallel by way of a series transformer and shunt transformer, respectively. A common capacitor connects the two converters' DC sides, providing the necessary DC voltage for the converters to function. To keep a constant voltage across the DC capacitor, the facility balancing between series and shunt converters is necessary. As a result of the UPFC's series branch injecting a voltage with changeable magnitude and phase angle, and the transmission line, can exchange actual power with it, which increases the line's capacity for power flow and its upper limit for transient stability. The facility system and shunt converter trade a current with a power factor angle and magnitude that can be controlled. By setting the DC bus voltage at a specific level, it is often controlled to balance the real power the series converter absorbs from or injects into the power supply with the losses. There have been several control methods proposed to adjust the magnitude and angle of the series voltage as well as the shunt current. According to the transmission line current, the series converter voltage phasor is frequently divided into in-phase and quadrature components. Quadrature-voltage components are further easily related to the reactive and actual power flows in the transmission system than are in-phase components, and vice versa. Controlling the quadrature component of the series converter voltage can frequently stop the decline in real power during short-circuit and transient circumstances, improving transient stability. The reactive power flow variation or voltage deviation at the inoculate bus where UPFC is located to control the series voltage in phase component, respectively.

The UPFC is set up as illustrated in and comprises two VSCs connected via a common DC terminal. Through a coupling transformer, one VSC converter 1 is linked to the road in a shunt; through an interface transformer, the other VSC converter 2 is connected to the transmission line in series. Both converters' DC voltage is provided by a typical source capacitor bank. The voltage phasor, V_{pq}, which can range from 0 to V_{pqmax}, is serially injected with the line by the series converter under supervision. In addition, the phase of V_{pq} is separately adjustable from 0 to 360. During this process, the series converter exchanges both actual and reactive power with the cable. The series converter generates and absorbs reactive power internally, but the capacitor, a DC energy storage device, makes it possible to generate and absorb real power. The primitive purpose of the shunt-connected converter 1 is to meet the real-power needs of converter 2, which are met by the cable itself. The shunt converter ensures a constant voltage of the DC bus. The losses of the two converters and their coupling transformers are therefore equivalent to the total actual power that is taken from the AC system. Additionally, the shunt converter performs a STATCOM-like function by producing or absorbing the necessary amount of reactive power to independently control the terminal voltage of the associated bus.

5 Power System Stability

A power system's ability to maintain a state of operating harmony under normal working conditions and to regain a sufficient state of balance after being subjected to stresses can be broadly characterized as having a stable power system. Depending on the design and operation mode of the Power framework, stability may be demonstrated in a variety of ways as the classification is shown in Fig. 3. Power system stability is the capacity of an electric Power framework, for a given starting working condition, to recover a condition of working balance after being exposed to a physical unsettling influence, with most framework factors limited, so basically the whole framework remains intact.

Rotor angle stability is the ability of the system to remain in synchronism when subjected to a disturbance. Small-disturbance rotor angle stability along with transient stability are labeled as short-term phenomena because the interruption due to the disturbances is not more than a minute.

6 ANN for the Control of the Power Systems

Nonlinear loads for instance rectifiers, computer equipment, air-conditioning units, and fluorescent lighting introduce harmonic distortions into electrical supplies, pos-

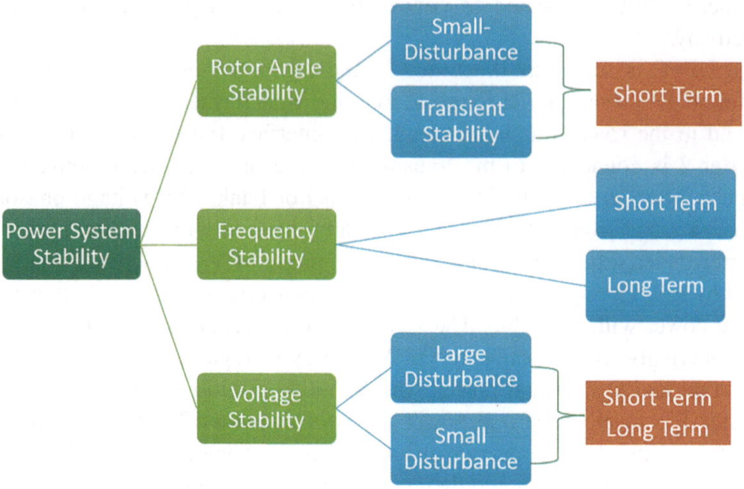

Fig. 3 Classification of power system stability

ing significant disturbances. These devices draw non-sinusoidal currents, thereby polluting voltages with harmonics that propagate through the electrical network. This can disrupt the performance of electrical devices and potentially damage them. Active Power Filters (APFs) have emerged as modern solutions for harmonic compensation, capable of effectively suppressing current harmonics and improving power factor, particularly under rapidly fluctuating loads, which sets them apart from other compensation devices. Moreover, APFs are versatile for integration into existing power systems, making them widely adopted in practical applications. In recent years, Artificial Neural Networks (ANNs) [7–11] have been successfully pertain to control APFs, showing promising results in power systems. ANNs excel in adapting online to changing parameters of electrical networks, such as nonlinear and time-varying loads, due to their robust training capabilities. One notable success of ANNs in power systems control is evident in the management of motor drives.

Reactive power/voltage regulation using ANN is also effective in power systems. First, it has been shown that neural networks can be used effectively to identify and manage nonlinear dynamical systems. This paper studies static and dynamic back-propagation learning for MLP (Multi-layer perceptron) architectures which is shown in Fig. 4 while introducing fundamental ideas and definitions. The recommended identification and adaptive control techniques are actually practicable, according to simulation results. To follow any desired reference signal, an MLP is created. In fact, the neural network connected to the reference model of the trajectory determines the load by identifying the nonlinear dynamics of the motor. Induction machines and inverter drives can both be controlled using ANNs as observers. A decentralized neural network control scheme of the reference compensation technique used to control an inverted pendulum with two degrees of freedom has recently been the subject of an experimental study. Two different neural network controllers are used

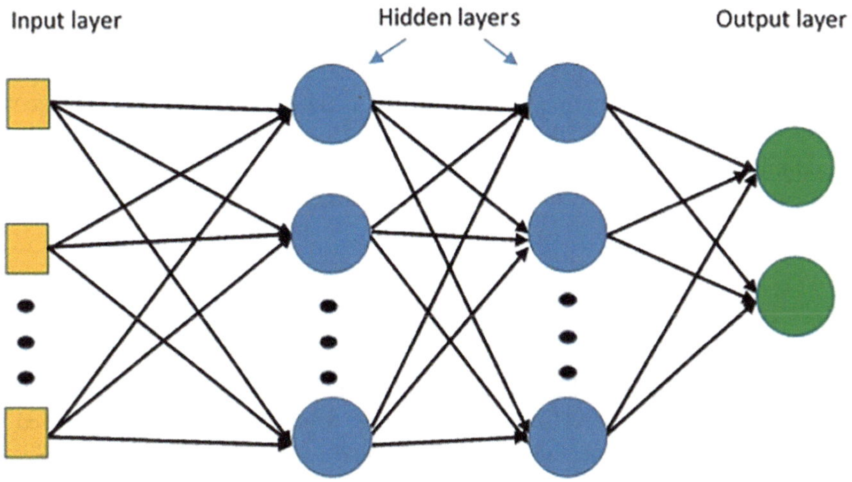

Fig. 4 Basic structure of Multilayer Perceptron Neural Network

to regulate each axis of the pendulum, resulting in a decoupled control structure. The decoupled control structure can eliminate coupling effects and capture current uncertainties. Experiments have been used to successfully apply this. Active power filtering methods based on ANNs are becoming increasingly effective. In fact, numerous studies demonstrate that ANNs are frequently utilized to implement the various APF components.

APF designs use inverters to inject reference currents into the electrical power lines. The identification of distortion harmonics from measurements of the voltages and currents of the power lines is a prerequisite for APF methods. This causes reference currents to need to be phase-oppositely injected into the electrical power systems. Accordingly, the reference currents regulate the inverter, which is often connected to an output filter built by analogue circuits. For this, many high-level control mechanisms are frequently employed. In real-world applications, pulse width modulation (PWM) implementation is necessary. It is even possible that PWM is a well-known and effective method for using a processor's digital outputs to control analogue systems and circuits. For instance, in an optimized PWM inverter control, the control is based on [12–15] and generates the best switching signals for the harmonic and voltage control of DC/AC bridge inverters. The approach's effectiveness is demonstrated by the results of an experimental implementation. We frequently compare various neural techniques. When compared to PID (Proportional Integral Derivative) regulators, these neuro-controllers—direct, inverse, and direct-inverse—perform the same functions. PID control systems are frequently employed as a fundamental control technology for industrial control systems, but PID responds slowly and requires fewer repetitions throughout the determined period. And another disadvantage of PID is, it deviates the output somewhat during the entire regulating procedure.

The PWM switching signals of an inverter are generated by ANNs when used to regulate an APF. To cancel the undesirable harmonics, reference currents are effectively injected into the power lines. Since the ANN has proven to be more effective than the conventional PI controller, it is thought to be more advantageous. The reaction time for the 5th and 7th harmonics in the normal PI, PID controller is determined to be 51 ms, while the Artificial Neural Network controller provides a response time of 31 ms for the same order of harmonics.

7 Circuit Simulation and Results

The proposed ANN-based UPFC configuration is illustrated in phasor mode using the Simulink tool in MATLAB. In this setup, a neural network is employed to regulate the shunt inverter and enhance damping signals for dynamic and reactive power management. The system employs two independent regulators to control the injected voltage, integrating both shunt and series techniques within a transformer-based power transmission model. The primary function of the UPFC is to control voltage and power flow, combining attributes of STATCOM (Static Synchronous Compensator) and SSSC (Static Synchronous Series Compensator). Gate pulse control of the shunt inverter and series inverter is achieved using a Pulse Generator with VSC (Voltage Source Converter) control, generating a Space Vector Duty Cycle. When power factor imbalance occurs due to nonlinear loads, the ANN controller manages both shunt and series operations to ensure a stable power factor close to unity, minimizing continuous oscillations.

This ANN-based system results can be compared with the PID and fuzzy-based controllers FIS (Fuzzy Inference System) but the neural network gives a better result. The layered architecture of an ANN (Artificial Neural Network) relies on the input, output, and hidden layers. In a neural network, basically predefine data for grid stabilization in the form of a duty cycle. The required data is then classified using a neural network classifier to compensate for the grid at the receiver end. In these data-based systems, a duty cycle is generated depending on the transmitting grid impedance. At this time, neural networks offer accurate impedance matching for sending and receiving ends.

The complete simulation circuit of an active power filter for a renewable energy system is shown in Fig. 5. By linking UPFC at the PCC (point of common coupling), reactive power compensation is accomplished. Figures 6 and 7 show the pulse generator circuits of the series and shunt device respectively. Results from the simulation are used to debate and analyze the Simulink diagram. Figures 8 and 9 show the ANN layers that regulate the number of neurons. The input layer is the first layer, followed by the hidden layer. Time is represented on the X axis in Fig. 10 while voltage and current amplitude are shown on the Y axis. These show the sources of grid power generation used by the UPFC system. The offset time and voltage THD fluctuation are illustrated on the X and Y axes, respectively, in

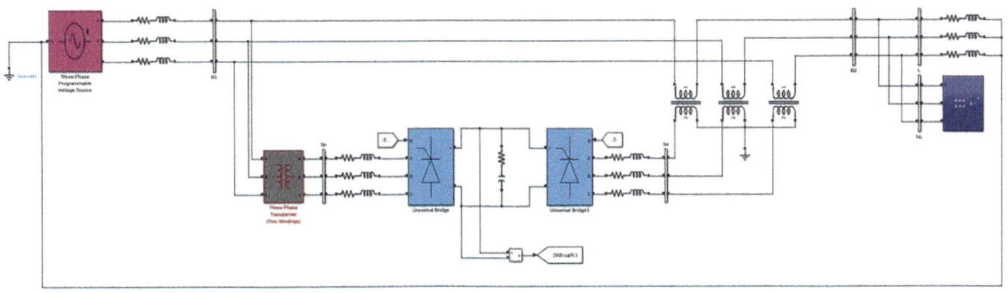

Fig. 5 Simulink diagram

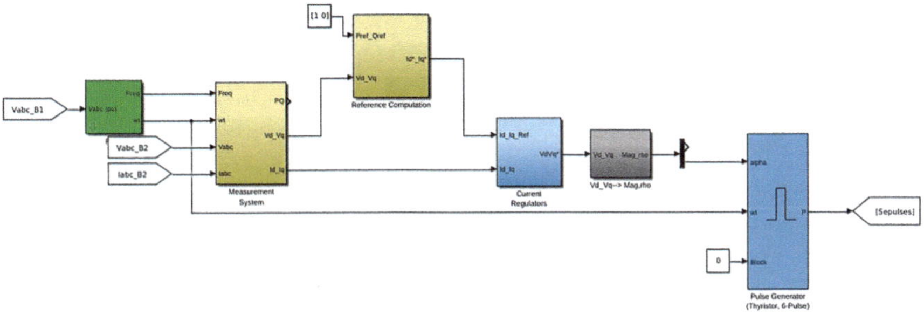

Fig. 6 Pulse generator circuit series device

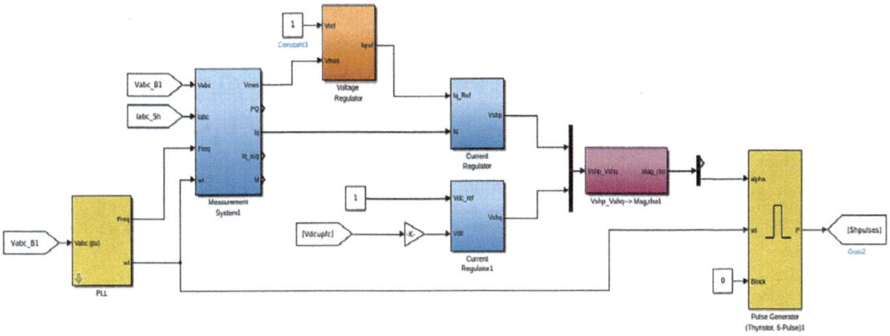

Fig. 7 Pulse generator circuit shunt device

Fig. 11. The offset time and voltage THD fluctuation are illustrated on X and Y axes, respectively, in Fig. 11. The offset time on the X axis and the linear load voltage THD stabilization on the Y axis are depicted in Fig. 12. The curves show that linear loads produce the most THD over a time span of >0.1 s off-set, and THD stabilization takes 0.4 s. In Fig. 13, the Y axis represents nonlinear load voltage THD stabilization while the X axis represents offset time. Nonlinear load maximal THD development spanning >0.1 s of offset time is shown by the curve, and THD stabilization takes 0.4 s.

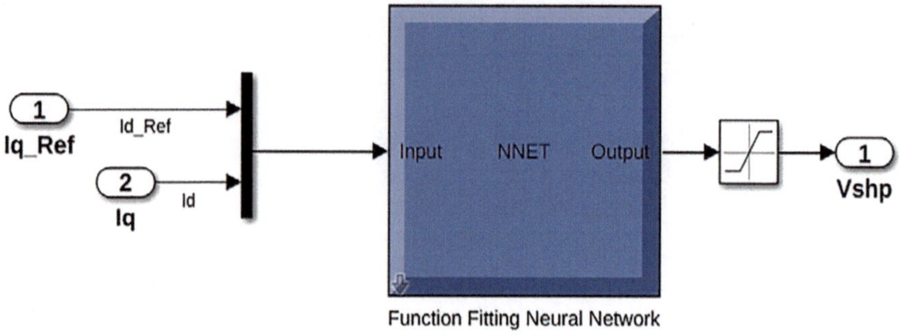

Fig. 8 Proposed ANN

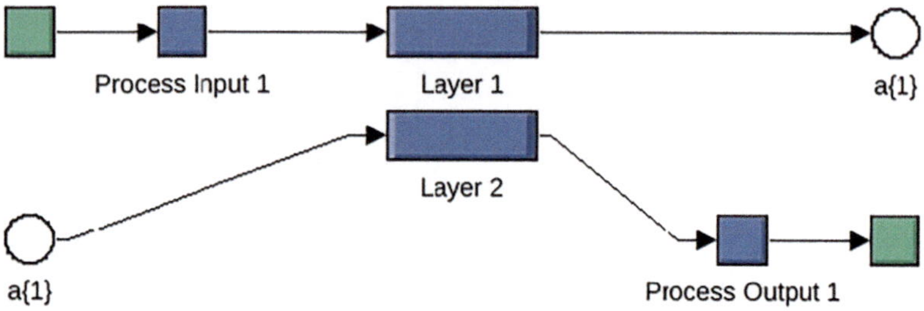

Fig. 9 Internal network of ANN

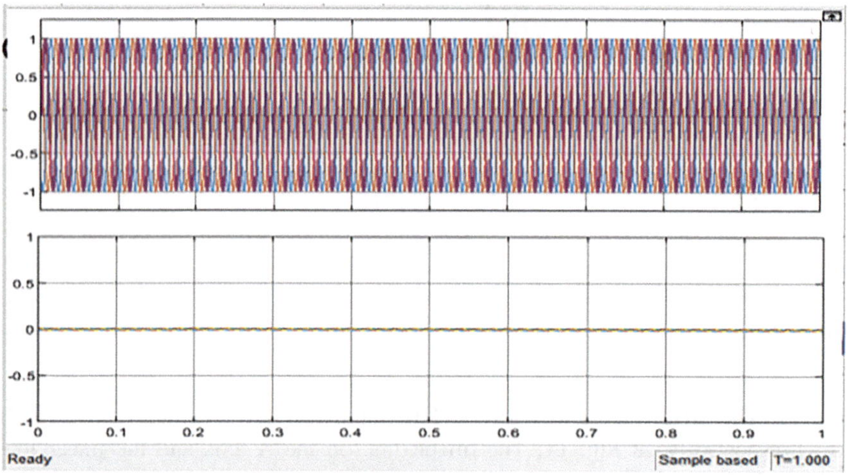

Fig. 10 Voltage and current outputs from power generation

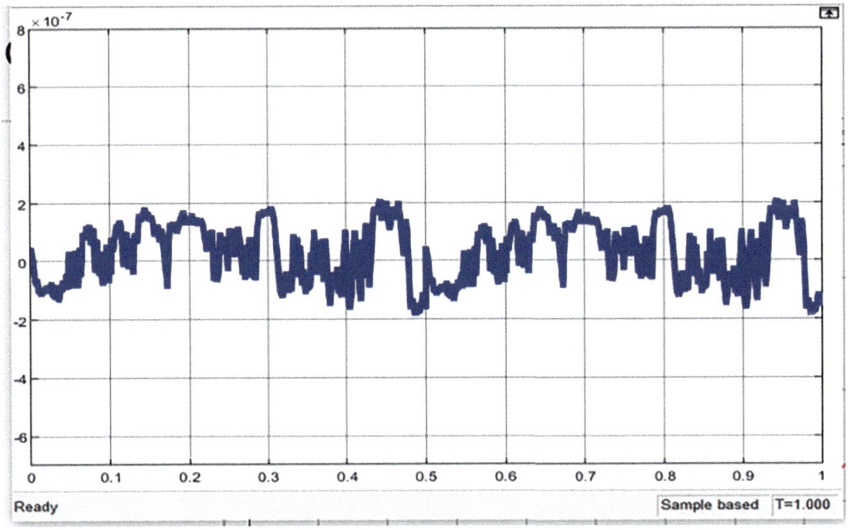

Fig. 11 Minimizing THD variation with shunt filters

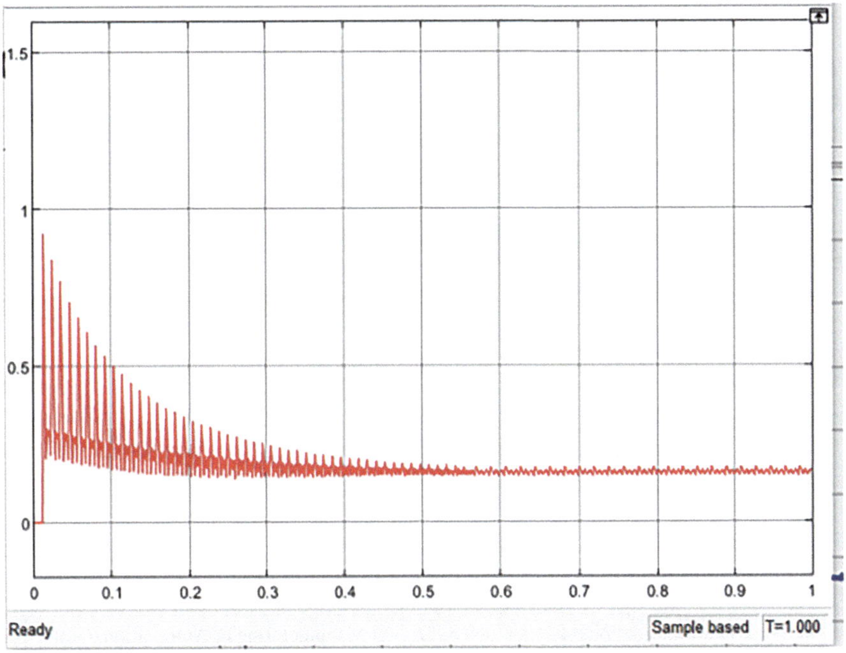

Fig. 12 THD of linear load

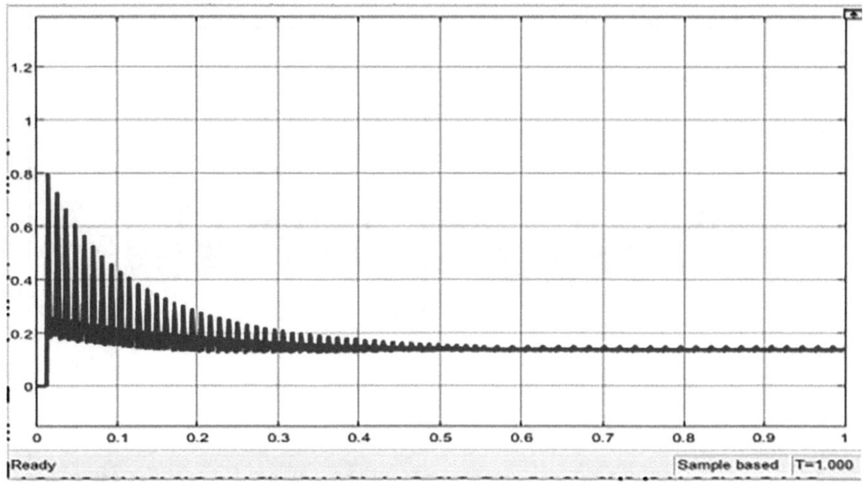

Fig. 13 THD of nonlinear load

8 Conclusion

Observation and analysis of ANN-based UPFC systems demonstrate improved stability, reduced transients, enhanced damping, and significantly lower Total Harmonic Distortion (THD) percentages. The system model achieves THD values of 0.3% for nonlinear loads and 0.25% for linear loads. The time for transitory stabilization is approximately the same as 0.8 s. Because of the reduced transmission loss and precise impedance matching provided by them, HVDC and HVDC power transmission systems can use them.

References

1. Joglekar, J., Nerkar, Y.: Application of UPFC for improving micro-grid voltage profile. In: 2010 IEEE International Conference on Sustainable Energy Technologies (ICSET), Kandy, pp. 1–5 (2010)
2. Khatami, S.V., Soleymani, H., Afsharnia, S.: Adaptive neuro-fuzzy damping controller design for a power system installed with UPFC. In: 2010 Conference Proceedings IPEC, Singapore, pp. 1046–1051 (2010)
3. Zhongshan, T., Yapeng, L., Daozhuo, J., Yuqing, Z. Design and test of UPFC- FCL prototype. In: 2012 Third International Conference on Digital Manufacturing & Automation, GuiLin, pp. 314–318 (2012)
4. Mortazavian, S., Shabestary, M.M., Hashemi Dezaki, H., Gharehpetian, G.B.: Voltage indices improvement using UPFC based on specific coefficients algorithm. In: Iranian Conference on Smart Grids, Tehran, pp. 1–5 (2012).
5. Murugan, A., Thamizmani, S.: A new approach for voltage control of IPFC and UPFC for power flow management. In: 2013 International Conference on Energy Efficient Technologies for Sustainability (2013)

6. Bhowmik, A.R., Chakraborty, A.K., Das, P.N.: Placement of UPFC for minimizing active power loss and total cost function by PSO algorithm. In: 2013 International Conference on Advanced Electronic Systems (ICAES), Pilani, pp. 217–220 (2013)
7. Kumar, P. Ganesh, T. Aruldoss Albert Victoire, P. Renukadevi, and Durairaj Devaraj. "Design of fuzzy expert system for microarray data classification using a novel genetic swarm algorithm." *Expert Systems with Applications* 39, no. 2 (2012): 1811–1821.
8. Shrawane, S.S., Diagavane, M., Bawane, N.: Concoction of UPFC for optimal reactive power dispatch using hybrid GAPSO approach for power loss minimisation. In: 2014 6th IEEE Power India International Conference (PIICON), Delhi, pp. 1–4 (2014)
9. Albatsh, F.M., Ahmad, S., Mekhilef, S., Mokhlis, H., Hassan, M.A.: D - Q model of fuzzy based UPFC to control power flow in transmission network. In: 7th IET International Conference on Power Electronics, Machines and Drives (PEMD 2014), Manchester, pp. 1–6 (2014)
10. Bonthu, Sridevi, Kompella Bhargava Kiran, Mamatha Deenakonda, VVR Maheswara Rao, and S. V. V. D. Jagadeesh. "Multi-label and Multi-class Classification on a Custom Dataset using Convolution Neural Networks." In 2023 7th International Conference on Intelligent Computing and Control Systems (ICICCS), pp. 576–579. IEEE, 2023.
11. Mallick, R.K., Nahak, N., Sinha, R.R.: Fuzzy sliding mode control for UPFC to improve transient stability of power system. In: 2015 Annual IEEE India Conference (INDICON), New Delhi, pp. 1–6 (2015)
12. Ferdosian, M., Abdi, H. Bazaei, A.: Improving the wind energy conversion system dynamics during fault ride through: UPFC versus STATCOM. In: 2015 IEEE International Conference on Industrial Technology (ICIT), Seville, pp. 2721–2726 (2015)
13. Venkata Subbarao, M., Chinimilli Pravallika, D. Ramesh Varma, and M. Prema Kumar. "Power quality event classification using wavelets, decision trees and SVM classifiers." In Innovations in Electronics and Communication Engineering: Proceedings of the 9th ICIECE 2021, pp. 245–251. Singapore: Springer Singapore, 2022.
14. Pragathi, Bellamkonda, Rajagopal Veramalla, Fazal Noorbasha, and Bangarraju Jampana. "Power quality improvement for grid interconnected solar PV system using neural network control algorithm." International Journal of Power and Energy Conversion 9, no. 2 (2018): 187–204.
15. Rupesh, M., and T. S. Vishwanath. "Fuzzy and ANFIS controllers to improve the power quality of grid connected PV system with cascaded multilevel inverter." IJEER 9, no. 4 (2021): 89–96.

Classification of Parkinson's Disease Using Machine Learning Techniques

Satish Dekka, K. Narasimha Raju, D. Manendra Sai, M. Pallavi, and Bosubabu Sambana

Abstract Emotional speech and adverse social effects, such as stigma, dehumanization, and loneliness, are the main issues with Parkinson's disease. This has a significant impact on the way of life of those who have Parkinson's disease. The slow death or destruction of brain neurons is one of the known causes of Parkinson's disease. As a result, dopamine is produced in the brain as a chemical messenger. Dopamine deficiency results in typical brain activity and a variety of modifications in how the body moves. Parkinson's disease (PD) is mostly recognized by its signs and symptoms, such as a little tremor. Body movement is slowed down by tremors. Rigid muscles in this body weaken and lose automatic movements like blinking and smiling, which are additional symptoms. Previous researchers are still working on this problem, but they are having trouble using the right techniques to solve it. The suggested work utilized the MATLAB environment to develop three machine learning algorithms. The outcome indicates that the KNN algorithm has the highest accuracy in detecting Parkinson's illness.

Keywords Parkinson's disease · Machine learning techniques · Decision tree · KNN · SVM

S. Dekka (✉)
Department of Computer Science and Engineering, Lendi Institute of Engineering and Technology (A), JNTU-GV, Vizianagaram, Andhra Pradesh, India

K. N. Raju
Department of Computer Science and Engineering, Gayatri Vidya Parishad College of Engineering (GVPCE), Andhra University, Visakhapatnam, Andhra Pradesh, India

D. M. Sai · M. Pallavi
Department of Computer Science and Engineering, Vignan's Institute of Engineering for Women. JNTU-GV, Visakhapatnam, Andhra Pradesh, India

B. Sambana
Department of Computer Science and Engineering (DataScience), School of Computing, Mohan Babu University, Tirupathi, Andhra Pradesh, India

A. Patel et al. (eds.), *Advances in Machine Learning and Big Data Analytics II*,
Springer Proceedings in Mathematics & Statistics 442,
https://doi.org/10.1007/978-3-031-51342-8_3

1 Introduction

Millions of seniors throughout the world suffer from Parkinson's disease (PD), a serious degenerative ailment of the central nervous system that lowers their quality of life [1]. Because the condition can vary greatly, each individual may have different symptoms from Parkinson's disease. Patients suffering from Parkinson's disease may also experience tremors, which typically occur when they are at rest. There are many different types of tremors, such as hand tremors, limb rigidity, and problems with balance and walking. Generally, there are two types of Parkinson's disease symptoms: non-motor and movement-related (motor). In reality, those who have non-motor symptoms endure greater suffering than those who have motor symptoms. Indigestion, trouble sleeping, olfactory loss, and cognitive decline are examples of non-motor symptoms. The Centers for Disease Control and Prevention (CDC) report that complications related to Parkinson's disease (PD) rank as the 14th most prevalent cause of death in the United States.

Parkinson's disease (PD) cannot yet be diagnosed [2]. However, several symptoms and diagnostic tests are employed in conjunction. Scientists have studied several indicators to detect Parkinson's disease early on and prevent its progression. Currently, all PD medications relieve symptoms but do not slow or stop the disease's progression. Several approaches for detecting Parkinson's disease have been proposed, based on various types of measurements such as speech data.

The neurodegenerative brain illness Parkinson's disease progresses slowly [3]. Brain cell loss is referred to as neurodegenerative disease. The human brain is normally composed of distinct areas that contain brain cells that produce dopamine. The brain's substantia nigra contains these cells. Communication between the substantia nigra and other brain regions that control movement is facilitated by the neurotransmitter dopamine [3]. People can move fluidly and in unison thanks to dopamine. The motor symptoms of Parkinson's disease (PD) appear when 60–80% of the cells that generate dopamine are damaged, resulting in inadequate dopamine production.

The enteric nervous system, lower brain stem, and olfactory tracts are where Parkinson's disease symptoms initially manifest. Higher brain regions, such as the substantia nigra and the brain shell, are affected by Parkinson's disease after spreading from these areas [3]. According to current theory, the disease starts years before motor symptoms including constipation, tremors, and slowing of movement, as well as smell loss or reduction. To make matters worse, vocal problems affect 90% of people with Parkinson's disease [4]. To halt the progression of the disease, the researchers are investigating methods for identifying these non-motor symptoms as early as feasible. Parkinson's disease affects numerous body parts. The brain is an integral part of the human body.

This condition affects the human brain and causes despair, anxiety, a reduction in internal functioning, and dizziness. Upper airway blockage affects the lungs. Muscles grew weaker. Eyes are also affected. Speaking and swallowing are difficult. Increased sweating is another sign. Due to its high accuracy and ease of implementation, machine learning (ML) has recently acquired favor for the identification of medical diseases [5]. Additionally, ML treatment for Parkinson's disease is mentioned in the literature. The development of feature selection (FS) for machine learning in brain surgery was studied by Wan and Maszczyk [6]. An ML-based method identifies the actual area to be operated on during brain surgery for Parkinson's disease. The research done once Parkinson's disease is discovered is the main emphasis of this effort [7]. Algorithms for machine learning are used to estimate the cognitive effects of Parkinson's disease. The degree of tremor in patients with Parkinson's disease was predicted by a machine learning method [8]. Additionally, the stage of Parkinson's disease was predicted using machine learning [9]. However, utilizing this widely used method of machine learning, the researchers mainly concentrate on the early detection of Parkinson's illness. Cavallo et al. [10] used motion data gathered from subjects' upper limbs to try and predict Parkinson's disease.

2 Review of Literature

ML approaches are utilized to estimate the cognitive implications of Parkinson's disease. A machine learning application predicted the level of tremor in Parkinson's patients [8]. ML was also used to forecast the stages of Parkinson's disease [9]. However, the researchers primarily focus on the early identification of Parkinson's disease using this common approach, ML. Cavallo et al. [10] attempted to predict PD based on motion data collected from people's upper limbs.

Machine learning is an emerging technology and it is used in many areas nowadays. In medicine, it is used to analyze different reports and identification of different diseases. Classification is one of the most popular and well-practiced techniques for knowledge discovery due to its simplicity and insightful properties [18]. In classification, there exist different model-building techniques for knowledge discovery such as decision trees, neural networks, support vector machines etc. Decision Tree models are very simple to understand and to interrupt for knowledge discovery. The decision tree formed contains root nodes and leaf nodes. Root nodes are the originating nodes of the decision tree. These root nodes descend downwards as branches lead to the terminating nodes known as leaf nodes. The branch leading from the root node to a leaf node will provide a decision for a particular case using a decision tree [19].

New samples are classified using the Support Vector Machine (SVM), a non-probability binary classifier. This method, which is supervised and kernel-based, categorizes unknown test samples after analyzing a set of known classes. The simplest kind of SVM classifier is the linear one, in which data are first mapped out as points in space and then split into two halves to minimize the gap width. For classification or regression, a Support Vector Machine (SVM) constructs a hyperplane in an infinite-dimensional space.

The hyperplane's maximum separation from the closest training data points occurs when the classifier's error is lowest. The data points with which this hyperplane was constructed are known as support vectors. Their proximity to the hyperplane makes them the closest. Both linear and nonlinear SVM processes are possible. Nonlinear Support Vector Machines operate better when the linear margin hyperplane is unable to produce a good match. To maximize hyperspace, nonlinear SVM manipulates the feature space through a kernel technique.

There are numerous prominent clustering algorithms, including k-means. K-means is one of the most extensively used clustering algorithms with real-world applications. Researchers have found various data properties that may have a significant impact on K-means clustering performance, such as data kinds and scales, attribute sparsity, and data noise and outliers [20]. In text clustering, K-Means has demonstrated comparable or superior performance to a range of other clustering algorithms and boasts an enticing computational efficiency [21–23]. Parkinson's disease symptoms include stiffness, poor balance, bradykinesia (slow movement), and movement abnormalities (tremors and gait) [24–28].

As previously noted, the lack of effective testing for Parkinson's disease has made diagnosis difficult [29–31]. However, a new study has shown that Parkinson's disease patients exhibit voice and speaking problems. Medical practitioners cannot notice these vocal problems in clinics. Consequently, early detection of Parkinson's disease and the identification of these vocal abnormalities require automated signal processing tools. According to recent studies, automated disease detection through risk factor extraction and classification can be achieved with machine learning and signal processing techniques [32–35]. In this study, we also try to create a PD detection approach based on machine learning and signal processing methods, inspired by these investigations.

This narrative review examines the present evidence surrounding some of the most frequent social symptoms of Parkinson's disease and their accompanying social repercussions, and suggests that proactively and appropriately treating these issues may improve disease outcomes [36]. In those with Parkinson's disease, exercise can improve motor function; nevertheless, depression makes it difficult to consistently work. The objective of this research was to design a new Therapy program that integrates peer-facilitated psychoeducation with manual-driven guided exercise for persons with Parkinson's disease and depression [37].

3 Methodology

In recent years, modern screening technologies and therapies have helped to reduce Parkinson's disease deaths. Different types of scientific research provide knowledge about the condition in unique but complementary ways. Parkinson's disease treatment has been made more effective by incorporating a variety of modern procedures and computational methodologies. As a result, the disease's fatality rate has fallen in recent years. Observational studies examine a population's

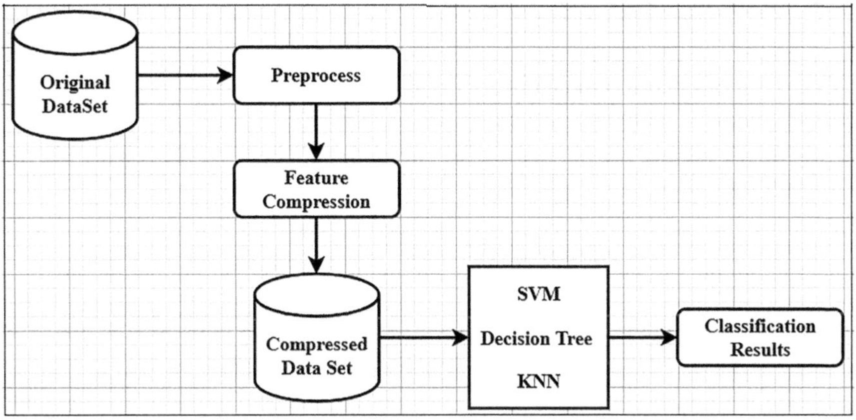

Fig. 1 Architecture

characteristics to identify the relationship between variables and outcomes. It does not always establish a link between cause and effect. This shows that there is a causal relationship between the variables and the outcomes.

Experiments are conducted using the Parkinson's disease dataset from UCI data repositories. For each data set, ID, Gender, PPE, DFA, NUMPULSES, MEANPERIOD PULSES... etc.

Figure 1 shows the process of architecture. First, we take the original data after pre-processing the data. Feature identification is crucial due to the factors which influence are mostly recognized. Various machine learning algorithms are applied to find the suitability for predicting the PD disease more accurately.

4 Experimental Setup and Assessment Criteria

The high-performance programming language for technical computing is MATLAB. Several experiments are run in MATLAB using various machine learning techniques. Data is extracted from a dataset made accessible via the UCI repository. Variables that influence Parkinson's disease identification were chosen, with a focus on symptoms and diagnosis and no consideration given to identification and treatment variables. This resulted in the selection of numerous factors connected with the symptoms or signs that a patient may exhibit, as well as the variable associated with the diagnosis, which allows for the identification of the kind of Parkinson's disease. The table compares the various ML algorithms on identified variables and describes them. In this work, three machine learning approaches, DT, KNN, and SVM, are utilized for early prevention and detection of patients to diagnose Parkinson's disease (PD). We discovered a system that performs better and provides more accurate findings, thereby saving human lives.

4.1 Performance Metrics

Different performance indicators are employed to evaluate the performance of our proposed model, including accuracy, AUC, and ROC.

Accuracy is simply defined as the ratio of properly predicted observations to total observations.

$$\textbf{Accuracy} = \frac{\text{TP} + \text{TN}}{\text{TP} + \text{TN} + \text{FP} + \text{FN}}$$

AUC The area under the ROC curve (AUC) indicates how well a parameter distinguishes between two diagnostic groups (diseased/normal).

$$\text{AUC} = \frac{\text{TP} + \text{TN}}{2 \times (\text{TP} + \text{TN} + \text{FP} + \text{FN})}$$

The dataset is separated into two sets: training and testing, with the former accounting for 80% of the data (320) and the latter for 20% (80).

5 Results

Experiments were carried out with various models—"Decision trees, support vector machines, and KNN"—and their comparative analysis was performed. The figures show the three models' confusion matrix and AUC curves. In parallel coordinates, the maximum accuracy for variables is achieved by utilizing a KNN, yielding an accuracy of 91.9%. This suggests that the KNN's classification of Parkinson's disease is consistent with that of the treating physician in 91.9% of the instances in the test set.

Here, Fig. 2 shows that we are generating the results of the confusion matrix for all the techniques, i.e., SVM, Decision Tree, and KNN.

In the above results, Fig. 3 shows the True Positive Rate (TPR) and False Positive Rate (FPR) of all ROC curves for all the techniques whereas KNN shows AUC as 0.87 among the three.

In the above, Fig. 4 shows the result X-axes represent features such as Gender, PPE, DFA, RPDE, Numpulses, Numperiodpulses, Nummeanpulses and StdDevperiodpulses, and Y-axes represent Standard deviation.

Figure 5 above shows the results representing the comparison of three algorithms in terms of accuracy for Decision Tree, SVM, and KNN as 78.7%, 86.5%, and 91.9%, respectively.

Table comparative analysis							
A complete model with 754 attributes							
S. No		Observations	Accuracy (%)	Precision	Training time	Predictors	Prediction speed
1	Decision trees	756	78.7	0.78	19.819 s	754	~ 2000 obs/s
2	Support vector machines	756	86.5	0.86	58.044 s	754	~ 2300 obs/s
3	KNN	756	91.9	0.91	82.552 s	754	~ 710 obs/s

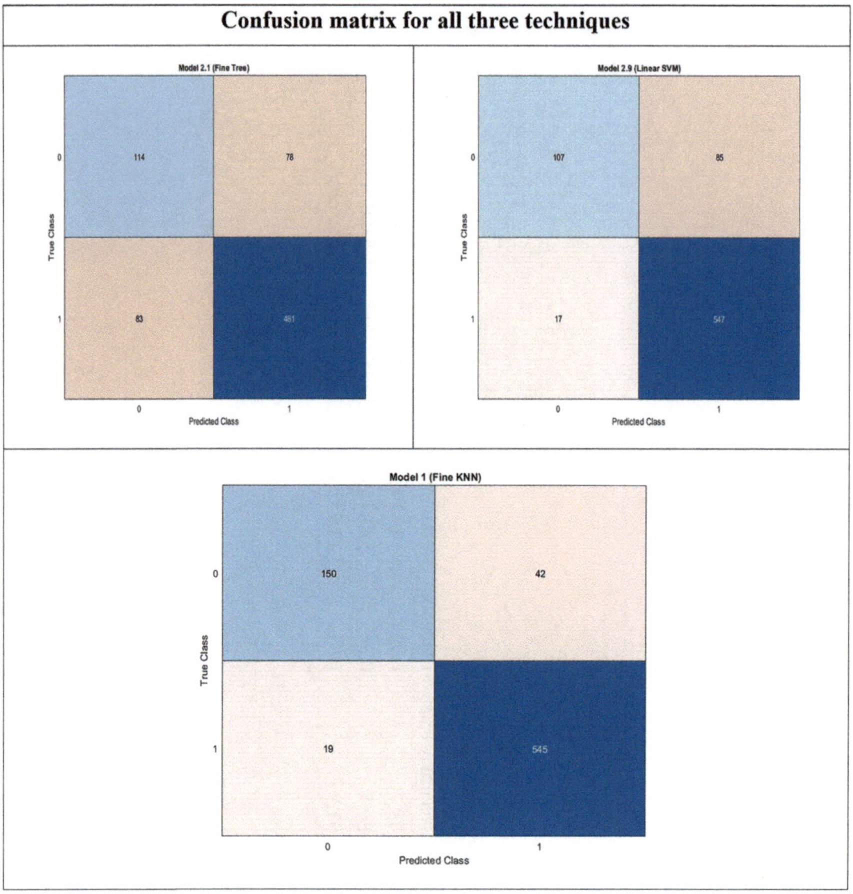

Fig. 2 Confusion matrix for all the techniques

The above table shows the comparative analysis across various attributes and the accuracy given for machine learning algorithms.

6 Conclusion

Parkinson's disease is a progressive neurological disorder characterized by a wide range of motor and non-motor characteristics. The natural progression of Parkinson's disease varies, but it is usually more rapid in patients with late onset and the PIGD variant. Future studies may identify disease-specific biomarkers,

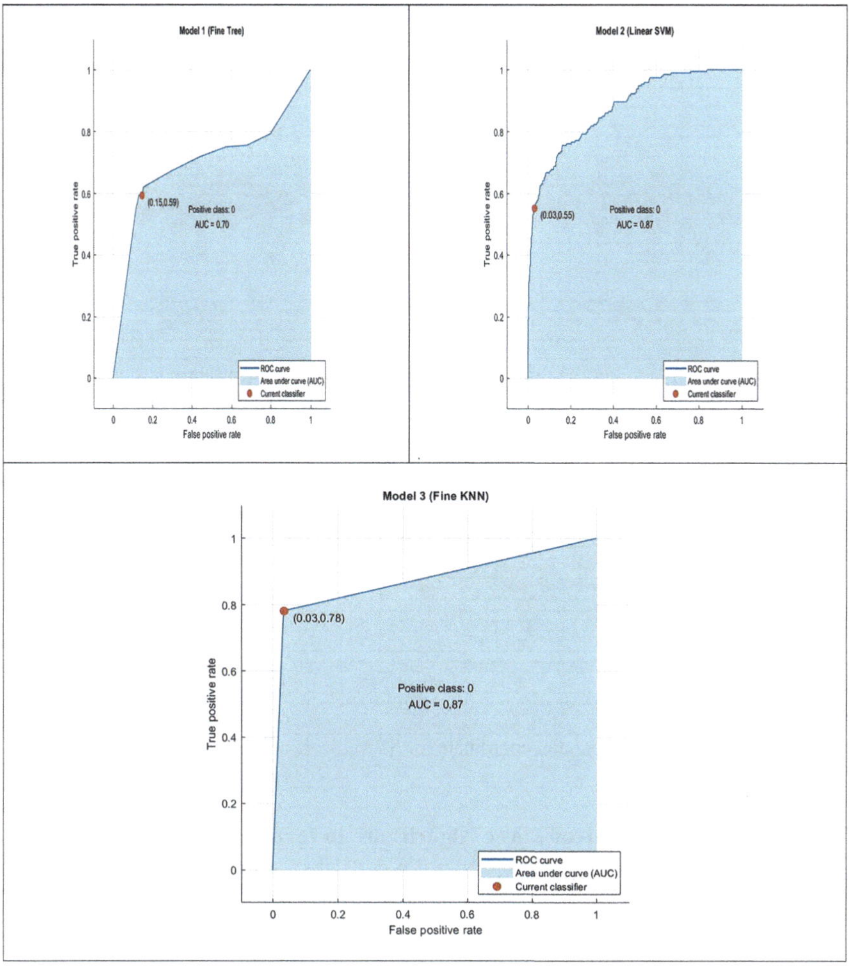

Fig. 3 ROC curve for all the techniques

allowing for distinction from other neurodegenerative illnesses. Such testing will be beneficial not just for diagnosing the disease in affected individuals, but also for identifying family members or communities at risk, allowing neuroprotective therapy to begin at an asymptomatic stage.

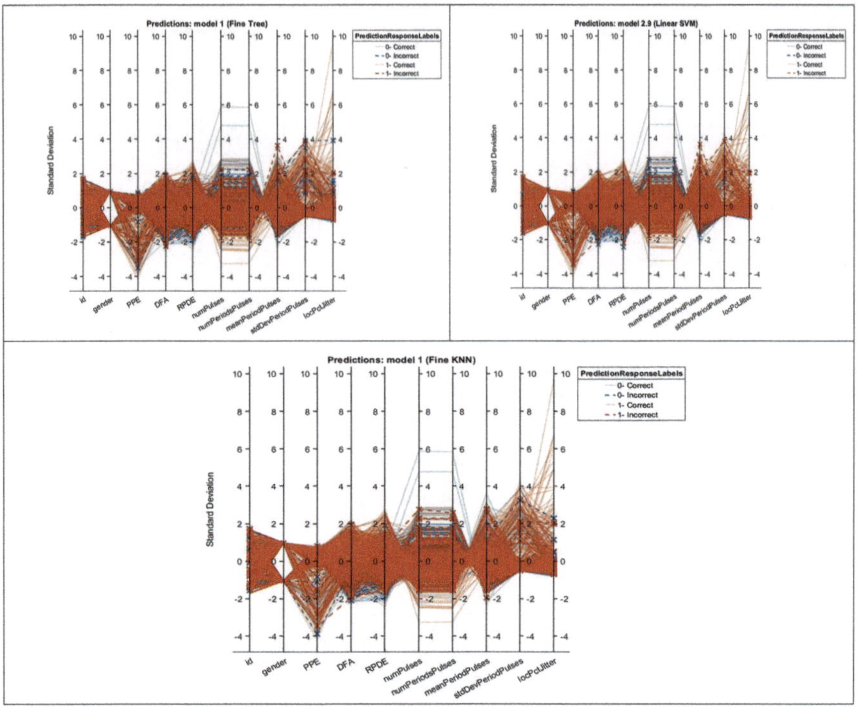

Fig. 4 Prediction model of parallel coordinate for SVM, decision tree, and KNN

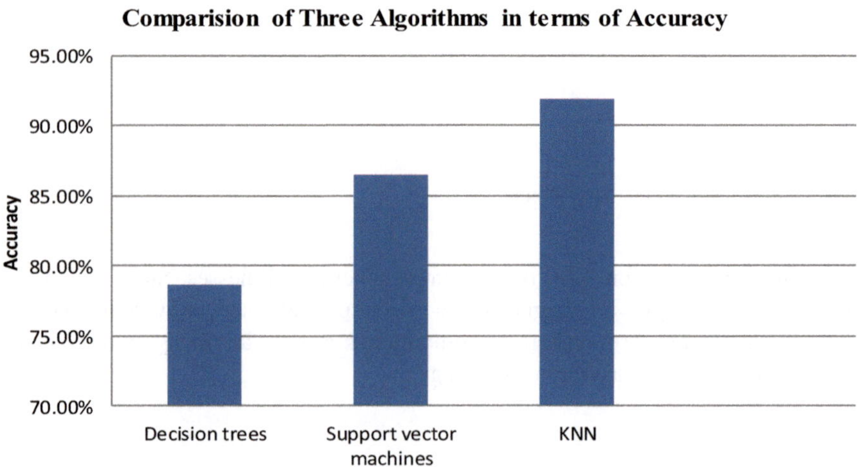

Fig. 5 Accuracy in decision tree, SVM, and KNN

References

1. R. Prashanth and S. D. Roy, "Early detection of Parkinson's disease through patient questionnaire and predictive modelling," International journal of medical informatics, vol. 119, pp. 75–87, 2018.
2. N. Singh, V. Pillay, and Y. E. Choonara, "Advances in the treatment of Parkinson's disease," Progress in neurobiology, vol. 81, no. 1, pp. 29–44, 2007.
3. Kara Budak R. Parkinson's disease. 2014.
4. Sakar CO, Kursun O. Telediagnosis of Parkinson's disease using measurements of dysphonia. J Med Syst 2010;34(4):591–9.
5. Rayan Z, Alfonse M, Salem A-BM. Machine learning approaches in smart health. Procedia Computer Sci 2019; 154:361–8. https://doi.org/10.1016/j.procs.2019.06.052.
6. Wan KR, Maszczyk T, See AAQ, Dauwels J, King NKK. A review on microelectrode recording selection of features for machine learning in deep brain stimulation surgery for Parkinson's disease. Clin Neurophysiol 2019;130(1):145–54.
7. Salman pour MR, et al. Optimized machine learning methods for prediction of cognitive outcome in Parkinson's disease. Computer Boil Med 2019;111.
8. Pedrosa T´i, Vasconcelos FF, Medeiros L, Silva LD. Machine learning application to quantify the tremor level for Parkinson's disease patients. Procedia Computer Sci 2018; 138:215–20.
9. Prashanth R, Dutta Roy S. Novel and improved stage estimation in Parkinson's disease using clinical scales and machine learning. Neurocomputing 2018; 305:78–103.
10. Cavallo F, Moschetti A, Esposito D, Maremmani C, Rovini E. Upper limb motor preclinical assessment in Parkinson's disease using machine learning. Park Relat Disorder 2019; 63:111–6
11. L. M. de Lau and M. M. Breteler, "Epidemiology of Parkinson's disease," Ae Lancet Neurology, vol. 5, no. 6, pp. 525–535, 2006.
12. L. Ali, C. Zhu, M. Zhou, and Y. Liu, "Early diagnosis of Parkinson's disease from multiple voice recordings by simultaneous sample and feature selection," Expert Systems with Applications, vol. 137, pp. 22–28, 2019.
13. L. Ali, C. Zhu, Z. Zhang, and Y. Liu, "Automated detection of Parkinson's disease based on multiple types of sustained phonation's using linear discriminant analysis and genetically optimized neural network," IEEE Journal of Translational Engineering in Health and Medicine, vol. 7, pp. 1–10, 2019.
14. S. Arora, F. Baig, C. Lo et al., "Smartphone motor testing to distinguish idiopathic rem sleep behaviour disorder, controls, and pd," Neurology, vol. 91, no. 16, pp. e1528–e1538, 2018.
15. J. Rusz, J. Hlavnicka, T. Tykalov et al., "Quantitative assessment of motor speech abnormalities in idiopathic rapid eye movement sleep behaviour disorder," Sleep Medicine, vol. 19, pp. 141–147, 2016.
16. J. Rusz, M. Novotny, J. Hlavnivcka, T. Tykalova, and E. Ruuvzivcka, "High-accuracy voice-based classification between patients with Parkinson's disease and other neurological diseases may be an easy task with inappropriate experimental design," IEEE Transactions on Neural Systems and Rehabilitation Engineering, vol. 25, no. 8, pp. 1319–1321, 2016.
17. J. Rusz, J. Hlavnicka, T. Tykalova et al., "Smartphone allows capture of speech abnormalities associated with high risk of developing Parkinson's disease," IEEE Transactions on Neural Systems and Rehabilitation Engineering, vol. 26, no. 8, pp. 1495–1507, 2018.
18. Juanli Hu, Jiabin Deng, Mingxiang Sui, A New Approach for Decision Tree Based on Principal Component Analysis, Proceedings of Conference on Computational Intelligence and Software Engineering, page no:1-4, 2009.
19. Shane Bergsma, Large-Scale Semi-Supervised Learning for Natural Language Processing, PhD Thesis, University of Alberta, 2010.
20. Tan, P.N., Steinbach, M., Kumar, V.: Introduction to Data Mining. Addison Wesley, Upper Saddle River (2005)

21. Krishna., Narasimha Murty, M.: Genetic k-means algorithm. IEEE Trans. Syst. Man Cybern. Part B 29(3), 433–439 (1999).
22. Steinbach, M., Karypis, G., Kumar, V.: A comparison of document clustering techniques. In: Proceedings of the KDD Workshop on Text Mining (2000)
23. Zhang, T., Ramakrishna, R., M. Livny: Birch: an efficient data clustering method for very large databases. In: Proceedings of 1996 ACM SIGMOD International Conference on Management of Data, pp. 103–114 (1996)
24. L. Cunningham, S. Mason, C. Nugent, G. Moore, D. Finlay, and D. Craig, "Home-based monitoring and assessment of Parkinson's disease," IEEE Transactions on Information Technology in Biomedicine, vol. 15, no. 1, pp. 47–53, 2011.
25. Z. A. Dastgheib, B. Lithgow, and Z. Moussavi, "Diagnosis of Parkinson's disease using electrovestibulography," Medical & Biological Engineering & Computing, vol. 50, no. 5, pp. 483–491, 2012.
26. G. Rigas, A. T. Tzallas, M. G. Tsipouras et al., "Assessment of tremor activity in the Parkinson's disease using a set of wearable sensors," IEEE Transactions on Information Technology in Biomedicine, vol. 16, no. 3, pp. 478–487, 2012.
27. M. A. Little, P. E. McSharry, E. J. Hunter, J. Spielman, L. O. Ramig et al., "Suitability of dysphonia measurements for telemonitoring of Parkinson's disease," IEEE Transactions on Biomedical Engineering, vol. 56, no. 4, pp. 1015–1022, 2009.
28. S. K. Van Den Eeden, C. M. Tanner, A. L. Bernstein et al., "Incidence of Parkinson's disease: variation by age, gender, and race/ethnicity," American Journal of Epidemiology, Vol. 157, no. 11, pp. 1015–1022, 2003.
29. R. Das, "A comparison of multiple classification methods for diagnosis of Parkinson disease," Expert Systems with Applications, vol. 37, no. 2, pp. 1568–1572, 2010.
30. L. Parisi, N. RaviChandran, and M. L. Manaog, "Feature driven machine learning to improve early diagnosis of Parkinson's disease," Expert Systems with Applications, vol. 110, pp. 182–190, 2018.
31. L. Naranjo, C. J. Perez, J. Martin, and Y. Campos-Roca, "A two-stage variable selection and classification approach for Parkinson's disease detection by using voice recording replications," Computer Methods and Programs in Biomedicine, vol. 142, pp. 147–156, 2017.
32. L. Ali, I. Wajahat, N. Amiri Golilarz, F. Keshtkar, and S. A. C. Bukhari, "LDA-GA-SVM: improved hepatocellular carcinoma prediction through dimensionality reduction and genetically optimized support vector machine," Neural Computing and Applications, 2020.
33. T. Meraj, A. Hassan, S. Zahoor et al., "Lung's nodule detection using semantic segmentation and classification with optimal features," Neural Computing and Applications Doi, 2019.
34. L. Ali and S. Bukhari, "An approach based on mutually informed neural networks to optimize the generalization capabilities of decision support systems developed for heart failure prediction," IRBM, 2020, In press.
35. L. Ali, S. U. Khan, N. A. Golilarz et al., "A feature-driven decision support system for heart failure prediction based on $\chi 2$ statistical model and Gaussian naive bayes," Computational and Mathematical Methods in Medicine, vol. 2019, Article ID 6314328, 8 pages, 2019
36. Juan Mario Haut, Mercedes Paoletti, Javier Plaza, Antonio Plaza, "Cloud implementation of the K-means algorithm for hyperspectral image analysis," J Supercomputer, 2017 73:514–529, DOI https://doi.org/10.1007/s11227-016-1896-3.
37. An update on the diagnosis and treatment of Parkinson disease, Philippe Rizek, MD, Niraj Kumar, MD, Mandar S. Jog, MD, cmaj Canadian medical association journal

A Survey on Skin Cancer Detection Using Artificial Intelligence

N. P. Patnaik M and Johan Jaidhan Beera

Abstract Skin cancer ranks among the worst types of disease. For skin cancer to occur, there must be a mutation in the skin's genetic material or a flaw in the DNA of skin cells. The importance of early skin cancer detection cannot be overstated since it can spread to other parts of the body and thus be less treatable later. It is important to recognize early warning signs of skin cancer because it is prevalent, has a high mortality rate, and is expensive to treat. Considering the gravity of these concerns, scientists have devised a number of methods for detecting skin cancer in its earliest stages. Skin cancer may be detected and classified as either benign or malignant based on several lesion factors, including but not limited to symmetry, color, size, form, etc. This study provides a comprehensive overview of AI methods for spotting skin cancer in its earliest stages. High-quality research publications on skin cancer diagnostics were reviewed. Tools, graphs, tables, methods, and frameworks portray research findings.

Keywords Skin cancer · AI · DNA · ML · KNN · CNN · Neural networks

1 Introduction

This efficient survey paper has examined different brain network methods for skin disease location and characterization. These methods are harmless. Skin disease recognition requires various stages, for example, preprocessing and picture division, trailed by include extraction and arrangement. This survey zeroed in on ANNs, CNNs, KNNs, and RBFNs for order of sore pictures. Every calculation enjoys its benefits and inconveniences. Appropriate choice of the grouping method is the center point for best outcomes. In any case, CNN gives improved results than

N. P. Patnaik M (✉) · J. J. Beera
Department of Computer Science & Engineering, GITAM Deemed to be University, Visakhapatnam, Andhra Pradesh, India
e-mail: jbeera@gitam.edu

A. Patel et al. (eds.), *Advances in Machine Learning and Big Data Analytics II*, Springer Proceedings in Mathematics & Statistics 442,
https://doi.org/10.1007/978-3-031-51342-8_4

different sorts of brain networks while characterizing picture information since it is more firmly connected with PC vision than others. The majority of the examination connected with skin malignant growth identification centers around whether a given sore picture is dangerous. Notwithstanding, when a patient inquires as to whether a specific skin malignant growth side effect shows up on any piece of their body, the momentum research cannot give a response. Hitherto, the exploration has zeroed in on the tight issue of order of the sign picture. Future exploration can incorporate full-body photography to look for the solution to the inquiry that ordinarily emerges. Independent full-body photography will mechanize and accelerate the picture-getting stage. The thought of auto-association has as of late arisen in the area of profound learning. Auto-association alludes to the course of solo realizing, which plans to distinguish highlights and find relations or examples in the picture tests of the dataset. Under the umbrella of convolutional brain organizations, auto-association methods increment the degree of element portrayal that is recovered by master frameworks [47]. At present, auto-association is a model that is still in innovative work. Nonetheless, its review can work on the exactness of picture handling frameworks later on, especially in the space of clinical imaging, where the littlest subtleties of elements are very critical for the right conclusion of sickness.

The point of convergence of this overview is on the use of reenacted knowledge as a gadget that guides during the time spent on skin infection diagnostics. Consequently, man-made reasoning-based skin illness demonstrative devices use either shallow or significant PC-based knowledge procedures. Both incorporate changing PC estimations through cooperation called getting ready to acquire data molded by predefined features. What is important is that shallow strategies keep an eye out for not utilizing multi-layer cerebrum networks using any and all means or using such associations confined to basic layers [6]. Strangely, significant methodologies incorporate planning enormous, significant multi-layer cerebrum networks with numerous mystery layers, typically going from small bunches to hundreds [7] (Fig. 1).

Fig. 1 Skin structure

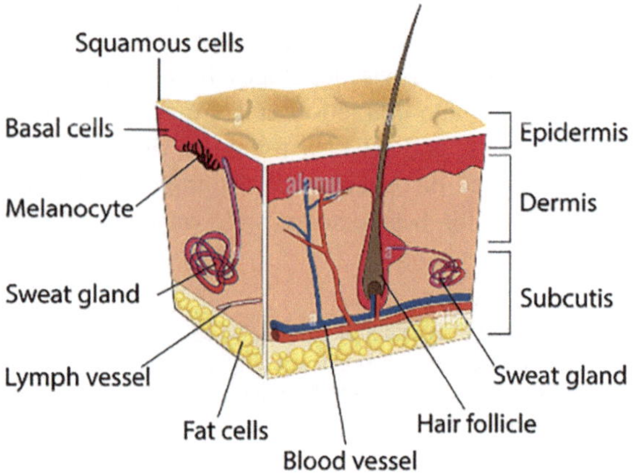

Most skin malignant growths are brought about by a lot of openness to bright (UV) beams. By shielding your skin from the sun's UV rays and counterfeit sources of skin cancer such as tanning beds and sunlamps, you can reduce your chances of contracting skin diseases.

Several factors can increase your chances of developing skin diseases including:

(a) Fair skin (anyone can develop skin diseases regardless of their skin tone).
(b) In the past, you have been burned by the sun.
(c) A tendency to excessively open the eyes to the sun.
(d) The environment must be sunny or high in elevation.
(e) Moles.
(f) Illnesses that cause cancer of the skin.
(g) The presence of malignant growths on the skin in the family.
(h) It is all diseases that prevent, trouble, or kindle your skin. A singular history of skin dangerous growth is included in skin diseases. The most common skin illnesses include rashes or changes to the appearance of your skin. Some skin illnesses are harmless, while others can cause severe side effects. There are several well-known skin illnesses, including:

    The presence of acne is caused by clogged follicles, which are a source of oil, microorganisms, and dead skin cells.
    Alopecia areata, a condition in which you lose hair gradually over time.
    There is also atopic dermatitis (inflammation of the skin), which causes dry, bothersome skin that expands, breaks, and flakes.

There are also psoriasis cases, where the skin feels hot and is layered.

(a) Raynaud's syndrome, in which your fingers, toes, or other body parts experience occasional decreases in blood circulation. This may lead to deadness or deterioration of the skin.
(b) Affected by Rosacea, flushing, a feeling of toughness, and pimples, usually on the face.
(c) A skin disease in which abnormal skin cells develop uncontrollably.
(d) Vitiligo is a condition in which patches of skin lose their color.

2 Literature Review

In the article "Computerized Division of Melanocytes in Skin Histopathological Pictures," C. Lu, M. Mahmood, N. Jha, and M. Mandal discuss how identifying melanocytes within the epidermis region is an important step in concluding skin melanoma by breaking down histopathological pictures. The identification of melanocytes in the epidermis, despite this, presents a problem, as other keratinocytes are essentially the same as the melanocytes. In this paper [1], we propose

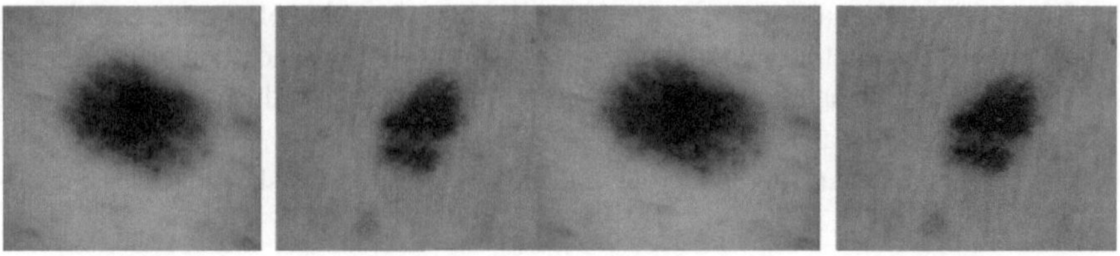

Fig. 2 Skin damage and basic identification of various parameters

a clever PC-aided strategy for dividing the melanocytes in skin histological pictures. This is achieved by applying a mean-shift calculation to the underlying division (Fig. 2).

A nearby locale recursive division calculation is then proposed to sift through the up-and-comer core districts given the space's earlier information. To recognize the melanocytes from other keratinocytes in the epidermis region, an original descriptor, named nearby twofold circle descriptor (LDED), is proposed to gauge the neighborhood elements of the competitor districts. The LDED utilizes two boundaries: district ellipticity and nearby example qualities to recognize the melanocytes from the applicant cores locales. Exploratory outcomes on 28 different histopathological pictures of skin tissue with various zooming factors show that the proposed method gives unrivaled execution. This study hopes to determine this issue by playing out a careful and direct review of the ongoing composition. We mean to address the ongoing man-made reasoning-based instruments used to perceive and bunch skin dangerous development (Fig. 3).

Hence, reenacted insight-based systems tend to be viewed as unobtrusive, easy to use, and open [5], thus avoiding the issues in the recently referenced existing disclosure procedures of skin-threatening development. Despite this, as the composition of the clinical use of PC-based insight rapidly creates and continues to announce revelations using inconsistent execution estimates, direct comparisons between previous sorts are extremely troublesome and hinder future investigations in the shortest amount of time possible.

Using joint Factual Surface Uniqueness to segment skin injuries from computerized pictures, J. Glaister, A. Wong, and D. A. Clausi have identified melanoma as the deadliest form of skin malignancy. It has been found that melanoma incidence rates are rising, particularly among white non-Hispanics, but when detected early, survival rates are high. As dermatologists incur substantial costs for screening every persistent, there is a need for a computerized framework to assess the chance of melanoma in a patient based on images taken using a standard computerized camera. Finding the skin sore in a computerized picture is one test in carrying out such a framework. We present a novel method for dividing skin injuries by the surface. An illumination-corrected photograph is used to learn representative texture distributions, and the distinctiveness metric for each distribution is calculated. Based on the occurrence of representative texture distributions in the image, normal skin and lesions are classified accordingly. We evaluate the proposed segmentation

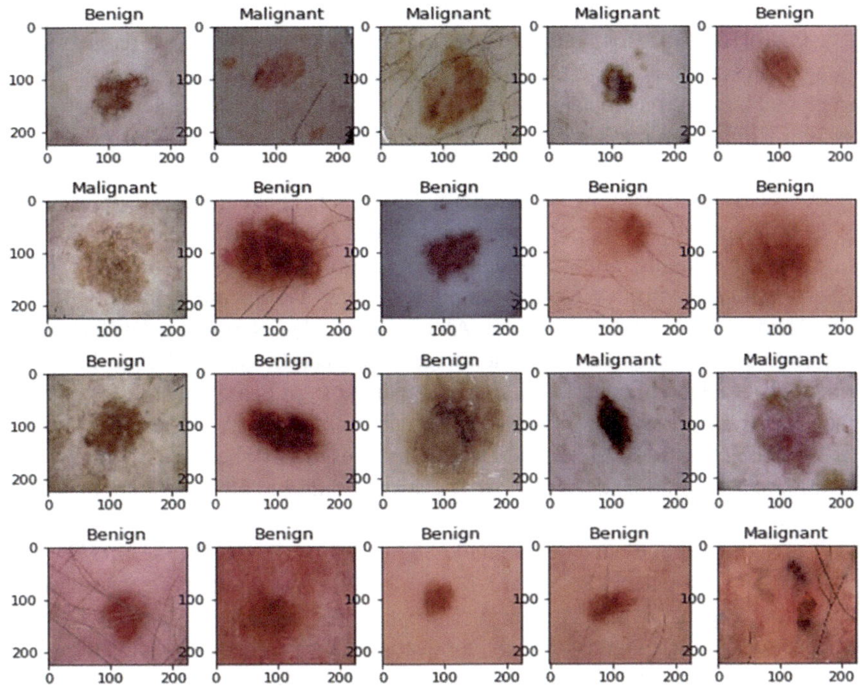

Fig. 3 Cancer image classification

framework by comparing lesion segmentation and melanoma classification results to the results obtained by other state-of-the-art algorithms.

The article "Noninvasive Real-Time Automated Skin Lesion Analysis System for Early Detection and Prevention of Melanoma" by O. Abuzaghleh, B. D. Barkana, and M. Faezipour shows how Melanoma spreads through metastasis, and as a result, has proven to be extremely fatal. A significant number of skin cancer deaths are caused by melanoma, according to statistical evidence. It has been found that the survival rate of melanoma depends on the stage of the cancer; early detection and intervention of the disease increase a patient's chances of survival. Although doctors' subjectivity makes it difficult to diagnose and prognosis melanoma accurately, the process is difficult to follow. To detect and prevent melanoma early, it is essential to analyze the shape, color, and texture of a skin lesion, since melanoma is asymmetrical, and has irregular borders, notched edges, and color variations. To detect and prevent melanoma early, two major components of a noninvasive and real-time automated skin lesion analysis system are presented in this paper. This system consists of two components: a real-time alert that warns users to be aware of sunburns; and a novel equation is used to determine how long it takes for the skin to burn. Among the features of the automated image analysis module are image acquisition, hair detection, exclusion, segmentation, feature extraction, and classification of images. To develop and test the proposed system [3], Pedro Hispano Hospital provided the PH2 Dermoscopy image database. The image database contains 200 dermoscopy images of lesions, ranging from benign to atypical, as well as melanoma. The experimental results show that the proposed system is efficient, classifying images with 96.3% accuracy, 95.7% accuracy, and 97.5% accuracy, respectively, for benign, atypical, and melanoma.

A system for classifying images dermoscopic for mobile devices is presented by F. E. S. Alencar, D. C. Lopes, and F. M. Mendes Neto in the article "Development of a System Classification of Images Dermoscopic for Mobile Devices." Several efforts were put into developing computer systems that facilitate the diagnosis of skin lesions, particularly melanoma. Alencar et al. [4] describe the development of a mobile device-based dermoscopic image classification system. Using the ABCD rule and its edge characteristics and color, the system classifies skin lesions, and then uses a backpropagation-trained MLP network to classify the lesions. The goal of the project is to develop a tool that can be used anywhere and by any health professional to conduct dermoscopy. The system achieved 66% sensitivity and 93% specificity, according to the results.

The authors of the paper "Automatic Segmentation of Wrist Bones in CT Using Statistical Wrist Shape $+$ Pose Model," published in IEEE Transactions on Medical Imaging, described how the wrist bones have been segmented in CT images for a wide range of clinical applications, such as arthritis evaluation, bone age assessment, and image-guided surgery. Non-uniformity and spongy texture of bone tissue, as well as tight space between bones, pose major challenges. A programmed wrist bone division method for CT pictures is presented in this paper [5] based on a factual model that captures the variations of wrist joint shapes and postures across 60 model wrists at nine different wrist positions. Based on a Gaussian Blend Model, a gathering-wise enrollment system is used to adjust the wrist bone surfaces at nonpartisan positions to lay out the correspondence between the preparation shapes. The significant methods of shape varieties are determined by examining the head part, and then integrating them into two posture models based on the variations present across the population and various wrist positions. By analyzing the similarity changes at all wrist positions across the population, an intra-subject posture model can be developed. A between-subject posture model is also used to show posture variations across wrist positions. In CT images, the created model is enrolled with the edge point cloud extracted from the CT volume using an assumption boost-based probabilistic method to divide the wrist bones. An enrollment strategy that is not inflexible is used to remedy lingering enrolment mistakes. We validate the proposed division technique by employing 66 concealed CT volumes of a normal voxel size of 0.38 mm to enlist the wrist model. According to our results, we recorded a mean surface distance error of 0.33 mm and a mean Jaccard list of 0.86 mm.

Y. Yuan, M. Chao, and Y. C. Lo's "Programmed Skin Sore Division Utilizing Profound Completely Convolutional Organizations With Jaccard Distance," which addressed programmed skin injury division in dermoscopic pictures is a moving undertaking because of the low differentiation among injury and the encompassing skin, the sporadic and fluffy sore lines, the presence of different curios, and different imaging obtaining conditions. In this paper [7], a completely programmed strategy for skin sore division by utilizing 19-layer profound convolutional brain networks that is prepared from start to finish and does not depend on earlier information. We propose a bunch of procedures to guarantee powerful and effective learning with restricted preparation of information. Besides, we plan a clever misfortune capability in view of Jaccard distance to dispose of the need for test re-weighting, a run-of-the-mill system while involving cross entropy as the misfortune capability for picture division because of the solid unevenness between the quantity of closer view and establishment pixels. We surveyed the proposed structure's ampleness, productivity, and speculation ability on two openly accessible information bases. One is from the ISBI 2016 skin injury examination towards the melanoma recognition challenge, and the other is the PH2 information base. Trial results showed that the proposed strategy beat other cutting-edge calculations on these two information bases. This strategy is sufficiently general and just necessities least pre-and post-handling, which permits its reception in an assortment of clinical picture division undertakings.

E. Ahn et al.'s "Saliency-Based Sore Division Through Foundation Discovery in Dermoscopic Pictures," which addressed the division of skin injuries in dermoscopic pictures is a major move toward robotized PC-helped conclusion of melanoma. Customary division techniques, in any case, experience issues when the sore lines are undefined and when the differentiation between the sore and the encompassing skin is low. They likewise perform ineffectively when there is a heterogeneous foundation or a sore that contacts the picture limits; this then, at that point, brings about under and over-division of the skin injury [8]. Here, it was proposed that saliency location utilizing the reproduction mistakes from a meager portrayal model combined with an original foundation identification can even more precisely segregate the injury from encompassing districts. Further proposed a Bayesian structure that better depicts the shape and limits of the sore. Here, it is likewise assessed the methodology on two public datasets including 1100 dermoscopic pictures, and contrasted with other customary and cutting-edge solo (i.e., no preparation required) injury division strategies, as well as the best-in-class unaided saliency discovery techniques. The outcomes show that this approach is more exact and stronger in dividing sores contrasted with different techniques.

A. F. Jerant, J. T. Johnson, C. Sheridan, T. J. Caffrey, et al.'s "Early identification and therapy of skin malignant growth," which addressed the division of skin sores is a significant stage in the robotized PC-helped finding of melanoma. Notwithstanding, existing division strategies tend to over-or under-fragment the sores and perform ineffectively when the injuries have fluffy limits, low differentiation with the foundation, inhomogeneous surfaces, or contain antiques. Besides, the exhibition of these strategies is intensely dependent on the suitable tuning of an enormous number of boundaries as well as the utilization of successful preprocessing methods, like light revision and hair evacuation. Bi et al. [9] proposed the use of completely convolutional networks (FCNs) to section the skin injuries naturally.

Using FCNs, brain networks consolidate low-level appearance data with undeniable level semantic data to locate objects. Developed a new equal reconciliation strategy to combine correlative data obtained from individual division stages to achieve a last division result with precise confinement and distinct injury limits, for even the most difficult wounds. A normal Dice coefficient of 9.18% was achieved for the ISBI 2016 Skin Sore Test dataset and a normal Dice coefficient of 9.66% for the PH2 dataset. A large number of exploratory results have been obtained comparing this method to other current methods for segmenting skin lesions, based on two publicly available benchmark datasets. In their paper "Noninvasive Real-Time Automated Skin Lesion Analysis System for Melanoma Early Detection and Prevention," O. Abuzaghleh, B. D. Barkana, and M. Faezipour indicated that melanoma is one of the deadliest types of skin cancer. Early detection may be enough to cure the disease, but highly trained specialists can identify it accurately. Since expertise is scarce, automated systems can identify diseases and save lives, reduce unnecessary biopsies, and reduce costs.

3 Research Gap Identified

According to the writing introduced, the importance and complexity of planning a continuous skin sore division method are determined by preprocessing and effective highlights, such as surface, variety, shape, and low-level elements. Various skin sore division models, including LDED, AI, FCRN, FCN, Bayesian Structure, and MLP, are used to counter the difficulty of identifying skin sore divisions in Melanoma. It is difficult to differentiate skin sores precisely in melanoma pictures. Customary division strategies can cause problems when the lines are virtually indistinguishable and when the contrast between the injured and surrounding skin is low. Furthermore, they fail to work effectively when there is a heterogeneous foundation or a sore that contacts the picture limits; this results in an under-division and over-division of the skin sore. To our knowledge, there are not a lot of methods that can provide mechanized skin sore division utilizing surface, variety, shape, and low-level highlights in melanoma pictures, which will give adequacy, proficiency, and speculation capability with minimal delay as far as we are aware.

4 Related Work

Despite being educated about skin disease risks and how to avoid them, the number of skin malignant growths/melanoma continues to grow at an alarming rate. Melanoma rates have been rising unexpectedly in Nordic nations since 1955, as skin melanoma is one of the fastest-growing diseases in Western nations. Since UV radiation damages the skin, this openness began practically stepwise a few decades ago, affecting people of all ages. A majority of melanoma cases were found on sun-exposed body parts, such as the head and feet, at the beginning of the twentieth century. After 1955, melanoma developed faster in areas that were rarely covered. Swedish studies have found a greater incidence of melanoma at lower scopes, whereas northern studies have found a lesser incidence.

This means that mechanized skin division plays an important role when determining Melanoma location. Various condition-of-at-strategies, however, degrade their presentation and prevent detection while degrading outcomes. During the fragmentation of melanoma, fundamental issues commonly occur: grouping, object discovery, object division, lack of preparation, covering issues, object location-related issues, and vascular division.

5 Conclusion

In general, the people groups are not aware of a few things about managing their skin health. It is only a matter of time before the people groups realize their concern is more serious than they think. Although the expert's findings are solid, it takes a great deal of time and effort. The division of skin sores in Melanoma dermoscopy images is an extremely difficult and time-consuming procedure that can be automated. As much as 90% of early analyses are reparable, as little as half of late ones are. Because of the low difference in skin sores, the enormous intraclass variation of melanomas, and the serious level of visual comparability between sores of melanoma and nonmelanoma, computerized melanoma recognition in dermoscopy pictures is a moving undertaking. Therefore, addressing these issues and providing better assistance for the conclusion of skin melanoma is crucial. Future research will focus on developing strategies to counter these issues during the early detection of skin malignancies.

References

1. C. Lu, M. Mahmood, N. Jha, and M. Mandal, "Automated Segmentation of the Melanocytes in Skin Histopathological Images," in IEEE Journal of Biomedical and Health Informatics, vol. 17, no. 2, pp. 284–296, March 2013.
2. J. Glaister, A. Wong, and D. A. Clausi, "Segmentation of Skin Lesions from Digital Images Using Joint Statistical Texture Distinctiveness," in IEEE Transactions on Biomedical Engineering, vol. 61, no. 4, pp. 1220–1230, April 2014.
3. O. Abuzaghleh, B. D. Barkana, and M. Faezipour, "Noninvasive Real-Time Automated Skin Lesion Analysis System for Melanoma Early Detection and Prevention," in IEEE Journal of Translational Engineering in Health and Medicine, vol. 3, pp. 1–12, 2015.
4. F. E. S. Alencar, D. C. Lopes, and F. M. Mendes Neto, "Development of a System Classification of Images Dermoscopic for Mobile Devices," in IEEE Latin America Transactions, vol. 14, no. 1, pp. 325–330, Jan. 2016.
5. E. M. A. Anas et al., "Automatic Segmentation of Wrist Bones in CT Using a Statistical Wrist Shape $+$ Pose Model," in IEEE Transactions on Medical Imaging, vol. 35, no. 8, pp. 1789–1801, Aug. 2016.

6. R. Kasmi and K. Mokrani, "Classification of malignant melanoma and benign skin lesions: implementation of automatic ABCD rule," in IET Image Processing, vol. 10, no. 6, pp. 448–455, 6 2016.
7. Y. Yuan, M. Chao, and Y. C. Lo, "Automatic Skin Lesion Segmentation Using Deep Fully Convolutional Networks With Jaccard Distance," in IEEE Transactions on Medical Imaging, vol. 36, no. 9, pp. 1876–1886, Sept. 2017.
8. E. Ahn et al., "Saliency-Based Lesion Segmentation Via Background Detection in Dermoscopic Images," in IEEE Journal of Biomedical and Health Informatics, vol. 21, no. 6, pp. 1685–1693, Nov. 2017.
9. L. Bi, J. Kim, E. Ahn, A. Kumar, M. Fulham, and D. Feng, "Dermoscopic Image Segmentation via Multistage Fully Convolutional Networks," in IEEE Transactions on Biomedical Engineering, vol. 64, no. 9, pp. 2065–2074, Sept. 2017.
10. L. Yu, H. Chen, Q. Dou, J. Qin, and P. A. Heng, "Automated Melanoma Recognition in Dermoscopy Images via Very Deep Residual Networks," in IEEE Transactions on Medical Imaging, vol. 36, no. 4, pp. 994–1004, April 2017.
11. N. C. F. Codella et al., "Deep learning ensembles for melanoma recognition in dermoscopy images," in IBM Journal of Research and Development, vol. 61, no. 4, pp. 5:1–5:15, July-Sept. 1, 2017.
12. Julian Sföttinger et al., Skin Paths for Contextual Flagging Adult Videos, in Advances in Visual Computing, Springer Berlin Heidelberg, vol. 5876, pp. 303–314, 2009.
13. Mohammad Saber I., and Ali Y., Skin Color Segmentation in Fuzz YCBCR Color Space with the Mamdani Inference, in American Journal of Scientific Research, July (2011), pp. 131–137, 2011.
14. F. A. Bahmer, P. Fritsch, J. Kreusch, H. Pehamberger, C. Rohrer, I. Schindera, et al., "Terminology in surface microscopy. Consensus meeting of the Committee on Analytical Morphology of the Arbeitsgemeinschaft Dermatologische For schung, Hamburg, Federal Republic of Germany, Nov. 17, 1989," J Am Acad Dermatol, vol. 23, pp. 1159–1162, Dec 1990.
15. A. F. Jerant, J. T. Johnson, C. Sheridan, T. J. Caffrey et al., "Early detection and treatment of skin cancer," American family physician, vol. 62, no. 2, pp. 357–386, 2000.
16. R. L. Siegel, K. D. Miller, and A. Jemal, "Cancer statistics, 2016," CA: A cancer journal for clinicians, 2015.
17. C. M. Balch, A. C. Buzaid, S.-J. Soong, M. B. Atkins, N. Cascinelli, D. G. Coit, I. D. Fleming, J. E. Gershenwald, A. Houghton, J. M. Kirkwood et al., "Final version of the American joint committee on cancer staging system for cutaneous melanoma," Journal of Clinical Oncology, vol. 19, no. 16, pp. 3635–3648, 2001.
18. R. Garnavi, M. Aldeen, M. E. Celebi, G. Varigos, and S. Finch, "Border detection in dermoscopy images using hybrid thresholding on optimized color channels," Comput. Med. Imag. Graph., vol. 35, no. 2, pp. 105–115, 2011.
19. C. Li, C. Y. Kao, J. C. Gore, and Z. Ding, "Minimization of region-scalable fitting energy for image segmentation," IEEE Trans. Image Process., vol. 17, no. 10, pp. 1940–1949, Oct. 2008.
20. B. Erkol, R. H. Moss, R. Stanley, W. V. Stoecker, and E. Hvatum, "Automatic lesion boundary detection in dermoscopy images using gradient vector flow snakes," Skin Res. Technol., vol. 11, no. 1, pp. 17–26, 2005.
21. J. Long, E. Shelhamer, and T. Darrell, "Fully convolutional networks for semantic segmentation," in Proc. IEEE Conf. Comput. Vis. Pattern Recognit., 2015, pp. 3431–3440.
22. D. S. Rigel, et al., "The incidence of malignant melanoma in the United States: issues as we approach the 21st century," Journal of the American Academy of Dermatology, vol. 34, pp. 839–847, 1996.
23. G. Argenziano, et al., "Dermoscopy, An Interactive Atlas," EDRA Medical Publishing, 2000.
24. M. Binder, et al., "Epiluminescence microscopy: A useful tool for the diagnosis of pigmented skin lesions for formally trained dermatologists," Archives of Dermatology, vol. 131, pp. 286–291, 1995.

25. P. Wighton, et al., "A fully automatic random walker segmentation for skin lesions in a supervised setting," in Medical Image Computing and Computer-Assisted Intervention–MICCAI 2009, ed: Springer, 2009, pp. 1108–1115.
26. American cancer society. Cancer facts & figures 2013; Available athttp://www.cancer.org/acs/groups/content/@epidemiologysurveilance/documents/document/acspc-036845.pdf.
27. J Nalepa, T Grzejszczak, and M Kawulok, Wrist Localization in Color Images for Hand Gesture Recognition, in Advances in Intelligent Systems and Computing, Springer International Publishing, vol. 242, pp. 79–86, 2014.
28. Erdem, C.E., Ulukaya, S., Karaali, A., and Erdem, A.T., Combining Haar Feature and skin color based classifiers for face detection, in ICASSP 2011, pp. 1497–1500, 2011.
29. Chan C.S., Liu H., and Brown, D.J., Recognition of human motion from qualitative normalised templates, J. Intell. Robot. Syst., vol. 48(1), pp. 79–95, 2007.
30. Zui Z, Gunes, H., and Piccardi, M., Head detection for video surveillance based on categorical hair and skin colour models, in Image Processing (ICIP) 2009, pp. 1137–1140, 2009.
31. H. Chen, et al., "DCAN: Deep Contour-Aware Networks for Accurate Gland Segmentation," in IEEE conference on Computer Vision and Pattern Recognition, 2016, pp. 2487–96.

A Review of Ranking Approach to Rank Interval-Valued Trapezoidal Intuitionistic Fuzzy Sets

S. N. Murty Kodukulla and V. Sireesha

Abstract An Interval-valued trapezoidal intuitionistic fuzzy set (IVTrIFS) is a powerful tool in modeling uncertainty. It is a special type of Intuitionistic Fuzzy Set (IFS) and interval-valued intuitionistic fuzzy set (IVIFS), which has a consecutive domain of real numbers. An IVTrIFSs is distinguished by using three characteristic functions namely membership degree, non-membership degree, and hesitancy degree. Ranking is a challenging task, and every ranking method has its own significance and applicability. Since their inception in 1965, several researchers have proposed various ranking methods by using concepts such as Score and Accuracy, Centroids, Centre of Gravity, Distance/Similarity measures, Value and Ambiguity, etc. yet there is no specific ranking approach that is appropriate for all applications or provides adequate results in all situations. The Centre of Gravity (COG) method is one of the most popular defuzzification techniques of fuzzy mathematics. By utilizing the notion of the COG, one can ascertain the central tendency or representative value of a given fuzzy set. As it is derived by using area and centroids of any geometric representation, this method is particularly useful in fuzzy logic systems, fuzzy control, decision-making processes, and fuzzy data analysis. In this paper, we derive a ranking method for IVTrIFS using the Centre of Gravity.

Keywords Intuitionistic Fuzzy Set (IFS) · Ranking of interval-valued Intuitionistic trapezoidal fuzzy set (IVTrIFS) · Centre of gravity

S. N. Murty Kodukulla (✉) · V. Sireesha
Department of Mathematics, GITAM Deemed to be University, Visakhapatnam, Andhra Pradesh, India
e-mail: sveerama@gitam.edu

A. Patel et al. (eds.), *Advances in Machine Learning and Big Data Analytics II*, Springer Proceedings in Mathematics & Statistics 442,
https://doi.org/10.1007/978-3-031-51342-8_5

1 Introduction and Review of Literature

The Centre of Gravity (COG) method is one of the popular defuzzification techniques of fuzzy theory. The COG in fuzzy sets is particularly useful in fuzzy logic systems, fuzzy control, decision-making processes, and fuzzy data analysis, where there is a need to interpret and make decisions based on uncertain or imprecise data.

The COG approach and the Trapezoidal Fuzzy Assessment Model (TrFAM) were first proposed by Subbotin [1], who also attempted to measure the quality performance of a student group using these models. They concluded that the COG approach is more precise than the existing conventional approaches. Additionally, Michael Gr. Voskoglou and Subbotin [2] worked on applying the COG approach in the Triangular Fuzzy Assessment Model (TFAM) and replaced isosceles triangles having common parts for the rectangles that appeared in the COG technique's graph. Subsequently, Michael Gr. Voskoglou [3] evaluated the TFAM and COG techniques by using them to assess students and Bridge players and contrasting them with other conventional approaches. In 2019 a new method to calculate the COG of Generalized Trapezoidal Fuzzy Numbers (GTFNs) is put forward by Peng et al. concluded that the new approach of computing the COGs of GTFNs is more flexible and reasonable as it adequately considers the definition of membership function.

In fuzzy theory, hesitancy is used to represent the degree of uncertainty in making a decision. It is a mathematical expression that quantifies the level of hesitation or reluctance in assigning a membership and non-membership value to a particular fuzzy set. In many fuzzy systems, particularly those involving decision-making processes, there might be situations where the input data is ambiguous or conflicting, leading to uncertainty in the output. The exact form of the hesitation can vary depending on the specific application and the nature of the uncertainty being modeled. It may be defined based on empirical data, expert knowledge, or through mathematical formulations tailored to the problem domain.

Since the hesitancy includes the membership and non-membership values of IVTrIFS. In the present study, a new method for ranking method is introduced by using the COG of hesitancy region.

2 Proposed Ranking Method

For any IVTrIFS, $\tilde{\alpha} = (0, 0, 0, 0)\,; \left[m_{\tilde{\alpha}}{}^{L}, m_{\tilde{\alpha}}^{U}\right], \left[n_{\tilde{\alpha}}{}^{L}, n_{\tilde{\alpha}}^{U}\right]\Big)$, the hesitancy region of IVTrIFS is divided into three parts namely S_1, S_2, and S_3 in which S_1 and S_3 are triangles and S_2 is a rectangle. The COG (x, y) is then calculated through the hesitancy region S_1, S_2, and S_3.

Here, the ranking function is defined as the Euclidian distance between x and y. The geometric representation of the hesitancy region is shown in Fig. 1.

The region S_1 is a Triangle with coordinate points: $(a, 0)$, $(b, 1 - m^U - n^U)$, $(b, 1 - m^L - n^L)$, S_2 is a Rectangle with coordinate points: $(b, 1 - m^L - n^L)$, (b,

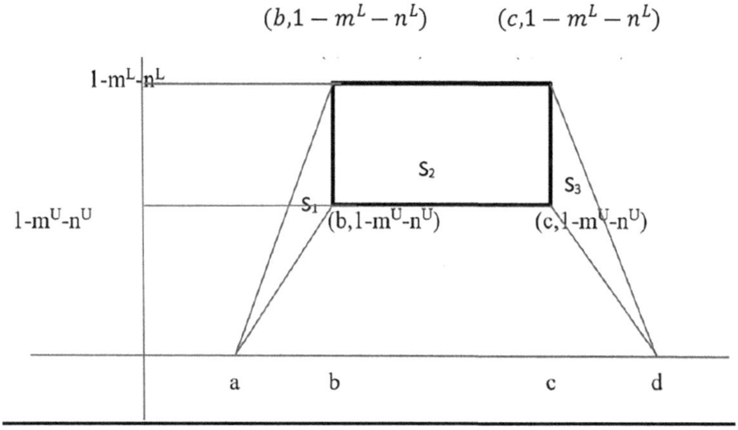

Fig. 1 Hesitancy region of IVTrIFS

$1 - m^U - n^U$), (c, $1 - m^U - n^U$), (c, $1 - m^L - n^L$) and S_3 is a Triangle with coordinate points: (c, $1 - m^U - n^U$), (c, $1 - m^L - n^L$), $(d, 0)$.

The area of S_1, S_2 and S_3 are respectively obtained as $(a - b)$ $(m^U + n^U - m^L - n^L)$, $2(b - c)$ $(m^U + n^U - m^L - n^L)$ and $(c - d)$ $(m^U + n^U - m^L - n^L)$

The Centre of gravities of S_1, S_2 and S_3 are respectively obtained as

$$(x_1, y_1) = \left[\frac{a + 2b}{3}, \frac{2 - m^U - n^U - m^L - n^L}{3}\right],$$

$$(x_2, y_2) = \left[\frac{b + c}{2}, \frac{2 - m^U - n^U - m^L - n^L}{2}\right]$$

and

$$(x_3, y_3) = \left[\frac{2c + d}{3}, \frac{2 - m^U - n^U - m^L - n^L}{3}\right]$$

The total area of the trapezoid is $S=$ the area of $(S_1 + S_2 + S_3)$

$$= (a\text{–}b)\left(m^U + n^U - m^L - n^L\right) + 2\,(b\text{–}c)\left(m^U + n^U - m^L - n^L\right) + (c\text{–}d)\left(m^U + n^U - m^L - n^L\right)$$

$$= (\,a + b - c - d)\,\left(m^U + n^U - m^L - n^L\right)$$

The COG of S is (x, y)

Where $x = \frac{1}{S}\sum(S_i x_i) = \frac{1}{S}(S_1x_1 + S_2x_2 + S_3x_3)$ and $y = \frac{1}{S}\sum(S_i y_i) = \frac{1}{S}(S_1y_1 + S_2y_2 + S_3y_3)$

By using the above equations, we have the COG

$$(x, y) = \left[\frac{a^2+b^2-c^2-d^2+ab-cd}{3(a+b-c-d)};\frac{(a+2b-2c-d)\left(2-m^U-n^U-m^L-n^L\right)}{3(a+b-c-d)}\right]$$

Where $a \neq b \neq c \neq d$, as in this case, there will be no hesitancy and it will be considered as a crisp number.

Note: If $a = b = c = d$ then its membership value will be 1 and hesitancy becomes 0. In such cases, the values will be treated as crisp numbers and the set with maximum crisp value will be preferred.

The ranking function is defined as: $H_d\left(\tilde{\alpha}\right) = \sqrt{x^2 + y^2}$

In the proposed approach, $H_d\left(\tilde{\alpha}\right)$ represents the distance from the origin at which there is no hesitancy to the COG of IVTrIFS. Hence, a set at a minimum distance from the origin means it is close to the set with minimum hesitancy and thus preferred the most.

For any two IVITFSs $\tilde{\alpha}_1, \tilde{\alpha}_2$ the ranking is defined as

(i) If $H_d\left(\tilde{\alpha}_1\right) > H_d\left(\tilde{\alpha}_2\right)$ then $\tilde{\alpha}_1 < \tilde{\alpha}_2$
(ii) If $H_d\left(\tilde{\alpha}_1\right) < H_d\left(\tilde{\alpha}_2\right)$ then $\tilde{\alpha}_1 > \tilde{\alpha}_2$
(iii) If $H_d\left(\tilde{\alpha}_1\right) = H_d\left(\tilde{\alpha}_2\right)$ then $\tilde{\alpha}_1 = \tilde{\alpha}_2$.

Proposition 1 The ranking function satisfies the following distance measure properties.

(i) $H_d\left(\tilde{\alpha}\right) \geq 0$ *for all* $\tilde{\alpha}$
(ii) $H_d\left(\tilde{\alpha}\right) = 0$ *iff* $\tilde{\alpha} = 0$
(iii) $H_d(\tilde{\alpha}_1) = H_d(\tilde{\alpha}_2)$ *iff* $\tilde{\alpha}_1 = \tilde{\alpha}_2$
(iv) $H_d(k\tilde{\alpha}\left.\right) = k.\ H_d(\tilde{\alpha})$ *for* $k \geq 0$

Proof:

The COG of hesitancy function of IVTrIFS given by

$$(x, y) = \left[\frac{a^2+b^2-c^2-d^2+ab-cd}{3(a+b-c-d)};\frac{(a+2b-2c-d)\left(2-m^U-n^U-m^L-n^L\right)}{3(a+b-c-d)}\right]$$

(i) Since $H_d(\tilde{\alpha})$ is the distance from the origin to the COG of hesitancy function of IVTrIFS and the distance of any set should not be negative. Hence by the definition $H_d\left(\tilde{\alpha}\right) \geq 0$ *for all* $\tilde{\alpha}$.

(ii) If $\tilde{\alpha} = (0, 0, 0, 0); \left[m_{\tilde{\alpha}}^{L}, m_{\tilde{\alpha}}^{U}\right], \left[n_{\tilde{\alpha}}^{L}, n_{\tilde{\alpha}}^{U}\right])$ then the value of $X = 0$ and $Y = 0$ and hence by the definition of $H_d\left(\tilde{\alpha}\right) = 0$.

(iii) Let $\tilde{\alpha}_1 = ([a_1, b_1, c_1, d_1]; [m_{\alpha_1}^{L}, m_{\alpha_1}^{U}]; \left[n_{\alpha_1}^{L}, n_{\alpha_1}^{U}\right])$ and $\tilde{\alpha}_2 = ([a_2, b_2, c_2, d_2]; [m_{\alpha_2}^{L}, m_{\alpha_2}^{U}]; \left[n_{\alpha_2}^{L}, n_{\alpha_2}^{U}\right])$

if $\tilde{\alpha}_1 = \tilde{\alpha}_2$ then $a_1 = a_2; b_1 = b_2; c_1 = c_2; d_1 = d_2$ and $m_{\tilde{\alpha}_1}^{L} = m_{\alpha_2}^{L}, m_{\alpha_1}^{U} = m_{\alpha_2}^{U}, n_{\alpha_1}^{L} = n_{\alpha_2}^{L}, n_{\alpha_1}^{U} = n_{\alpha_2}^{U}$. Clearly by the definition of $H_d(\alpha)$ we have $H_d(\tilde{\alpha}_1) = H_d(\tilde{\alpha}_2)$.

(iv) $H_d\left(k\tilde{\alpha}\right) = \sqrt{(k.X)^2 + (k.Y)^2} = \sqrt{k^2.X^2 + k^2.Y^2} = \sqrt{k^2.\left(X^2 + Y^2\right)}$
$= k\sqrt{X^2 + Y^2} = k.\,H_d\left(\tilde{\alpha}\right)$

Numerical Example: 1 Let $\tilde{\alpha}_1 = ([0.5, 0.6, 0.7, 0.75]; [1, 1]; [0, 0])$ and $\tilde{\alpha}_2 = ([0.45, 0.65, 0.7, 0.75]; [1, 1; [0, 0])$ be two IVTrIFS.

By using Wan [4]

$H\left(\tilde{\alpha}_1\right) = 0.6375$, $H\left(\tilde{\alpha}_2\right) = 0.6375$ which implies $\tilde{\alpha}_1 = \tilde{\alpha}_2$

By using Wu and Liu [5]

$I\,(Sx\,(\tilde{\alpha}_1)) = 0.6375, (Sx\,(\tilde{\alpha}_2)) = 0.6375,$
$I\,(Hx\,(\tilde{\alpha}_1)) = 0.6375, (Hx\,(\tilde{\alpha}_2)) = 0.6375$
Which implies $\tilde{\alpha}_1 = \tilde{\alpha}_2$

By using Dong and Wan [6]

$S\left(\tilde{\alpha}_1\right) = 0.26$ and $S\left(\tilde{\alpha}_2\right) = 0.21$ which implies that $\tilde{\alpha}_1 > \tilde{\alpha}_2$

By using Sireesha and Himabindu [7]

$V\left(\tilde{\alpha}_1\right) = 0.3208$, $V\left(\tilde{\alpha}_2\right) = 0.325$, which implies that $\tilde{\alpha}_2 > \tilde{\alpha}_1$

By using the proposed method

Consider $\tilde{\alpha}_1 = ([0.5, 0.6, 0.7, 0.75]; [1, 1]; [0, 0])$

$x = 0.6358$ and $y = 0$ then $H_d\,(\tilde{\alpha}_1) = \surd\,((0.6357)^2 + (0)^2) = 0.6357$.

Consider $\tilde{\alpha}_2 = ([0.45, 0.65, 0.7, 0.75]; [1, 1; [0, 0])$

$x = 0.6285$ and $y = 0$ then $H_d(\widetilde{\alpha_2}) = \sqrt{((0.6285)^2 + (0)^2)} = 0.6285$.

Clearly $H_d\left(\widetilde{\alpha_1}\right) > H_d\left(\widetilde{\alpha_2}\right)$ which implies that $\widetilde{\alpha_1} < \widetilde{\alpha_2}$

3 Conclusion

Ranking is a challenging task, and every ranking method has its own significance and applicability. Numerous researchers proposed various ranking methods, and stated that there is no unique ranking method that is suitable for all kinds of applications. COG is a well-proven approach for ranking in many areas and several researchers proposed various ranking methods in fuzzy sets, IFS, IVIFS, and in many other domains. In this chapter, we attempted to rank IVTrIFSs using COG of the Hesitancy region. The proposed ranking method satisfied the distance measure properties. The comparison of ranking methods depicts that the proposed ranking approach exactly coincides with an existing ranking method based on Score and Accuracy defined by Jiang and Wang (2014) and Sireesha and Himabindu [7]. Whereas, it is almost consistent and performs better than the ranking methods defined by Wan [4], Vu and Liu (2013) and Dong and Wan [6] in specific cases. Therefore, the proposed ranking method offers an alternative approach to ranking the IVTrIFSs.

References

1. Subbotin, I., & Badkoobehi, H. (2004). Application of fuzzy logic to learning assessment.
2. Voskoglou, M. Gr, Subbotin, I. Ya, Dealing with the Fuzziness of Human Reasoning, International Journal of Applications of Fuzzy Sets and Artificial Intelligence, 3, 91–106, 2013.
3. Voskoglou, M. G. (2015). An application of triangular fuzzy numbers to learning assessment.
4. Wan S P (2011), Multi-attribute decision making method based on interval-valued intuitionistic trapezoidal fuzzy number, Control and Decision, vol. 26, no. 6, pp. 857–860, 2011.
5. Wu J and Liu Y (2013), An approach for multiple attribute group decision making problems with interval-valued intuitionistic trapezoidal fuzzy numbers, Computers & Industrial Engineering, vol. 66, no. 2, pp. 311–324.
6. Dong J and Wan S (2015), Interval-valued trapezoidal intuitionistic fuzzy generalized aggregation operators and application to multi attribute group decision making, Scientia Iranica E, vol. 22(6), pp. 2702–2715.
7. Sireesha V. and Himabindu K. An ELECTRE Approach for Multi-criteria Interval-Valued Intuitionistic Trapezoidal Fuzzy Group Decision Making Problems Vol. 2016, Article ID:1956303, http://dx.doi.org/10.1155/2016/1956303. Volume 156, Issue 2, 16 July 2004, Pages 445–455.

An Empirical Evaluation of ResNet-SE-16 for Accurate Classification of Lung Cancer Using Histopathological Images

P. Vishal, D. Kishore Kalyan Kumar, K. Venu Kiran Raju, K. Jayalaxmi Manojna, and Mohan Mahanty

Abstract Lung cancer is a leading cause of cancer-related deaths worldwide, accounting for approximately 25% of all cancer-related deaths. It is much more deadly than colon, breast, and prostate cancers combined. Early detection and treatment of cancer is crucial for a patient's recovery. Radiologists use histopathological images to diagnose potentially infected lung regions, but this process can be time-consuming. Deep learning, a type of machine learning, can hasten this process by simulating how the human brain functions. Convolutional Neural Networks (CNNs) can quickly and accurately classify and identify the various kinds of lung cancer, improving patient treatment and survival chances. While many CNN models have been developed by researchers, they often require significant computational power and time. In this project, we developed an improved CNN model that classifies lung cancer types with high accuracy while requiring less computational power than other models. Using metrics like accuracy, precision, recall, and F1 score, our model obtained an accuracy of 98% during training and 97% during validation.

Keywords Deep learning · Convolution neural network · Lung cancer · Resnet · Squeeze and excitation · Histopathological images

1 Introduction

Lung cancer is a highly malignant form of cancer affecting both males and females. It can be classified into two primary types: non-small cell lung cancer (NSCLC) and small cell lung cancer (SCLC). NSCLC, accounting for approximately 85% of cases, has subtypes including adenocarcinoma, squamous cell carcinoma, and large cell carcinoma. Adenocarcinoma is the most common subtype and is often

P. Vishal · D. K. K. Kumar · K. V. K. Raju · K. J. Manojna · M. Mahanty (✉)
Department of Computer Science and Engineering, Vignan's Institute of Information Technology (A), Visakhapatnam, Andhra Pradesh, India

A. Patel et al. (eds.), *Advances in Machine Learning and Big Data Analytics II*,
Springer Proceedings in Mathematics & Statistics 442,
https://doi.org/10.1007/978-3-031-51342-8_6

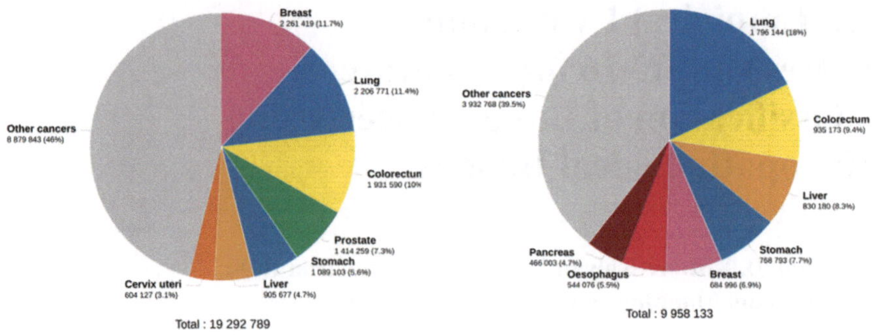

a. Incidence [1] **b.** Mortality [1]

Fig. 1 Incidence and mortality rate of lung cancer. (**a**). Incidence [1]. (**b**). Mortality [1]

found in the outer regions of the lungs. Squamous cell carcinoma, linked to smoking, typically occurs in the middle of the lungs. Large cell carcinoma, a less common subtype, can develop anywhere in the lung. Lung cancer, responsible for 18% of cancer-related deaths, is primarily caused by smoking but can also be triggered by factors such as radon exposure, air pollution, and asbestos exposure. A mortality rate of 18% is observed based on recent studies as shown in Fig. 1.

The objectives of the study are listed below.

1. The main objective of this project is to create a lung cancer classification model that can identify lung cancer with accuracy.
2. To accurately classify lung cancer based on histopathological images.
3. To build a model that requires less computational power and less time.

2 Literature Review

In their paper, Sanidhya Mangal et al. [2] presented a model that used the LC25000 dataset of 5000 images. To categorize lung cancer, they employed a shallow neural network design. They discovered that the system they created had a diagnostic accuracy of 96% for recognizing colon adenocarcinomas and more than 97% for detecting lung squamous cell carcinomas and adenocarcinomas.

In their paper, Bijaya Kumar Hatuwal et al. [3] presented a CNN deep learning model that classified histopathological images into 3 categories using the LC25000 dataset. The model they developed attained a training accuracy of 96.11% and a validation accuracy of 97.2%. M. Abbas et al. [4] presented a model that utilized multiple pre-trained convolutional neural networks, which had been previously trained on the ImageNet dataset, to classify histopathological slides into three categories: benign lung tissue, squamous cell carcinoma-lung, and adenocarcinoma-lung. On the test dataset, the F-1 scores achieved by the various networks, including AlexNet, VGG-19, ResNet-18, ResNet-34, ResNet-50, and ResNet-101, were 0.973, 0.997, 0.986, 0.992, 0.999, and 0.999, respectively. In their paper, Satvik Garg et al. utilized eight well-known pre-trained CNN models [5] (namely, VGG16, ResNet50, MobileNet, InceptionResNetV2, NASNetMobile, Xception, InceptionV3, and DenseNet169) and achieved remarkable results, with accuracy ranging from 97.5% to 100%. In particular, the InceptionResNetV2, InceptionV3, and MobileNet models achieved perfect scores of 100% for precision, accuracy, AUROC score, recall, and F1 score across all classification tasks.

In their paper, Mehedi Masud et al. [6] proposed a classification system that employed contemporary techniques in both Deep Learning (DL) and Digital Image Processing (DIP) to distinguish between five different types of lung and colon tissues (including two benign and three malignant types) by examining their histopathological images. Their findings demonstrated that their system could detect cancerous tissues with an accuracy of up to 96.33%. In their paper, Mumtaz Ali and Riaz Ali et al. [7] introduced a new type of multi-input dual-stream capsule network that utilized the feature learning abilities of both conventional and separable convolutional layers to categorize histopathological images of lung and colon cancer into five different classes, comprising three malignant and two benign categories. According to their findings, their model attained an overall accuracy of 99.58% and an F1 score of 99.04%.

In their paper, Liu et al. [8] suggested a new model, where they integrated CroReLU into the SeNet network (SeNet_CroReLU). According to their findings, the diagnostic accuracy of their proposed model reached 98.33%, which was significantly superior to the accuracy achieved by conventional neural network models at the same stage. In their paper, Setiawan et al. [9] constructed a CNN model that was composed of one gamma correction layer, three convolution layers, three max-pooling layers, and two fully connected layers. They gathered 3000 images from the public LC25000 dataset and assessed their model through a five-fold cross-validation technique using gamma values of 0.8, 1.0, and 1.2. Their findings demonstrated that their model attained its highest accuracy of 87.16% when a gamma value of 1.2 was used.

In their paper, Kwabena Adu et al. [10] demonstrated the efficacy of DHS-CapsNet through empirical evidence. They utilized this approach on histopathological images obtained from the LC25000 dataset and reported improved results of 99.23%, surpassing the performance of traditional CapsNet which obtained an accuracy of 85.55%. Furthermore, DHS-CapsNet was observed to have a top-1 classification error of only 0.77% compared to 14.45% for traditional CapsNet. In their paper, Aya Hage Chehade et al. [11] introduced the XGBoost model as the top-performing method for differentiating between subtypes of lung and colon cancers based on accuracy, precision, and recall measures. Their XGBoost model achieved a 99% accuracy rate and an F1 score of 98.8%.

The drawback of other models is that they require more computational power and take more time. Our model which is an enhanced version of resnet is ResNet16-SE which has fewer layers but uses SE block to enhance the performance and produce results on par with other models. Compared to other models, our model has the advantage of requiring less computational power and time.

3 Proposed Model

Deep learning is a form of machine learning that employs artificial neural networks to enable computers to learn and make decisions using input data. It has revolutionized AI, particularly in areas such as speech and image recognition. Deep learning models excel at complex tasks, but they require substantial computing power, and abundant labeled data, and pose challenges in interpretability and explaining decisions.

3.1 Convolutional Neural Network

CNNs are popular deep neural networks used for image identification and computer vision tasks. They consist of convolutional, pooling, and fully connected layers. Convolutional layers extract features from images using learnable filters, while pooling layers reduce the dimensionality of feature maps. Fully connected layers classify images based on the extracted features. CNNs offer advantages like automated feature learning, scalability, and solving various computer vision problems.

3.2 Proposed Method

In our proposed system we have followed the subsequent steps: Data acquisition, Data pre-processing, Model, Training, and Testing. Our proposed model is based on the Convolution Neural Network. We have used ResNet as our base model. We have developed a ResNet model with 16 layers and we have integrated it with Squeeze-and-Excitation Networks (SENets). The architecture of our proposed model is shown in Fig. 2.

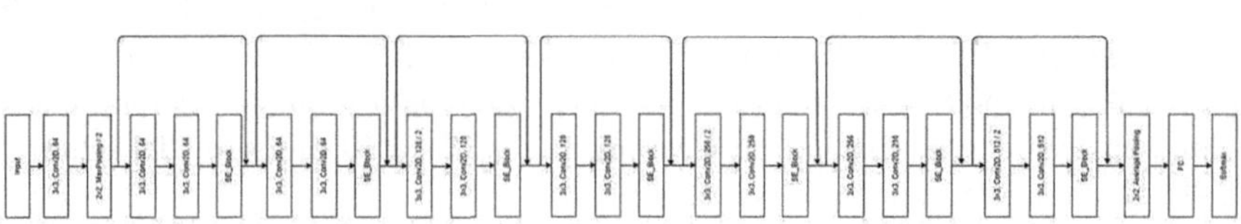

Fig. 2 Proposed model architecture

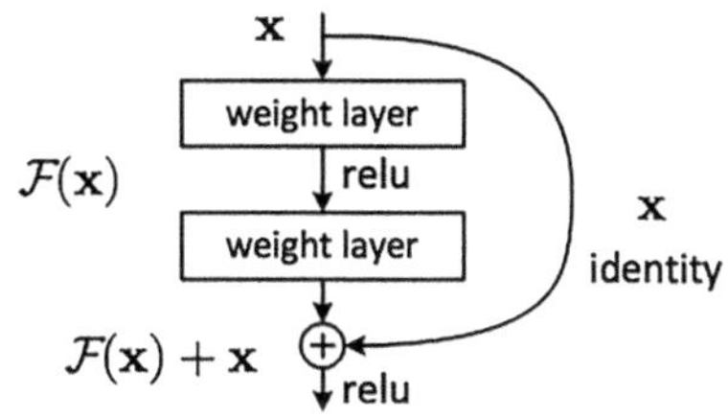

Fig. 3 Residual block [12]

3.3 Residual Block

ResNet's residual block consists of two convolutional layers, batch normalization, and ReLU activation function. It employs a shortcut connection to merge the input and second output, which are then passed through a non-linear activation function. Figure 3 illustrates the working of a ResNet's residual block [12], showcasing how the input and second output are merged through a shortcut connection and passed through a non-linear activation function.

The output of the residual block can be expressed as shown in Eq. 1.

$$y = F\left(x, \{W_i\}\right) + W_s x \tag{1}$$

3.4 Squeeze and Excitation Block

The SE block can be included in any convolutional neural network (CNN) architecture. The basic idea is to add a SE block after each convolutional layer. The SE block creates a set of scaled feature maps as output from the convolutional layers' output, which are the feature maps. The architecture of the SE block is shown in Fig. 4.

In our proposed model we have further extended the SE block by adding two additional convolution layers which improve the performance of the SE block. The Squeeze operation can be expressed as shown in Eq. 2.

$$F_{\text{sq}}\left(u_c\right) = \frac{1}{H \times W} \sum_{i=1}^{H} \sum_{j=1}^{W} u_c\left(i, j\right) \tag{2}$$

The Excitation operation can be expressed as shown in Eq. 3.

$$F_{\text{ex}}\left(z, W\right) = \sigma\left(W_2 ReLU\left(W_1 z\right)\right) \tag{3}$$

Finally, for compiling the model we have used the Adam optimizer, and the loss function used is the categorical cross-entropy (CE). It is calculated as depicted in Eq. 4.

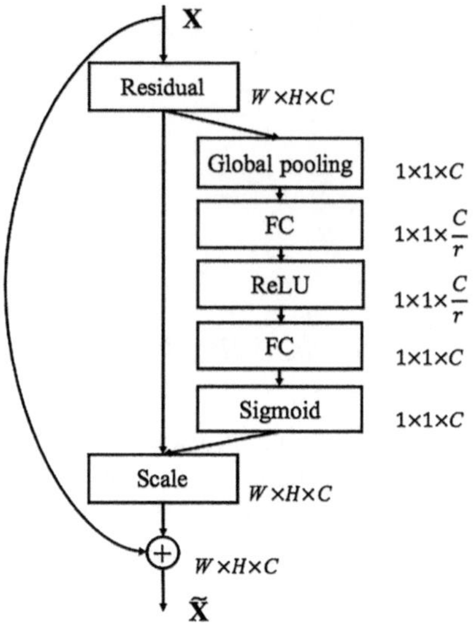

Fig. 4 SE block architecture [13]

$$\mathrm{CE} = -\log\left(\frac{e^{S_p}}{\sum_j^C e^{S_j}}\right) \tag{4}$$

4 Comparative Result and Analysis

4.1 Data Acquisition

Our model uses the LC25000 dataset consisting of histopathological images of lung and colon cancer. It focuses on three subtypes of lung cancer: Squamous Cell Carcinoma, Benign Tissue, and Adenocarcinoma, with 5000 images per class. Figure 5 illustrates each class of lung cancer in our model's LC25000 dataset, including Squamous Cell Carcinoma, Benign Tissue, and Adenocarcinoma. The model performs down-sampling operations on the images and uses a decoder block to generate feature vectors. Ultimately, the model generates segmented output image masks.

4.2 Data Augmentation

The dataset contained RGB images in .jpeg format. Initially, the dataset contained 250 images of each class, and using the augmenter package the images were augmented to 5000 for each class. The images were resized to (256, 256) pixel

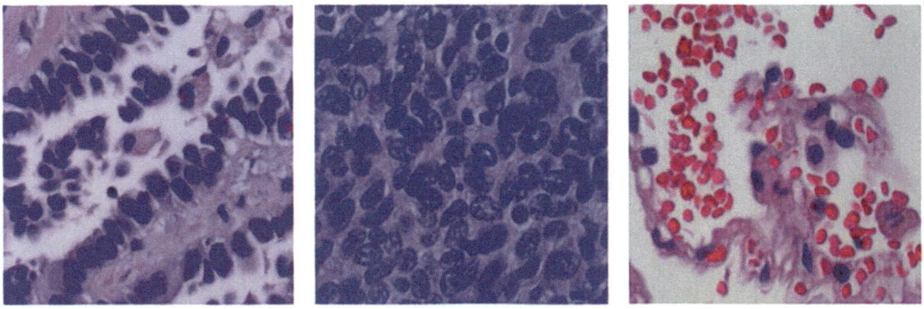

a. Adenocarcinoma   b. Squamous cell carcinoma   c. Benign Tissue

Fig. 5 Histopathological images of lung cancer [14]. (**a**) Adenocarcinoma. (**b**) Squamous cell carcinoma. (**c**) Benign tissue

size and since the images were RGB they have been normalized to maintain values between (0, 1). Then each class is identified with 0, 1, and 2 where 0 is for Adenocarcinoma, 1 is for Squamous Cell Carcinoma, and 2 is for Benign Tissue.

4.3 Training and Testing

The dataset is split into 80% for training and 20% for testing. The model undergoes 10 epochs with 188 steps per epoch and a batch size of 64. Early Stopping and Reduce LROn Plateau callbacks are employed to prevent over-fitting and dynamically adjust the learning rate during training.

4.4 Experimental Setup

All the experiments were performed on a computer with I7 PROCESSOR- 12700 (16 CORE 25 MB CACHE), nVIDIA CHIPSET GEFORCE GTX1630 4GB GDDR6 GPU, CPU with 32 GB of RAM. The model execution has been conducted with Python 2.7 on TensorFlow and Keras. The proposed architecture is trained by using the augmented images, conducted over 50 epochs.

4.5 Metrics

The accuracy metric determines the proportion of clearly identifiable samples out of the total sample count. However, relying solely on accuracy may not be sufficient,

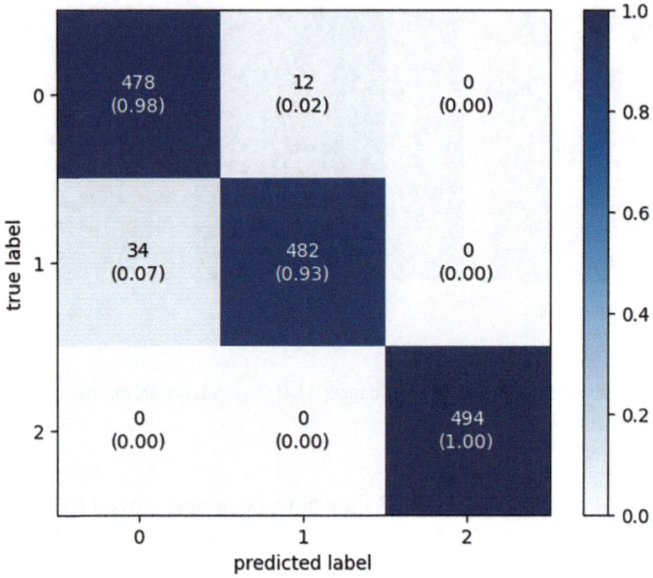

Fig. 6 Confusion matrix

especially with uneven classes or varying misclassification impacts. Therefore, employing diverse evaluation criteria is recommended for a comprehensive assessment of model performance.

4.6 *Confusion Matrix*

A confusion matrix is a table used to assess the effectiveness of a classification model. True positives, true negatives, false positives, and false negatives are represented in the matrix. It helps determine the accuracy, precision, recall, and F1 score of the model and identifies areas of difficulty for specific classes. Figure 6 displays the confusion matrix of our project, presenting the results of the classification model by showcasing the true positives, true negatives, false positives, and false negatives.

4.6.1 Accuracy

Accuracy measures the proportion of correctly classified instances in a dataset. As shown in Eq. 5, it indicates the overall correctness of a classification model by calculating the ratio of correct predictions to the total number of observations. Higher accuracy reflects a higher number of true positive and true negative values in the confusion matrix.

$$\text{Accuracy} = \frac{(\text{TP} + \text{TN})}{(\text{TP} + \text{FP} + \text{FN} + \text{TN})} \tag{5}$$

4.6.2 Recall

In a confusion matrix, recall measures a model's ability to correctly identify positive instances. As shown in Eq. 6, it is calculated as the ratio of true positives to the sum of true positives and false negatives. High recall is important in scenarios where missing a positive instance carries significant consequences, like in medical diagnosis.

$$\text{Recall} = \frac{\text{TP}}{(\text{TP} + \text{FN})} \tag{6}$$

4.6.3 Precision

Precision measures the accuracy of positive predictions by the model, representing the proportion of correctly predicted positive instances out of all predicted positives. A high precision score indicates low false positives, while a low precision score indicates high false positives. Equation 7 represents the process of calculating the precision.

$$\text{Precision} = \frac{\text{TP}}{(\text{TP} + \text{FP})} \tag{7}$$

4.6.4 F1 Score

The F1 score is a single metric that combines precision and recall, representing a model's correctness. It is the harmonic mean of precision and recall, with a range of 0 to 1. A higher F1 score indicates better overall performance, considering false positives and false negatives. Equation 8 represents the process of calculating the F1 score.

$$\text{F1} - \text{Score} = \frac{2 \times (\text{Recall} \times \text{Precision})}{(\text{Recall} + \text{Precision})} \tag{8}$$

4.7 Result Comparison

As shown in Table 1, our model surpasses all the existing models by means of F1 score, precision, recall, and accuracy.

Table 1 Result comparison table

References	F1 score	Precision	Recall	Accuracy
[2]	–	–	–	97
[3]	95	97	96	96.11
[6]	96.38	96.39	96.37	96.33
[9]	87	86.67	87	87.16
Resnet 18	93	95	92	95
Proposed model	97.21	97.18	97.13	97.04

5 Conclusion

In this work, we have performed lung cancer classification using histopathological images. For this task, an enhanced convolutional neural network has been implemented based on ResNet architecture. Our model can perform better than many existing models at low computational cost. The proposed model can classify lung cancer into three different categories. Our model, with 97% accuracy, can identify adenocarcinoma, benign conditions, and squamous cell carcinoma. A confusion matrix was plotted after the metrics such as the F1 score, precision, and recall were calculated. The effectiveness of the model can be evaluated using these metrics.

References

1. F. Bray, J. Ferlay, I. Soerjomataram, R. L. Siegel, L. A. Torre, and A. Jemal, "Global cancer statistics 2018: GLOBOCAN estimates of incidence and mortality worldwide for 36 cancers in 185 countries," CA. Cancer J. Clin., vol. 68, no. 6, pp. 394–424, Nov. 2018, https://doi.org/10.3322/caac.21492.
2. S. Mangal Engineerbabu, A. Chaurasia Engineerbabu, and A. Khajanchi, "Convolution Neural Networks for diagnosing colon and lung cancer histopathological images," September 2020 (accessed on March 24, 2023). Available online at: http://arxiv.org/abs/2009.03878
3. B. K. Hatuwal and H. C. Thapa, "Lung Cancer Detection Using Convolutional Neural Network on Histopathological Images," Int. J. Comput. Trends Technol., vol. 68, no. 10, pp. 21–24, Oct. 2020, https://doi.org/10.14445/22312803/IJCTT-V68I10P104.
4. M. A. Abbas, S. U. K. Bukhari, A. Syed, and S. S. H. Shah, "The Histopathological Diagnosis of Adenocarcinoma & Squamous Cells Carcinoma of Lungs by Artificial intelligence: A comparative study of convolutional neural networks," medRxiv, p. 2020.05.02.20044602, May 2020, https://doi.org/10.1101/2020.05.02.20044602.
5. S. Garg and S. Garg, "Prediction of lung and colon cancer through analysis of histopathological images by utilizing Pre-trained CNN models with visualization of class activation and saliency maps," ACM Int. Conf. Proceeding Ser., pp. 38–45, Dec. 2020, https://doi.org/10.1145/3442536.3442543.
6. M. Masud, N. Sikder, A. Al Nahid, A. K. Bairagi, and M. A. Alzain, "A Machine Learning Approach to Diagnosing Lung and Colon Cancer Using a Deep Learning-Based Classification Framework," Sensors 2021, Vol. 21, Page 748, vol. 21, no. 3, p. 748, Jan. 2021, https://doi.org/10.3390/S21030748.
7. M. Ali and R. Ali, "Multi-Input Dual-Stream Capsule Network for Improved Lung and Colon Cancer Classification," Diagnostics 2021, Vol. 11, Page 1485, vol. 11, no. 8, p. 1485, Aug. 2021, https://doi.org/10.3390/DIAGNOSTICS11081485.

8. Y. Liu et al., "CroReLU: Cross-Crossing Space-Based Visual Activation Function for Lung Cancer Pathology Image Recognition," Cancers (Basel)., vol. 14, no. 21, Nov. 2022, https://doi.org/10.3390/CANCERS14215181
9. W. Setiawan, M. M. Suhadi, Husni, and Y. D. Pramudita, "Histopathology of lung cancer classification using convolutional neural network with gamma correction," Commun. Math. Biol. Neurosci., vol. 2022, no. 0, p. Article ID 81, 2022, https://doi.org/10.28919/CMBN/7611.
10. K. Adu, Y. Yu, J. Cai, K. Owusu-Agyemang, B. A. Twumasi, and X. Wang, "DHS-CapsNet: Dual horizontal squash capsule networks for lung and colon cancer classification from whole slide histopathological images," Int. J. Imaging Syst. Technol., vol. 31, no. 4, pp. 2075–2092, Dec. 2021, https://doi.org/10.1002/IMA.22569.
11. A. Hage Chehade, N. Abdallah, J. M. Marion, M. Oueidat, and P. Chauvet, "Lung and colon cancer classification using medical imaging: a feature engineering approach," Phys. Eng. Sci. Med., vol. 45, no. 3, pp. 729–746, 2022, https://doi.org/10.1007/S13246-022-01139-X/METRICS.
12. K. He, X. Zhang, S. Ren, and J. Sun, "Deep Residual Learning for Image Recognition," (accessed on March 24, 2023). Available online at: http://image-net.org/challenges/LSVRC/2015/
13. J. Hu, L. Shen, and G. Sun, "Squeeze-and-Excitation Networks," (accessed on March 24, 2023). Available online at: http://image-net.org/challenges/LSVRC/2017/results
14. A. A. Borkowski, M. M. Bui, L. B. Thomas, C. P. Wilson, L. A. DeLand, and S. M. Mastorides, "Lung and Colon Cancer Histopathological Image Dataset (LC25000)," December 2019 (accessed on March 24, 2023). Available online at: https://www.google.com/search?q=%2Fdatasets%2Fandrewmvd%2Flung-and-colon-cancer-histopathological-images&rlz=1C1GCEU_enIN1080IN1080&oq=%2Fdatasets%2Fandrewmvd%2Flung-and-colon-cancer-histopathological-images&gs_lcrp=EgZjaHJvbWUyBggAEEUYOTIGCAEQRRg60gEIMjQyMmowajmoAgCwAgE&sourceid=chrome&ie=UTF-8

Design of EV Battery System Using Grid-Interface Solar PV Power with Novel Adaptive Digital Control Algorithm

C. Sudhakar, K. Keerthi Reddy, D. Gopal Krishna, P. Pavan, N. Sreelatha, and K. Rohini

Abstract This inquiry aims to show a solar photovoltaic (PV) system that is linked to the grid and includes characteristics that allow for adjustments to be made to the power quality. Three major phases may be identified within the system. This system is employed to compensate for a range of power quality (PQ) issues, including harmonics, redundant reactive power, and load unbalancing, in addition to transferring power that is produced by a photovoltaic (PV) array to feed linear and nonlinear loads. This system also feeds linear and nonlinear loads. These are only some of the problems that our technology could be able to solve in the future. In addition to this, it can provide linear power to loads when this kind of power is required. A three-phase voltage source converter (VSC) is used so that the direct current (DC) electricity that is produced by the PV array may be converted into alternating current (AC). Alternating current, or AC, is the kind that is most often used. It is essential for the solar PV system that is connected to the grid to have an effective control strategy in order to simplify the transmission of active energy and to reduce the potential that power quality issues will occur. This research aims to illustrate how an adaptive generalized maximum Versoria criteria (AGMVC) controller may be used for a variable speed drive (VSC), which is a component of a solar photovoltaic energy conversion system. To guarantee effective utilization of the solar photovoltaic array, the maximum power point tracking (MPPT) method, which is founded on the perturb and observe algorithm, is used. The grid-integrated PV system's experimental environment is first fabricated in the laboratory using an IGBT-based VSC and DSP (dSPACE DS-1202), and then the system itself is put together. Experiments are carried out to determine how effective the AGMVC control approach is. These experiments make use of a prototype that was developed inside the laboratory. This control method is evaluated in comparison with a variety of different conventional controllers, such as synchronous reference frame theory

C. Sudhakar (✉) · K. K. Reddy · D. G. Krishna · P. Pavan · N. Sreelatha · K. Rohini
Department of Electrical and Electronics Engineering, Mother Theresa Institute of Engineering and Technology, Palamaner, Andhra Pradesh, India
e-mail: sudhakareee365@mtieat.org

A. Patel et al. (eds.), *Advances in Machine Learning and Big Data Analytics II*,
Springer Proceedings in Mathematics & Statistics 442,
https://doi.org/10.1007/978-3-031-51342-8_7

(SRFT) and instantaneous reactive power theory (IRPT), in addition to recently developed weight-based controllers, such as least mean square (LMS), least mean mixed norm (LMMN), and normalized kernel least mean fourth-neural network (NKLMFNN). To make a comparison between AGMVC and the control methods that have been discussed in the past, a number of criteria, such as fundamental weight convergence, steady state error, computational complexity, the requirement for phase lock loop (PLL), and the potential for providing harmonic compensation, are taken into consideration. The purpose of this comparison is to establish whether the AGMVC is more effective than the control approaches that were discussed before. In accordance with the IEEE-519 standard, the functionality of the system is evaluated and ranked in accordance with its performance during the testing. Harmonics, the maximum Versoria criteria, power factor correction (PFC), and solar photovoltaic are just some of the index terms that may be discovered in this section.

Keywords AGMVC · VSV · MPPT · PLL · PV · EVB · THD

1 Introduction

Because of the rise in the price of petrol and the fall in the price of PV modules, which has paved the way for the quick development of the electric vehicle (EV), as the future mode of transport, there is a need for the grid integration of a solar photovoltaic (PV) array with an electric vehicle battery (EVB) charging system. This is because increasing petrol prices have opened the door for the rapid expansion of the EV. This desire is being fueled by the fact that there is a demand for this kind of integration. The utilization of renewable energy sources like solar, wind, and other similar sources in the growth of electrical energy production in the form of distributed generation leads to reduced levels of greenhouse gas emission and carbon footprints. Other such sources include: solar, geothermal, and hydroelectric. Solar power offers a low-hassle, pollution-free, and clean alternative to more conventional kinds of electric energy. Tuttle and Baldrics [1] present a diagrammatic illustration of the grid-interfaced electric vehicle concept. The intermittent nature of solar photovoltaic (PV) power results in the need for maximum power point tracking (MPPT) solutions, as shown in Brito et al. and Debnath et al.'s papers [2, 3]. This problem is brought on by variations in the quantity of solar insolation and the temperature of the environment, respectively. To get the highest amount of solar photovoltaic (PV) electricity, this research makes use of the perturb and observe (P&O) method. This technology has a low overall cost, is simple to put into action, and provides reliable tracking capabilities. These are all reasons why it is useful to use. The most important factor that contributes to concerns about power quality is a significant rise in the total number of non-linear loads that are connected to the system. The incorporation of non-conventional sources increases the amount of strain that is imposed on the electrical system. Xiaodong [4] gives a full overview of the many power quality challenges that occur because of grid-integrated renewable energy sources.

Xiaodong's paper [4] is a good resource to look at if you want to learn more about this topic. Harmonics injection, which is induced by power converters, frequency and voltage fluctuation, and other related issues are some of the problems. Killi and Samanta's study [5] is an example of the improved P&O strategy, which is one factor that leads to the overall increase in the system's level of complexity. The innovative converter that is utilized in the solar photovoltaic system that is integrated with the utility grid employs a single-stage design that is both inexpensive and efficient. This setup, which is shown in Kim et al. [6], is used for the DC-to-DC conversion as well as the DC-to-AC conversion. Both conversions require direct current. The battery that combines plug-in electric cars and the vehicle-to-grid concept for the grid may be able to counteract the effects of solar photovoltaic arrays, such as an increase in voltage or a decrease in peak load demand. These effects include the potential for an increase in voltage. Alam et al. [7] delve into these risk-reduction methods in further detail. The results of testing a PV-EV system that was connected to the electric grid are shown in Montiero et al.'s study [8]. This performance consists of sinusoidal grid currents and a low total harmonic distortion (THD) in a number of different operating modes. EV to grid, utility grid to EV, solar PV power to utility grid, and solar power to charge EVB are the many operating modes that may be used. Sechilariu et al. [9] give information on solar photovoltaic battery-integrated buildings that are working in partnership with the grid to moderate the peak load demand in metropolitan areas. These buildings are located in large urban centers. Tabari and Yazdani, and Traube et al. [10, 11] provide a report on the stability analysis and elimination of fluctuation in solar PV array power of a DC distribution network that incorporates electric cars into the grid and makes use of a DC-DC buck-boost converter as a battery charger. Battery charging is accomplished using a DC-DC buck-boost converter by this network. According to Thale et al. [12], the energy storage that is offered by batteries may be utilized as a power backup in a microgrid that is based on renewable energy sources and batteries and can function in a variety of different operational situations. This is according to a microgrid that was designed to run in a number of different operational conditions. The reliability and steadiness of the system are both improved because of this factor. Wu et al. [13] give a complete overview of the criteria and rules that must be followed by grid-integrated solar PV systems to provide a reliable and secure connection. These criteria and regulations must be met for grid-integrated solar PV systems to deliver a reliable and secure link. The grid feeding converter makes use of a number of different control techniques, including synchronous reference frame phase-locked loop (SRF-PLL), dual second-order generalized integrator (DSOGIFLL), and stationary reference frame-based control with proportional and resonant controller. When the grid voltages are unbalanced and distorted, the SRF technique does not work as well as it does when the grid voltages are underbalanced. However, the SRF approach performs extremely well when the grid voltages are underbalanced. Traditional controllers can fix the grid frequency by using PLL to achieve this goal.

A phase shift will occur between the input and output of the controller as a direct consequence of the DSOGIFLL control's production of transition delay. Robert et al. [14] explain a wide variety of power converters, each of which incorporates a different technique of control and a distinct mode of operation. The control strategies that are used in grid-integrated solar PV structures are discussed in further detail by Adefarati and Bansal [15], Yang et al. [16], and Krithiga and Gounden [17]. Battery storage is essential in order to circumvent the problems that are cropping up because of the variable nature of renewable energy. In this study, the EVB is merged with a bidirectional converter at the DC link of the three-phase voltage source converter (VSC). As a result, a battery with a low-voltage rating is used. This allows the battery to store the excess power and deliver power during times of reduced load demand and peak load periods, respectively. Rallabandi et al. [18], Yang et al. [19], Mahmud et al. [20], and Hollinger et al. [21] detail the control strategies as well as the energy management of the grid that is merged with the PV battery system. For a solar PV-based EVB system that is connected to the utility grid, a control mechanism that is based on recursive digital adaptive filters is used. The following is a description of the primary contributions that this study has made: (1) The recursive digital adaptive filter is straightforward to build, and its response is satisfactory in terms of the speed at which it operates. (2) The polluted load current is sent through the filter to extract a fundamental active current component that is devoid of harmonics. Because of the filter, there is no discernible phase shift between these two waves at any time throughout the experiment.

Overall system efficiency is increased and costs are reduced due to the elimination of a second DC-DC boost converter in the single-stage design.

A simple and basic technique known as P&O is utilized to extract the maximum amount of energy from the solar PV array. Connecting the EVB with the DC-DC buck-boost converter at the DC link eliminates the second harmonic current constituent from the EV's battery current, extending the battery life of the EV, as opposed to a system in which the EVB is connected directly across the DC link, in which the oscillations at the DC link are replicated directly on the EVB. As a result, the EV's battery will last longer.

The recursive digital filter-based controller has fewer computations for each step and a lighter computing load as a result of the utilization of basic addition and subtraction blocks, as well as multiplication blocks, rather than the blocks for coordinate translation, amplitude evaluation, and phase estimation. To properly compensate for non-linear load changes and account for a variety of atmospheric conditions, the appropriate filter coefficients must be chosen. The controller-self is responsible for locking the grid frequency in place.

The EVB will automatically charge and discharge itself in response to changes in load. The current controller additionally regulates the charging and discharging of the EVB based on whether the system is experiencing a low unit price of electric energy/ultimate load state or a high unit price of electric energy. When solar energy is scarce, the electric vehicle's battery assists the VSC by acting as an active power filter and reactive power compensator.

A portion of the system's active power is contributed to the grid. In addition to this, if there is a rapid spike or drop in the load, the grid currents will fall or increase in a smooth manner respectively. By using the feed-forward component of solar PV power (FFSPV), the dynamic responsiveness of the system is increased under adaptive climatic circumstances. Additionally, this improves the system's ability to lessen its dependency on the proportional and integral (PI) controller. The created system's control is verified using test results for changes in solar insolation and variations in load.

The utility grid benefits from an increase in power quality thanks to the power factor correction and harmonic mitigation provided by this technology. In line with the IEEE- 519 standard, the THD for grid currents is measured and recorded [22].

The word "photovoltaic," or "PV," refers to a method of generating energy by converting solar radiation into direct current electricity utilizing semiconductors that exhibit the photovoltaic effect. This method is sometimes referred to as "photovoltaic" or "PV." Producing power using photovoltaics requires solar panels, each of which consists of a collection of individual cells that house a photovoltaic material. Materials used to produce solar cells range from monocrystalline silicon and polycrystalline silicon to amorphous silicon and cadmium telluride and copper indium selenite/sulfide [1]. The manufacturing of solar cells and photovoltaic arrays has advanced greatly in recent years in response to the rising demand for these technologies.

As of the year 2010, solar photovoltaic is the technology that produces energy in more than one hundred nations, and even though it only accounts for a minuscule portion of the total power-generating capacity of 4800 gigawatts (GW) worldwide, it is the technology that is expanding at the quickest pace of any in the world.

2 Literature Survey

At a time when electricity consumption is at a peak, the centralized power grid is under a significant strain. Transmission and distribution line losses are a major drag on the grid's efficiency. There are both environmental and non-environmental causes of grid-wide power outages, such as storms and faulty equipment. Most of India's big central power plants run on fossil fuel, resulting in annual emissions of around 1.026 billion tons of CO2 [1]. More than half of India's rural population lives in areas without access to consistent electricity, which limits opportunities for economic growth. Renewable energy sources are required to combat these problems [2], and the United States has a great deal of untapped potential for harnessing the power of the sun to create electricity.

Solar photovoltaic cells and converters based on power electronics make up a solar energy conversion system. The system may operate while linked to the grid or independently. The upfront and ongoing expenditures of a solar PV system are higher because of the need for battery storage for a standalone system. Since battery storage is not necessary for a solar PV system that is grid-interfaced, these systems are more financially viable [3]. Algorithms known as maximum power point tracking (MPPT) are used for solar photovoltaic (PV) arrays to extract the maximum amount of usable energy from them. A large number of MPPT algorithms, such as incremental conductance (InC), fuzzy logic control, hill climbing, perturb and observe (P&O), a numerical method, and many more, have been described and documented in the scientific literature [4, 5].

For connecting a solar PV array to the grid, either a single-stage or a two-stage topology may be used. In a single-stage configuration, there is no intermediary stage between the PV array and the DC link of the VSC. The single-stage system has the benefit of increased efficiency, while the two-stage system has the advantage of greater stability. As more and more nonlinear loads—like LED bulbs, battery chargers, electronic ballasts, adjustable speed drives, uninterruptible power supplies, etc.—are being plugged into the grid, harmonics are being introduced. This leads to voltage distortion on the grid, malfunction of associated electrical equipment, and overheating of motor loads. As a result, the solar PV array's VSC must be able to both provide grid-based active power and protect against power quality issues [6]. The development of efficient, low-complexity, and reliable methods of control might make this a reality.

Mastromauro et al. [7] designed a single-phase low-power PV system that can sustain the grid voltage and rectify harmonics with the help of a repeating controller. For the four-legged VSC, Singh and Jain [8] propose a decoupled adaptive noise detection-based control. Gonzalez-Esp'n et al. [9] have created an adaptive controller for observing a reference signal through a Schur-lattice IIR filter construction. Singh et al. [10] have developed a control strategy for the two-stage, three-phase grid-integrated solar PV system based on the fast zero-attractive normalized least-mean-squared-error algorithm. Sliding mode control and Lyapunov function-based control method for maximum power tracking and a DC-AC converter were presented by Rezkallah et al. [11] for active power injection and harmonic correction in a solar PV grid-interfaced system. Flota et al.'s [12] double-loop control strategy allows the converter to take on additional reactive power adjustment responsibilities. A DC/AC inverter coupled in series with thyristor-controlled LC filters is a key component of the hybrid coupling inverters disclosed by Wang et al. [13]. Artificial neural networks [14], adaptive neuro-fuzzy inference systems [15], Fuzzy logic [16], and recurrent neuro-control algorithms [17] are only some of the AI approaches that have been reported in controllers for grid-coupled PV systems.

The error rate is low and the algorithms work well under dynamic loading situations, but they are more difficult to implement and need some training. More and more people are turning to control strategies based on adaptive filters because adaptive controllers can adjust their settings in response to real-time data, making them more effective under both static and fluctuating loads. Several different types of adaptive control are described in the scientific literature. This category includes the least mean mixed norm [19], adaptive notch filter [20], adaptive improved phase lock loop [21], and multistage adaptive filter (MAF) [22]. According to the aforementioned sources, PV systems may improve power quality by supplying active power to the grid. However, the system only makes use of a small fraction of the features available in active power filters. There are few controllers in the aforementioned systems that can balance loads and suppress harmonics all at once, and those that can often suffer from complexity, lackluster dynamic performance, or the disadvantage of incorporating more passive components into the system to achieve this goal.

3 Conclusion

The proof was shown that the solar PV-powered EVB grid-intertied system could achieve satisfactory performance with adaptive recursive digital filter control. This technique not only improves electricity quality at PCC, but also adds active power to the grid. The EVB stores more energy while the load's demand is low, and it discharges that energy in response to a high load demand. The DC-DC converter in the integrated EVB has allowed for maximum power point tracking voltage to be reached at the DC connection. The vehicle's battery is kept on a continuous cycle of charging and draining thanks to the buck-boost converter. The recursive filter method eliminates any visible phase delay between the basic current component and the load current. Results from experiments have shown unity power factor functioning, which prevents grid current harmonics. Therefore, the total harmonic distortion (THD) of the grid current is kept below 5% even when the load is non-linear, as specified by the IEEE-519 standard. The system's steady-state responsiveness and dynamic behavior under changing loads, fluctuating quantities of solar insolation, and the interruption of solar PV arrays have been shown as a result of its successful implementation. The VSC has adjusted for the load's reactive power and behaved in a manner similar to that of an active power filter when EVB was used but there was no solar power production. The findings of the tests have established the control method's credibility.

References

1. Tuttle, D.P., Baldrics, R.: 'The evolution of plug-in electric vehicle-grid interactions', IEEE Trans. Smart Grid, 2012, 3, (1), pp. 500–505
2. S. Brito, M.A.G. de Galotto, Sampaio, L.P., et al.: 'Evaluation of the main MPPT techniques for photovoltaic applications', IEEE Trans. Ind. Electron., 2013, 60, (3), pp. 1156–1167
3. Debnath, D., De, P., Chatterjee, K.: 'Simple scheme to extract maximum power from series connected photovoltaic modules experiencing mismatched operating conditions', IET Power Electron., 2016, 9, (3), pp. 408–416
4. Xiaodong, L.: 'Emerging power quality challenges due to integration of renewable energy sources', IEEE Trans. Ind. Appl., 2017, 53, (2), pp. 855– 866
5. Killi, M., Samanta, S.: 'Modified perturb and observe MPPT algorithm for drift avoidance in photovoltaic systems', IEEE Trans. Ind. Electron., 2015, 62, (9), pp. 5549–5559

6. Kim, H., Parkhideh, B., Bongers, T.D., et al.: 'Reconfigurable solar converter: a single-stage power conversion PV-battery system', IEEE Trans. Power Electron., 2013, 28, (8), pp. 3788–3797
7. Alam, M.J.E., Muttaqi, K.M., Sutanto, D.: 'Effective utilization of available PEV battery capacity for mitigation of solar PV impact and grid support with integrated V2G functionality', IEEE Trans. Smart Grid, 2016, 7, (3), pp. 1562–1571
8. Monteiro, V., Pinto, J.G., Afonso, J.L.: 'Experimental validation of a three port integrated topology to interface electric vehicles and renewables with the electrical grid', IEEE Trans. Ind. Inf., 2018, 14, (6), pp. 23642374
9. Sechilariu, M., Wang, B., Locment, F.: 'Building integrated photovoltaic system with energy storage and smart grid communication', IEEE Trans. Ind. Electron., 2013, 60, (4), pp. 1607–1618
10. Tabari, M., Yazdani, A.: 'Stability of a dc distribution system for power system integration of plug-in hybrid electric vehicles', IEEE Trans. Smart Grid, 2014, 5, (5), pp. 2564–2573
11. Traube, J., Lu, F., Maksimovic, D., et al.: 'Mitigation of solar irradiance intermittency in photovoltaic power systems with integrated electric-vehicle charging functionality', IEEE Trans. Power Electron., 2013, 28, (6), pp. 3058–3067
12. Thale, S.S., Wandhare, R., Agarwal, V.: 'A novel reconfigurable micro grid architecture with renewable energy sources and storage', IEEE Trans. Ind. Appl., 2015, 51, (2), pp. 1805–1816
13. Wu, Y.K., Lin, H.J., Lin, J.H.: 'Standards and guidelines for grid-connected photovoltaic generation systems: a review and comparison', IEEE Trans Ind. Appl., 2017, 53, (4), pp. 3205–3216
14. Robert, J., Rodriguez, P., Blaabjerg, F., et al.: 'Control of power converters in ac micro grids', IEEE Trans. Power Electron., 2012, 27, (11), pp. 4734– 4749
15. Adefarati, T., Bansal, R.C.: 'Integration of renewable distributed generators into the distribution system: a review', IET Renew. Power Genre. 2016, 10, (7), pp. 873–884
16. Yang, Y., Blaabjerg, F., Wang, H., et al.: 'Power control flexibilities for grid connected multi-functional photovoltaic inverters', IET Renew. Power Gener. 2016, 10, (4), pp. 504–513
17. Krithiga, S., Gounden, N.G.A.: 'Power electronic configuration for the operation of PV system in combined grid-connected and stand-alone modes', IET Power Electron., 2014, 7, (3), pp. 640–647
18. Rallabandi, V., Akeyo, O.M., Jewell, N., et al.: 'Incorporating battery energy storage systems into multi-MW grid connected PV systems', IEEE Trans. Ind. Appl., 2019, 55, (1), pp. 638–647
19. Yang, Y., Ye, Q., Tung, L.J., et al.: 'Integrated size and energy management design of battery storage to enhance grid integration of large-scale PV power plants', IEEE Trans. Ind. Electron., 2018, 65, (1), pp. 394–402
20. Mahmud, N., Zahedi, A., Mahmud, A.: 'A cooperative operation of novel PV inverter control scheme and storage energy management system based on ANFIS for voltage regulation of grid-tied PV system', IEEE Trans. Ind. Inf., 2017, 13, (5), pp. 2657–2668
21. Hollinger, R., Diazgranados, L.M., Braam, F., et al.: 'Distributed solar battery systems providing primary control reserve', IET Renew. Power Gener. 2016, 10, (1), pp. 63–70
22. IEEE Recommended Practices and Requirements for Harmonic Control in Electrical Power Systems, IEEE Std. 519, 2014
23. Singh, B., Chandra, A., Al-Haddad, K.: 'Power quality: problems and mitigation techniques' (John Wiley & Sons Ltd., United Kingdom, 2015

A Decentralized and Intelligent Approach for Suspicious Event Detection in Surveillance Range

K. U. V. Padma and E. Neelima

Abstract As the world becomes progressively more concerned about security, a gigantic amount of observational data is being generated by security systems. This data, produced on a daily basis, presents challenges for companies looking to store and manage it efficiently. The scattered nature of the data presents a critical challenge to data storage and analysis. One major concern is the security of data exchanged over devices, as any adversary can capture or alter the data, deceiving the reconnaissance system. Users can exchange files and data over the globe due to the scattered nature of the arrangement. Nevertheless, large files consuming a large amount of transmission bandwidth to upload and download over the web can benefit from the use of IPFS, which has rapidly gained notoriety due to its capacity to run the best of multiple protocols, including FTP and HTTP.

Despite IPFS's numerous focal points, there are security and access control issues, including a lack of traceability in how files are accessed. To address these issues, this article proposes a novel procedure that enhances IPFS with blockchain technology to give a straightforward audit trail. By leveraging blockchain as an asset, data reliability and source security can be improved, offering a clear way to trace all activity related to a specific file. This approach can enhance the security and integrity of the data, increasing confidence in the surveillance system's ability to work effectively and securely.

Keywords IPFS · Blockchain · Cryptography · Surveillance · Smart contract

1 Introduction

Over the past few years, methods for shielding data and avoiding unauthorized access to critical data have advanced significantly. In today's advanced world, there

K. U. V. Padma (✉) · E. Neelima
Department of CSE, GITAM Deemed to Be University, Visakhapatnam, Andhra Pradesh, India
e-mail: neadha@gitam.edu

A. Patel et al. (eds.), *Advances in Machine Learning and Big Data Analytics II*,
Springer Proceedings in Mathematics & Statistics 442,
https://doi.org/10.1007/978-3-031-51342-8_8

is an explosion of data, making personal, vital, and sensitive data more vulnerable to security threats such as data management or misuse. Centralized monitoring systems are the primary source of such data, but they need flexibility, unwavering reliability, and computational efficiency in handling essential data.

To address centralization concerns, blockchain technology, which uses a distributed ledger system, aims to decentralize the storage of data. The Inter-Planetary File System (IPFS) [1] focuses on data storage and management, and blockchain and IPFS work together to resolve centralization issues. A smart contract is used to enroll and grant permission to users, and it digitally validates and enforces the protection of stored data.

This paper discusses decentralized storage solutions that integrate with blockchain and certificate authority. The smart contract is responsible for the initial and required verification of user credentials, and certificate authority is established with the configured system to monitor the behavioral designs of different users. This includes performing operations such as policy verification, certificate issuing, certificate authority and cancellation, enrollment specialist maintenance, and servicesubstantiation. By employing these techniques, the security and judgment of imperative data can be upgraded, mitigating the risks of unauthorized access or data manipulation.

2 Literature Survey

In this section, we present the related knowledge and background to better understand our proposed system.

2.1 Blockchain Technology

Bitcoin, a decentralized cryptocurrency, was originally presented to the public in 2009, along with its fundamental blockchain technology, which generated significant interest [4]. In recent years, the scholarly academia and businesses have conducted extensive research and found that it can be applied in various industries, such as finance, healthcare, public utilities, identity management, government agencies, asset registration, and more, playing a significant role [39]. The blockchain is a distributed database shared among all individuals in a peer-to-peer network [4]. There is no central authority in this network, and no single node controls the whole network.

As shown in Fig. 1, the blockchain is composed of an array of interconnected data blocks. Blocks are added to the blockchain through consensus among the majority of nodes in the system. Each block comprises of a block header and an arrangement of exchanges, which incorporate pointers to the piece headers of the going before block, a Merkle root that organizes the exchange data, and a timestamp.

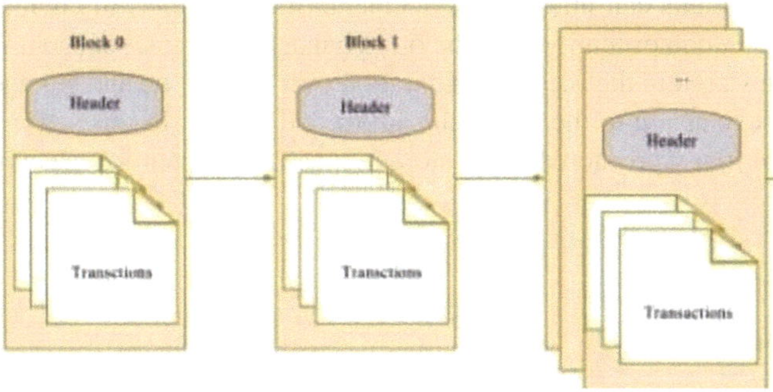

Fig. 1 Blockchain structure

The blocks are sequentially linked together. The cryptographic hash algorithm ensures the permanence of transaction data in each block, and the associated blocks in the blockchain cannot be changed. This technology has potential applications in an assortment of industries, including finance, healthcare, public utilities, asset registration, and government agencies [1].

2.2 *Ethereum*

Ethereum, which is based on Bitcoin, is a decentralized application platform that utilizes smart contracts [5, 40, 41]. In contrast to Bitcoin's stack-based non-Turing complete programming language, Ethereum utilizes a Turing total programming language, making it possible to develop more complex logic and expand its use in various industries. Ethereum is a programmable blockchain, which means that users can create, implement, and execute smart contracts on the platform.

Once a contract is deployed, it can be automatically executed according to the agreed-upon smart contract logic, providing security against downtime, censorship, fraud, and third-party interference.

The Ethereum platform's key components are briefly laid out below.

2.3 *Ethereum Account*

The Ethereum platform supports two major types of accounts: Externally Owned Accounts (EOA) and Contract Accounts, which are identified by a 20-byte hexadecimal string, such as 0xe4874f5a6077b6793de633f52b5ff112aec012de [5]. EOAs are controlled by the external user's private key and have corresponding Nonce fields and balances. These accounts can initiate transactions to transfer funds to another address or start contract code execution. On the other hand, the contract account is controlled by the code stored in the account, and it moreover has a code and storage space related to it, in expansion to the adjust the adjust and Nonce. In this setting, the encrypted keyword files were stored in the storage space of the contract account [40].

2.4 *Ether and Gas*

The decentralized network, Ethereum, utilizes the cryptocurrency Ether. The Ethereum Virtual Machine (EVM) is a smart contract execution environment that works within the Ethereum network. EVM is executed by each mining node as part of the block validation protocol when the contract code is activated by a transaction or message. The transaction is validated and executed by miner nodes in the Ethereum network. They approve each transaction in the piece and execute the transaction-triggered code in EVM, performing the same computations and storing the same data. Each operation in EVM consumes a unique amount of gas that is tracked. Gas can be obtained by acquiring Ether, and the transaction initiator must pay Ether for the activities he intends to perform (computation, data storage, etc.). The transaction cost is determined as follows: Ether = gas used * gas price [5].

2.5 *Transactions and Messages*

An Ethereum transaction is a signed packet that enables the transfer of ether from one account to another as well as the execution of smart contract code [5]. The transaction contains several fields, including the account number, the recipient's address, the gas price, the gas limit, the amount of ether being transferred, the sender's signature, and any other optional data fields [40].

To make an immutable record, the Information column in an Ethereum transaction can be filled with any information desired by the sender. For instance, we can include vital data such as encrypted client characteristics, file location, and secret key in JSON format, which is then encoded as hexadecimal and stored in the Information field of the transaction on the Ethereum blockchain. Ethereum's official JSON API interface, eth_getTransactionByHash, can be utilized to get and prepare this information. The difference between an Ethereum message and a transaction is that messages can only be transmitted between contracts, and by sending a message, the recipient's contract code can also be executed.

3 Related Work

3.1 Smart Contract for Secure Surveillance Storage Model

The proposed system involves two primary entities: the Data Owner Group (DOG) and the Data User Group (DUG). The DOG, an individual or network, holds a set of files to share, while the DUG consists of authorized data users who can access these files. The system framework, illustrated separately, shows the deployment of the smart contract on the blockchain. The following algorithms are involved in our system model:

Illustration, it can be drawn separately. The system framework is shown in Fig. 2. In our work, the system model consists of algorithms given below.

- Setup(1) (PK, MK): The system setup algorithm is run by DO. It takes as input security parameter.

A. It outputs system public parameter PK and system master key MK.

The public key (PK) can be shared freely by the data owner (DO) through various platforms, such as a website or public database. The system master key is encrypted by DO and is integrated into an Ethereum transaction. To store encrypted keyword files and offer system services for data users (DU), DO makes a smart contract on the Ethereum blockchain, as outlined in the to begin with two steps of the system.

- Upon receiving a registration request from the DU, the DO first confirms the identity of the user. If the identity is confirmed, the corresponding attribute set is assigned to DU. The Ethereum account address of DU is added to the authorized

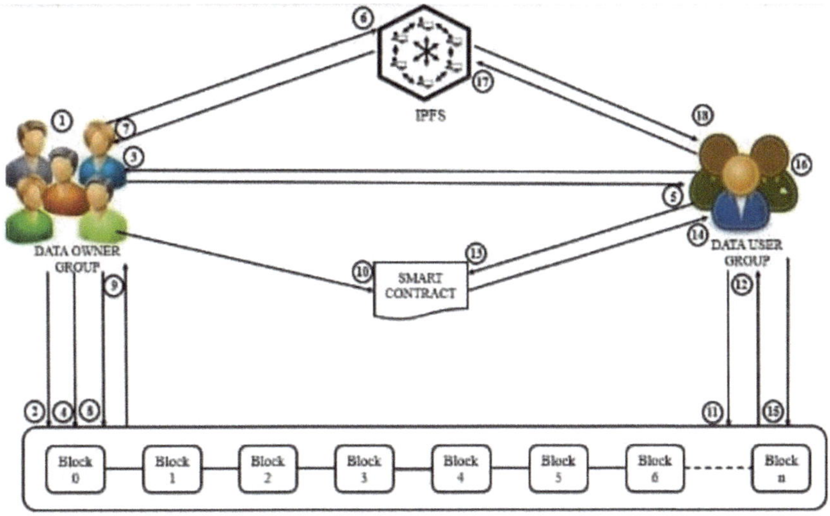

Fig. 2 System framework

user list in the smart contract. Subsequently, DO creates a secret key for DU, taking after the steps outlined below:

- UserRegistration(MK, S) → (SKS, SKD): The process of registering a user is conducted by DO and uses the system master key MK and the DU's attribute set as inputs to the user registration algorithm. The result of this algorithm is the generation of a secret key, SK, which is specific to the DU. To protect this key, it is encrypted using the AES algorithm and integrated into an Ethereum transaction. The shared key created through the Diffie-Hellman key exchange protocol serves as the encryption key. DO then transmits the transaction ID, smart contract address, smart contract ABI, and smart contract source code to the DU through a secure channel. These activities are depicted in steps 3, 4, and 5 in Fig. 2 of the system framework [42].
- Encrypt: The encrypt algorithm is run by DO. It consists of the following three sub-algorithms.

1. FileEncrypt(F) (CTF, K, kw): To encrypt a file F, DO utilizes the file encryption algorithm. The algorithm takes the file F as input and produces a file ciphertext CTF, a file encryption key K, and a keyword set kw as yield. From the file F, DO selects a set of keywords kw, selects an AES key K from the key space, encrypts the file F using the selected key, and uploads the resulting ciphertext CTF to IPFS. The file location returned by IPFS is recorded by DO for future retrieval. As shown in steps 6–7 of Fig. 2 of the system framework.
2. KeyEncrypt(PK, K, hlocation,) CTmd: To encrypt the file encryption key K, the key encryption algorithm requires the system public parameters PK, file location, and access policy as inputs. The algorithm produces the ciphertext CTmd as output. DO employs K to encrypt the location as CTI and encrypts the file encryption key K using a chosen ABE algorithm. To finish this, DO uses the system public parameters PK and access policy. DO randomly selects AES key K1 from the key space and uses it to encrypt CTI and CTk as CTmd. After the transaction is confirmed, DO files the transaction ID and K1. As shown in step 9 of Fig. 2 of the system framework IndexGen(kw, MK) index—to generate an encrypted keyword index, the index generation algorithm requires the system master key MK and a set of keywords kw as input. After processing, it outputs an encrypted keyword index. DO is responsible for constructing the encrypted keyword index based on the provided keyword set kw and subsequently storing the index on the smart contract. As shown in step 10 of the Fig. 2 of system framework.

   - tokenGen(kw, SKs) token: To generate a search token, DU runs the token generation algorithm, which takes a keyword kw and secret key SKS as input and produces the token as output. DU retrieves the transaction data associated with its secret key from the Ethereum blockchain and decrypts it to obtain both SKS and SKd. Utilizing kw, DU creates the search token and utilizes the smart contract to perform the search, as shown in steps 11, 12, and 13 of Fig. 2 of the system framework,

- Test (token, file) result: To perform the matching, the smart contract executes the match test algorithm using the search token and the encrypted keyword file as inputs. Upon successful matching, the smart contract returns the set of relevant transaction IDs and their corresponding keys, as shown in step 14 of Fig. 2 of the system framework.

4 Existing System

For a long time, decentralized cryptocurrencies such as Bitcoin [4], Ethereum [5], and Zcash [6] have been prevalent, with blockchain technology being the fundamental innovation of cryptocurrency [4–6]. Blockchain is currently being widely used in the financial sector [7] and has been found to be advantageous in different non-financial industries, including decentralized supply chains [8], identity-based PKI [9], decentralized verification of archive presence [10], decentralized Internet of Things [11], and decentralized storage [12–14].

To improve the security of users' data, a personal data management system based on blockchain technology has been created, permitting data owners to control their data [15]. A blockchain architecture system has been developed for IoT [16], which utilizes attribute-based encryption (ABE) technology to make data access control and address the issue of data protection in IoT systems. Moreover, a blockchain-based access control architecture for enhancing the security of big data platforms has been presented [17] to address the security and privacy challenges that are hindering the development of big data.

Decentralized storage systems such as IPFS [13], Storj [12], and Sia [18] work without the requirement for a central service provider and allow users to store data on storage nodes that rent out free storage space. The blockchain is indispensable to these platforms, and IPFS employs Filecoin [14] as an incentive layer to encourage nodes to provide storage and retrieval services as a content-addressed decentralized storage platform. However, IPFS lacks a robust privacy cryptographic algorithm interface for user-uploaded files [13]. On the other hand, Storj technology employs end-to-end encryption and stores the file's cryptographic hash unique fingerprint on the blockchain while allowing for file integrity verification. Sia integrates blockchain technology with a peer-to-peer storage format, encrypting the uploaded content into multiple file parts and using smart contracts to send the file ciphertext to the nodes that provide storage services. Users pay Siacoin for storage, and the storage nodes provide file verification of storage periodically to prevent the storage node from deleting the file.

5 Proposed System

The Interplanetary File System (IPFS) is a decentralized, peer-to-peer hypermedia system that aims to make the web more open, faster, and more secure [1]. Unlike conventional web systems, where files are location-addressed, IPFS employs content-addressing, which creates a unique multihash address for each uploaded file based on its content and node ID [1]. This means that any node with the content's hash can search the network for the correct file, and any node with a matching file will serve it, making the P2P network reliable and safe from server attacks.

Moreover, IPFS ensures that identical files are only saved once on the network to prevent content duplication [1]. If a certain node transfers a file, IPFS creates a single instance of the file, regardless of the number of times it is uploaded. The IPFS gateway allows users to transfer files on a private format or over the Internet, and nodes can access local content offline by downloading it from the network.

IPFS's decentralized architecture and content-addressing system make it a promising technology for file storage and transfer. It is anticipated to have applications in areas such as cloud storage, content, delivery, and peer-to-peer communication.

IPFS (Interplanetary File System) [13] is a distributed file system that works in a peer-to-peer network with the objective of connecting all computing devices to a shared file system. IPFS is designed as a content-addressed block storage system with content-addressed hyperlinks, and it is capable of handling high traffic. The technology incorporates various components such as distributed hash tables (DHT), a financial block exchange, and self-certifying namespaces. IPFS is designed to avoid from any single point of failure and does not require trust between nodes. One of the advantages of IPFS is that data is distributed and stored globally without the requirement for a central server, which separates it from conventional cloud storage.

Using a combination of blockchain technology and IPFS, the created reconnaissance data can be secured with extra encryption and decoding strategies to give additional layers of security for the data [1]. Furthermore, various mechanisms were implemented using smart contracts to validate all credentials [2]. As a result, nodes have become independent servers capable of serving content to others, and to bring the IPFS network down, all of the nodes must be down at the same time, which is highly unlikable [4]. The advantage of using IPFS is that it reduces the bandwidth required for sharing large files over the Internet, and users can also host websites using it [4]. Peers with the hash address can view and download the file, and they can view it locally or through the author's node [4].

6 Conclusion

IPFS works in a similar way to the current use of the web in several respects. When a file is uploaded to the IPFS framework, a unique cryptographic hash string is generated, which can be used to retrieve the file. The hash string is similar to a Uniform Resource Locator (URL) for the web. In this instance, we refer to the hash string as the location. Due to blockchain's block bloat and transaction costs, it is not practical to store large amounts of data, such as videos and music, on the blockchain in real-world applications. As a result, in this scheme, the encrypted file is stored on IPFS.

When the smart contract is deployed to EVM byte code and sent on the Ethereum blockchain, it is essential to record the contract address and Application Binary Interface (ABI). Users may interact with the contract using the contract address and ABI. In this scheme, smart contracts are used to store encrypted keyword lists and related data, as well as to ensure verifiability. The service fee is only deducted from the contract when the desired result is achieved. This addresses the issue of searchers intended to fail producing a result or providing an incorrect result to save resources in traditional cloud storage.

References

1. C. Gray (2014) Stoo) Vk. Dropbox Why Decentralized Storage is the Future Online]. Available: https://bitcoinmagazine.com/articles/storsdropboxdecentralizedstorage-future-1408177107
2. A. Sahai and B. Waters, Fuzzy identity based encryption in Proe Annu Int. Conf. Theory Appl Cryptograph Techin Berlin, Germany Springer, 2005, pp. 457–473.
3. Zhang, X. A. Wang, and J. Ma. Data owner based attribute based encryption" in Proc. Int. Conf. intellNetw. Collaborative Syst. (INCOS), Sep. 2015, pp. 144–148
4. S. Nakamoto (2008) Bitcoin A Peer to Peer Electronic: Cash System: [Online] Available: https://bitco.in/pdf/bitcoin.pdf
5. G. Wood, Ethereum: A secure decentralised generalised transaction ledger, Yellow Paper Accessed Mar 25, 2018 [Online] Available https://ethereum.gitnode.io/yellowpaper/paper.pdf
6. D. Hopwood, S. Bown, T. Homby, and N. Wilcox Zcash protocol specification
7. Zerocain Electric Coin Company, Oakland, CA, USA, Tech Rep. 2016-1.10.2016.
8. Blockchain for Financial Services Accessed: Mar 25, 2018. (Online) Available: https://www.ibm.com/blockchain/financial-services
9. Blockchain for Supply Chain. Accessed Mar 25, 2018 Online]. Available: https://www.ibm.com/blockchain/supply-chain
10. C. Fromknecht and D. Velicana (2014) A Decentralized Public Kay Infrastructure With Identity Retention [Online]. Available: https://eprint.acs.org/2014/803.pdf
11. Proof of Existence Accessed Mar 25, 2018 [Online] Available: https://proofofexistence.com
12. A Decentralized Network for Internet of Things. Accessed: Mar. 25, 2018. [Online]. Available: https://iotex.io
13. S. Wilkinson, T. Boshevski, J. Brandoff, and V. Buterin, "Storj a peer- to-peer cloud storage network," White Paper. Accessed: Mar. 25, 2018. [Online]. Available: https://storj.io/stofj.pdf
14. J. Benet. (2014). "IPFS-content addressed, versioned, P2P file system." [Online], Available: https://arxiv.org/abs/1407.3561
15. P. Labs. (2018). Filecoin: A Decentralized Storage Network. [Online]. Available: https://filecoin.io/filecoin.pdf
16. G. Zyskind, O. Nathan, and A. S. Pentland, "Decentralizing privacy: Using blockchain to protect personal data," in Proc. Secur. Privacy Work- shops (SPW), May 2015, pp. 180–184.

17. Y. Rahulamathavan, R. C.-W. Phan, M. Rajarajan, S. Misra, and Kondoz, "Privacy-preserving blockchain based lOT ecosystem using attributebased encryption," in Proc. IEEE Int. Conf. Adv. Netw. Telecom- mun. Syst., Dec. 2017, pp. 1–6.
18. H. Es-Samaali, A. Outchakoucht, and J. P. Leroy, "A blockchain-based access control for big data," Int. J. Comput. Netw. Common. Secur., vol. 5, no. 7, pp. 137–147,2017.
19. D. Vorick and L Champine. (2014). Sia: Simple Decentralized Stor-age. [Online]. Available: https://prod.coss.io/documents/white-papers/ siacoin.pdf
20. L Zu, Z. Liu, and J. Li, "New ciphertext-policy attribute-based encryption with efficient revocation," in Proc. IEEE Int. Conf. Comput. Inf. Tech- nol. (CIT), Sep. 2014, pp. 281–287.
21. P. Zhang, Z. Chen, K. Liang, S. Wang, and T. Wang, "A cloud-based access control scheme with user revocation and attribute update," in Proc. Australas. Conf. Inf. Secur. Privacy. Berlin, Germany: Springer, 2016, pp. 525–540.
22. J. Li, Y. Shi, and Y. Zhang, "Searchable ciphertext-policy attribute-based encryption with revocation in cloud storage," Int. J. Commun. Syst., vol. 30, no. 1, p. e2942,2017.
23. J. Li, W. Yao, J. Han, Y. Zhang, and J. Shen, "User collusion avoidance CP-ABE with efficient attribute revocation for cloud storage," IEEE Syst. J., vol. 12, no. 2, pp.1767–1777, 2018.
24. A. Kapadia, P. P. Tsang, and S. W. Smith, "Attribute-based publishing with hidden credentials and hidden policies," in Proc. NDSS, vol. 7,2007, pp. 179–192.
25. Y. Zhang, X. Chen, J. Li, D. S. Wong, H. Li, and I. You, "Ensuring attribute privacy protection and fast decryption for outsourced data security in mobile cloud computing " inf. Sci., vol. 379, pp. 42–61, Feb. 2017.
26. M. Chase, "Multi-authority attribute based encryption," in Proc. Theory Cryptogr. Conf. Berlin, Germany: Springer, 2007, pp. 515–534.
27. H. Qian, J. Li, Y. Zhang, and J. Han, "Privacy-preserving personal health record using multi-authority attribute-based encryption with revocation," Int. J. inf. Secur., vol. 14, no. 6, pp. 487–497,2015.
28. H. S. G. Pussewalage and V. A. Oleshchuk, "A distributed multi-authority attribute based encryption scheme for secure sharing of personal health records," in Proc. 22nd ACM Symp. Access Control Models Techno/., 2017, pp. 255–262.
29. J. Li, Y. Wang, Y. Zhang, and J. Han, "Full verifiability for outsourced decryption in attribute based encryption," IEEE Trans. Services Comput, to be published.
30. J. Li, X. Lin, Y. Zhang, and J. Han, "KSF-OABE: Outsourced attribute- based encryption with keyword search function for cloud storage," IEEE Trans. Services Comput., vol. 10, no. 5, pp. 715–725, Sep./Oct. 2017.
31. D. X. Song, D. Wagner, and A. Perrig, "Practical techniques for searches on encrypted data," in Proc. IEEE Symp. Secur. Privacy, May 2000, pp. 44–55.
32. D. Boneh, G. Di Crescenzo, R. Ostrovsky, and G. Persiano, "Public key encryption with keyword search," in Proc. Int. Conf. Theory Appl. Cryptograph. Techn. Berlin, Germany: Springer, 2004, pp. 506–522.
33. D. Boneh and B. Waters, "Conjunctive, subset, and range queries on encrypted data," in Proc. Theory Cryptogr. Conf. Berlin, Germany: Springer, 2007, pp. 535–554.
34. Z. Wan and R. H. Deng, "VPSearch: Achieving verifiability for privacy- preserving multi-keyword search over encrypted cloud data," IEEE Trans. Dependable Secure Comput., to be published.
35. H. Li, F. Zhang, J. He, and H. Tian. (2017). *'A searchable sym- metric encryption scheme using blockchain." [Online!. Available: https://arxiv.Org/abs/1711.01030
36. C. Cai, X. Yuan, and C. Wang, "Towards trustworthy and private keyword search in encrypted decentralized storage," in Proc. lEEEInt. Conf. Com-mun. (ICC), May 2017, PP 1–7.
37. H. G. Do and W. K. Ng, "Blockchain-based system for secure data storage with private keyword search," in Proc. IEEE World Congr. Services (SERVICES), Jun. 2017, pp. 90–93.
38. P. Jiang, F. Guo, K. Liang, J. Lai, and Q. Wen, "Searchain: Blockchain-based private keyword search in decentralized storage," Future Gener. Comput. Syst., to be published.[Online]. Available: http://www.sciencedirect.com/science/article/pii/S0167739X17318630, https://doi.org/10.1016/j.future.2O17.08.036.

39. Y. Zhang, D. Zheng, X. Chen, J. U, and H. Li, "Computationally efficient ciphertextpolicy attribute-based encryption with constant-size cipher- texts," in Proc. Int. Conf. Provable Secur, vol. 8782. Springer. 2014, pp. 259–273, https://doi.org/10.1007/978-3-319-12475-918.
40. M. Crosby, P. Pattanayak, S. Verma, and V. Kalyanaraman, "Blockchain technology: Beyond bitcoin," Appt. Innov., vol. 2, pp. 6–10, Jun. 2016.
41. Ethereum Homestead Documentation. Accessed: Mar. 25, 2018. [Online]. Available: https://readthedocs.org/projects/ethereum-homestead
42. Ethereum Blockchain App Platform. Accessed: Mar. 25, 2018. [Online], Available: https://www.ethereum.org
43. Diffie-Hellman Key Exchange. Accessed: Mar. 25, 2018. [Online]. Available: https://en.wikipedia.org/wiki/Diffie%E2%80% 93Hellman_key_exchange
44. S. Hu, C. Cai, Q. Wang, C. Wang, X. Luo, and K. Ren. (2018). Searching an Encrypted Cloud Meets Blockchain: A Decentralized. Reliable and Fair Realization.[Online].Available: http://nisplab.whu.edu.cn/paper/infocom_2018_1.pdf
45. Verify Contract Code. Accessed: Mar. 25, 2018. [Online]. Available: https://etherscan.io/verifyContract
46. Units and Globally Available Variables. Accessed: Mar. 25, 2018. [Online]. Available: http://solidity.readthedocs.io/en/latest/units-and-global-variables.html
47. MIRACL Accessed; Mar. 25, 2018. [Online]. Available: https://www.miracl.com

Image Selection for Graphical Password Authentication

P. Anjaneyulu, D. Priyanka, T. Chalapathi, B. Samanvi, and B. Mounika

Abstract One of the most crucial elements of information security is user authentication. User authentication is the primary strategy for ensuring the utmost factors for computer security. It gives the base for access control and user liability. Over time, among numerous types of user authentication systems, alphanumeric passwords have remained the most common. A graphical password is a type of authentication system that relies on the user's selection of images presented in a graphical user interface (GUI), in a specific order. This approach, known as graphical user authentication (GUA), differs from the traditional system of using alphanumeric usernames and passwords, which is presently the most common way of authenticating computer users. This system has been shown to have significant disadvantages. For example, users tend to choose passwords that can be fluently guessed. On the other hand, if a password is delicate to guess, then it is frequently difficult to remember. To overcome this problem of low security, authentication styles are developed by experimenters, using images as a password. This explorative paper is a detailed examination of current graphical passwords, proposing a new proposition. Graphical passwords have been proposed as a volition to textbook-grounded schemes, based on the idea that humans tend to remember pictures more fluently than text. Also, pictures are considered to be more user-friendly.

Keywords Graphical password · Image selection · Authentication · Graphical user authentication (GUA) · Security

1 Introduction

Within information security, user authentication, and data security, there are something known as abecedarian factors. Computer, network, data, and information

P. Anjaneyulu (✉) · D. Priyanka · T. Chalapathi · B. Samanvi · B. Mounika
Department of Computer Science & Engineering, Aditya Institute of Technology and Management, Tekkali, Andhra Pradesh, India

A. Patel et al. (eds.), *Advances in Machine Learning and Big Data Analytics II*,
Springer Proceedings in Mathematics & Statistics 442,
https://doi.org/10.1007/978-3-031-51342-8_9

security have been calculated as the major specialized problems currently faced. Every association, social network, or other platform tries to provide better security for their users, which is accurate and more secure for users. User authentication is a system that keeps unauthorized users from penetrating sensitive information. Research has shown that alphanumeric passwords are filled with both security and usability problems that make them significantly less effective. Old security methods that have been used for a long time give lower security for authentication than advanced security methods [1]. There are some common types of user authentication systems, namely password-based authentication, multi-factor authentication, certificate-based authentication, biometric authentication, and token based authentication.

1.1 Password-Based Authentication

Passwords are the most common style of authentication. They can be in the form of a string of letters, figures, or special characters. To protect yourself, you need to produce strong passwords that include a combination of all three of these considerations. The reality is that there are a lot of passwords to remember. As a result, numerous people choose convenience over security. Most people use simple passwords rather than creating dependable passwords because they are easier to remember. Passwords have a lot of issues and are not sufficient for guarding online information. Hackers can fluently guess user credentials by running through all possible combinations until they find a match.

1.2 Multi-Factor Authentication

Multi-factor authentication (MFA) is an authentication system that requires two or more independent ways to identify a user. Examples include codes generated from the users':

- Smartphone
- Captcha tests
- Fingerprints
- Voice biometrics
- Facial recognition

MFA authentication styles and technologies increase the confidence of users by adding multiple layers of security. MFA can effectively protect against cyber attacks. However, it has its risks: users may lose their phones or SIM cards, preventing them from completing authentication.

1.3 Certificate Authentication

Certificate-based authentication identifies users by using digital instruments. A digital instrument is an electronic document based on the design of a driver's license or passport. The certificate contains the digital identity of a user including a public key and the digital hand of a certification authority. Digital instruments prove the power of a public key and are issued only by a certification authority. Users give their digital certificates when they subscribe to a server, which then verifies the credibility of the digital hand and the certificate authority. The server also uses cryptography to confirm that the user has a correct private key associated with the certificate.

1.4 Biometric Authentication

Biometrics authentication is a security process that relies on the unique biological characteristics of an individual. It is used by consumers, governments, and private entities—including airfields, military bases, and border agencies. This technology increasingly adopted due to the ability to achieve a high position of security without causing friction for the user. Common biometric authentication styles include facial recognition, fingerprint scanners, speaker recognition, and eye scanners.

1.5 Token-Based Authentication

Token-based authentication enables users to enter their credentials once and receive a unique string of characters, known as a token. This token can then be used to access protected systems without reentering credentials each time. The digital token serves as formal proof of authorization. Use cases of token-based authentication include RESTful APIs, which support multiple frameworks and clients.

2 Literature Survey

- Susan Wiedenbeck, Jim Waters, Jean-Camille Birget, Alex Brodskiy, and Nasir Memon. Passpoints: design and longitudinal evaluation of a graphical password system. International Journal of Human-Computer Studies, 63:102–127, July 2005 [2].

Computer security depends largely on passwords to authenticate human users. However, users have difficulty remembering passwords over time if they choose a secure password, i.e., one that is long and random. Therefore, they tend to choose short and insecure passwords. Graphical passwords, which consist of clicking on images rather than typing alphanumeric strings, may help to overcome the problem of creating secure and memorable passwords. In this paper we describe PassPoints, a new and more secure graphical password system. We report an empirical study comparing the use of PassPoints to alphanumeric passwords. Participants created and practiced either an alphanumeric or graphical password. The participants subsequently carried out three longitudinal trials to input their password over the course of six weeks. The results show that the graphical password users created a valid password with fewer difficulties than the alphanumeric users. However, the graphical users took longer and made more invalid password inputs than the alphanumeric users while practicing their passwords. In the longitudinal trials the two groups performed similarly in memory of their password, but the graphical group took more time to input a password.

- Xiaoyuan Suo, Ying Zhu, and G. Scott Owen. Graphical passwords: A survey. In Proceedings of Annual Computer Security Applications Conference, pages 463–472, 2005 [3].

The most common computer authentication method is to use alphanumerical usernames and passwords. This method has been shown to have significant drawbacks. For example, users tend to pick passwords that can be easily guessed. On the other hand, if a password is hard to guess, then it is often hard to remember. To address this problem, some researchers have developed authentication methods that use pictures as passwords. In this paper, we conduct a comprehensive survey of the existing graphical password techniques. We classify these techniques into two categories: recognition-based and recall-based approaches. We discuss the strengths and limitations of each method and point out the future research directions in this area. We also try to answer two important questions:

1. "Are graphical passwords as secure as text-based passwords?";
2. "What are the major design and implementation issues for graphical passwords?" This survey will be useful for information security researchers and practitioners who are interested in finding an alternative to text-based authentication methods.

- Antonella De Angeli, Lynne Coventry, Graham Johnson, and Karen Renaud. Is a picture really worth a thousand words? exploring the feasibility of graphical authentication systems. International Journal of Human-Computer Studies, 63:128–152, July 2005 [4].

The weakness of knowledge-based authentication systems, such as passwords and personal identification numbers (PINs), is well known, and reflects an uneasy compromise between security and human memory constraints. Research has been undertaken for some years now into the feasibility of graphical authentication mechanisms in the hope that these will provide a more secure and memorable alternative. The graphical approach substitutes the exact recall of alphanumeric codes with the recognition of previously learnt pictures, a skill at which humans are remarkably proficient. So far, little attention has been devoted to usability, and initial research has failed to conclusively establish significant memory improvement.

This paper reports on two user studies comparing several implementations of the graphical approach with PINs. Results demonstrate that pictures can be a solution to some problems relating to traditional knowledge-based authentication but are not a simple panacea, since a poor design can eliminate the picture superiority effect in memory. The paper concludes by discussing the potential of the graphical approach and providing guidelines for developers contemplating using these mechanisms.

- E. Blonder. Graphical password. U.S. Patent 5559961, Lucent Technologies, Inc. (Murray Hill, NJ), August 1995 [5].

A graphical password arrangement displays a predetermined graphical image and requires a user to "touch" predetermined areas of the image in a predetermined sequence, as a means of entering a password. The password is set by allowing the arrangement to display the predetermined areas, or *tap regions*, to a user, and requiring the user to position these tap regions in a location and sequence within the graphical image, with which the user desires the password to be set at. These tap regions are then removed from the display, leaving the original image by itself. The arrangement then waits for an entry device (user) to select the tap regions, as described above, for access to a protected resource.

Blonder [1] proposed a graphical password authentication technique for the first time. According to the introduced process, a user can select some click points to choose a password from a predefined image in the registration phase. At the time of login, the user selects points which were chosen in the registration phase. If those points were matched, then the user is identified as an authorized user Feature Extraction.

3 Existing System

The existing system also uses traditional mechanisms like alphanumeric passwords that are hard to remember, and it also uses fingerprints and recognition systems that are complex to use. The existing system uses captcha. A captcha test is used to identify whether a user is a human or a bot [6].

And by using a captcha and an alphanumeric password, anyone can be authenticated easily as a user. There can be a chance for unauthorized access. Strong passwords are often difficult to remember, which can lead users to write them down or store them in insecure locations. A weakness of captcha is that some patterns could be hard to read for older human users. Alphanumeric passwords and captchas only protect against a limited set of threats and may not be effective against more sophisticated attacks, such as social engineering. The paper emphasizes the need for strong passwords and effective defenses against password-cracking techniques [7].

Disadvantage

- Less secure.
- Vulnerable to brute force attacks.

Factors techniques	Installation cost	Security	Required devices	Reliability	User acceptance
Textual password	–	Low	–	High	Very high
Fingerprint identification	Medium	High	Fingerprint scanner	High	High
Face recognition	Medium	High	Camera	High	Medium
Hand geometry	Medium	High	Scanner	High	High
Voice analysis	Medium	Medium	Microphone	High	Low
Signature recognition	Low	Low	Touch panel, optic pen	High	Very high

4 Proposed System

In the proposed solution, we want to use another mechanism such as a graphical password. In a graphical password authentication system, the user has to select from images among a lot of images, presented to them in a graphical user interface (GUI). A graphical password system is a method of authentication that uses images rather than alphanumeric characters to verify the identity of a user. Instead of typing a password, the user selects a series of images among groups of images to unlock their account [8].

In a graphical password system, users are presented with a set of images or pictures from which they choose the right sequence [9].

This is more user-friendly and provides higher security than other traditional password schemes. It provides strong security against bot attacks.

Graphical password systems offer a number of advantages over traditional alphanumeric passwords, particularly in terms of ease of use, enhanced security, and accessibility.

Advantages

- More secure
- Infinite search space

5 Problem Description

The problem at hand is to design and implement an image selection system for graphical password authentication. Graphical password authentication is an alternative to traditional text-based passwords that utilize images or graphical elements as a means of authentication. In this system, users are required to select specific images from a set of options to create a password. However, there are several challenges and issues that need to be addressed to ensure the effectiveness and security of the graphical password authentication system.

User Memorability One of the primary concerns with graphical passwords is the ability of users to remember their chosen images. Unlike text-based passwords that can be easily memorized, remembering specific images or graphical patterns can be more challenging. The problem lies in finding a balance between selecting images that are meaningful and memorable to the user while being unique and not easily guessed by attackers.

Image Set Selection Choosing the appropriate set of images for users to select from is crucial. The images should be diverse, visually distinct, and universally recognizable to accommodate users from different cultural backgrounds. The challenge lies in creating an image set that is large enough to provide sufficient options for users while ensuring that the images are relevant and meaningful to a wide range of users.

Security and "Guessability" The security of graphical passwords depends on the ability of users to select images that are difficult to guess or predict. The system must mitigate the risk of attackers guessing or brute-forcing the password by analyzing the patterns or preferences of users. The challenge is to develop algorithms or techniques that can detect and prevent common vulnerabilities, such as users consistently selecting images from the same region of the image grid or following predictable patterns [10].

Usability and User Experience The graphical password authentication system should be intuitive and easy to use for users. The challenge lies in designing a user-friendly interface that allows users to easily select and remember their chosen images without experiencing frustration or confusion. Considerations such as image size, grid layout, and feedback mechanisms need to be considered to enhance usability and user experience.

System Performance and Efficiency As graphical password authentication involves image processing and comparison; the system's performance and efficiency are important factors. The system should be able to handle a large number of users concurrently, perform quick and accurate image comparisons, and ensure low latency during the authentication process. The challenge is to optimize the system's algorithms and infrastructure to achieve a balance between security and performance.

Addressing these challenges will contribute to the development of a robust and secure image selection system for graphical password authentication, enhancing the overall security of user accounts and data.

6 Implementation

- *User registration*: Django provides built-in authentication views and forms, which you can use to handle user registration. You can create a custom registration form with the required fields and validation and use the built-in UserCreationForm to handle the actual creation of the user. Once the user is registered, you can redirect them to a page where they can select images and sequences.
- *Image selection*: To allow users to select images and sequences, you can create a page with a list of all available images and sequences. You can use Django's built-in templates and views to render this page and handle user interactions. You can provide checkboxes or other input elements to allow users to select images and sequences. Once the user has made their selection, you can save their choices in a database.

  - *Login*: Django provides built-in authentication views and forms for handling user login. You can create a custom login form with the required fields and validation and use the built-in Authentication Form to handle the actual authentication. Once the user is logged in, you can redirect them to a page where they can choose the same sequence as before.
  - *Same sequence selection*: To allow users to choose the same sequence as before, you can create a page that displays their previously selected images and sequences. You can use Django's built-in templates and views to render this page and handle user interactions. Once the user has made their selection, you can save their choices in a database.
  - *Login status*: After the user attempts to login, you can display a message indicating whether the login was successful or not. You can use Django's built-in authentication views and forms to handle the login and display the appropriate message.

Implementing these requirements using Django should be relatively straightforward, given its built-in authentication views and forms, as well as its powerful template engine and database ORM as shown in Fig. 1.

Email Verification

In a graphical password system, email verification can be used to confirm the identity of the user when they forget their password. Here's how it could work:

- The user clicks the Forgot Password button on the login page. The system prompts the user to enter the email address associated with their account.
- The system sends a verification email to the provided email address with a unique code or link that the user needs to click to confirm that they initiated the password reset.

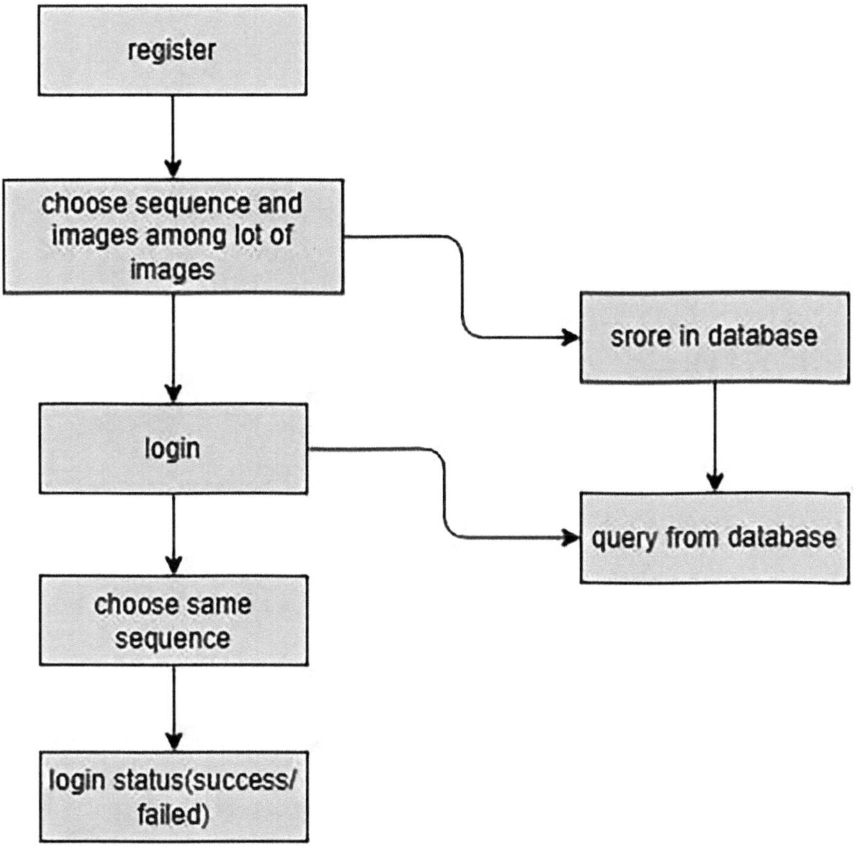

Fig. 1 Flow chart of graphical password authentication system for web-based applications

- Once the user clicks the link or enters the code, the system confirms that the email address belongs to the user and allows the user to reset their password.
- The user is then directed to a new page where they can create a new password for their account. After the user successfully creates a new password, they can log in to their account using their new password.
- This process ensures that the user is the legitimate owner of the email address associated with the account and prevents unauthorized access to the user's account. It also allows the user to securely reset their password without the need for human intervention or assistance.

Software used: HTML, PHP, PyCharm, Django

7 Result and Analysis

Figures 2, 3, 4, and 5 show the result screens.

Fig. 2 Front page

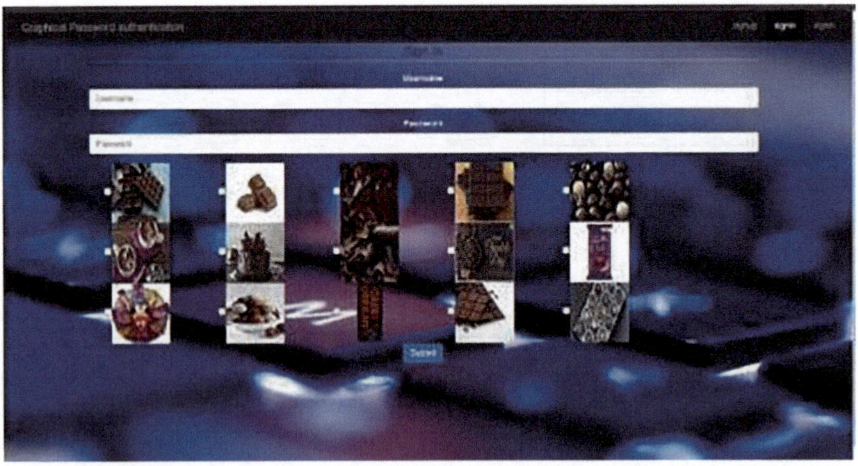

Fig. 3 Signup page

Fig. 4 Forgot page

8 Conclusion

Implementing image selection and sequence choice for password creation and login could provide several advantages over traditional alphanumeric passwords.

- *Ease of use*: Users may find it easier to remember images and sequences compared to random alphanumeric passwords, which can be more difficult to remember and require frequent resetting.
- *Increased security*: Image selection and sequence choice may provide increased security over traditional passwords, as they are less vulnerable to brute force attacks and more difficult to guess.
- *Reduced risk of password reuse*: Users often reuse passwords across different accounts, which can increase the risk of security breaches. With image selection and sequence choice, users are less likely to use the same password for different accounts, reducing the risk of password reuse.

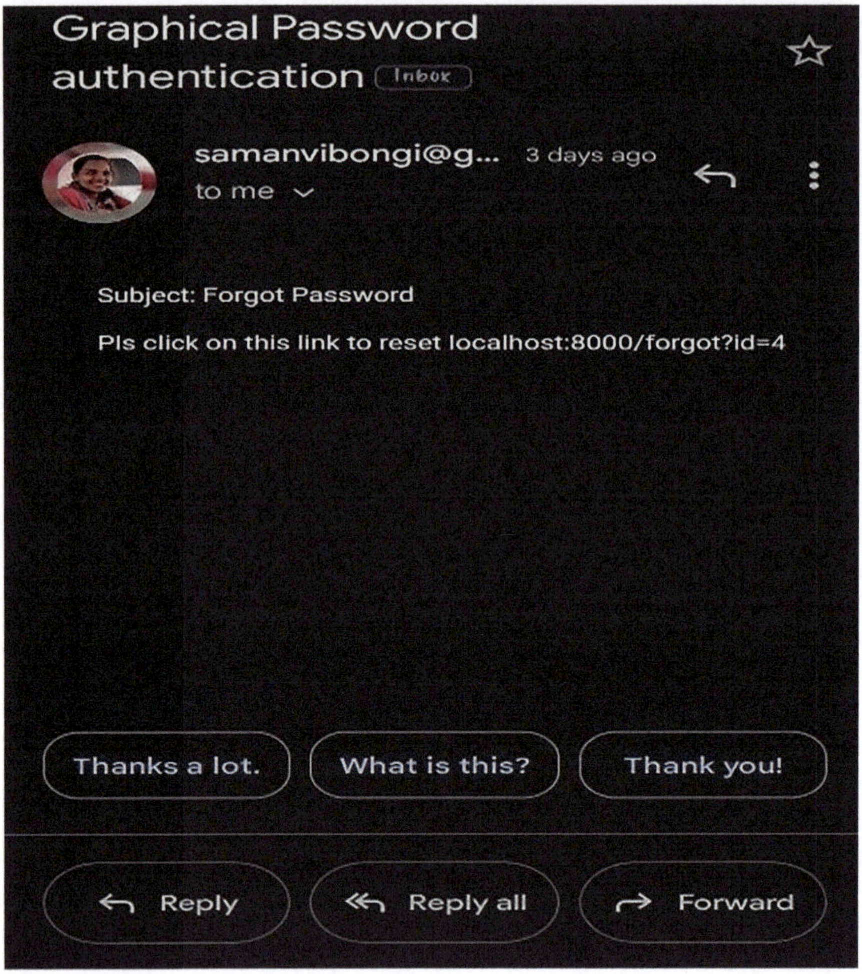

Fig. 5 Email validation for forgotten password

References

1. Real User Corporation. The science behind passfaces, June 2004.
2. Susan Wiedenbeck, Jim Waters, Jean-Camille Birget, Alex Brodskiy, and Nasir Memon. Passpoints: design and longitudinal evaluation of a graphical password system. International Journal of Human-Computer Studies, 63:102–127, July 2005.
3. Xiaoyuan Suo, Ying Zhu, and G. Scott Owen. Graphical passwords: A survey. In Proceedings of Annual Computer Security Applications Conference, pages 463– 472, 2005.
4. Antonella De Angeli, Lynne Coventry, Graham Johnson, and Karen Renaud. Is a picture really worth a thousand words? exploring the feasibility of graphical authentication systems. International Journal of Human- Computer Studies, 63:128–152, July 2005

5. G. E. Blonder. Graphical password. U.S. Patent 5559961, Lucent Technologies, Inc. (Murray Hill, NJ), August 1995.
6. Robert Morris and Ken Thompson. Password security: a case history. Communications of the ACM, 22:594– 597, November 1979.
7. Daniel V. Klein. Foiling the Cracker: A Survey of, and Improvements to, Password Security. In Proceedings of the 2nd USENIX UNIX Security Workshop, 1990.
8. Eugene H. Spafford. Observing reusable password choices. In Proceedings of the 3rd Security Symposium. Usenix, pages 299–312, 1992.
9. William Stallings and Lawrie Brown. Computer Security: Principle and Practices. Pearson Education, 2008.
10. Sigmund N. Porter. A password extension for improved human factors. Computers & Security, 1(1):54—56, 1982.

Soft Computing for Visual Recognition Through Audio for the Visually Impaired

S. R. K. L. Amulya, Vishakha Singh, Tummalapalli Sandeep, Surya Sasidhar, and Rita Roy

Abstract Visually impaired people are frequently unaware of outside obstructions and require assistance to avoid dangerous collisions. The sense of sight is the most fundamental aspect of human perception and plays. We propose an object recognition algorithm and assistance system that is very useful for their safety and quality of life. This project aims to use an available mobile device as a talking guide dog to aid those with vision impairments in outdoor navigation and inform them about obstacles. The proposed technology will lower the risks of obstacle contact by allowing users to move outside without stumbling, potentially informing the user what the objects are. The approach proposed uses algorithms of deep learning. CNN recognizes salient functions and captions of photos and converts written text to speech by detecting features through the broadcasted picture alongside its relevant caption, while the Gated Recurrent Unit (GRU) network acts as a tool that captions and describes the text detected from photographs. The predicted caption is transformed into an audio message. The proposed Convolution Neural Networks [CNNs] Gated Recurrent Units [GRUs] model has investigated the usage of numerous network architectures: Inception Net, Mobile Net, Xception Net, ResNet, and VGG16.

Keywords Convolution Neural Network · Gated Recurrent Unit · Inception Net · Mobile Net · Xception Net

1 Introduction

Vision is a sensory modality. It is the eye and brain's acute and bidirectional response to various means. The eyes act as light receptors, sending signals to the brain as a starting point for visual functions such as, for example, contrast and

---

S. R. K. L. Amulya · V. Singh · T. Sandeep · S. Sasidhar · R. Roy (✉)
GITAM(Deemed to be University), Visakhapatnam, India

A. Patel et al. (eds.), *Advances in Machine Learning and Big Data Analytics II*,
Springer Proceedings in Mathematics & Statistics 442,
https://doi.org/10.1007/978-3-031-51342-8_10

color perception. According to the International Classification of Diseases 11, vision impairment is classified into distance and near-presenting. These disabilities restrict the ability of visually impaired individuals to navigate and complete their daily tasks, consequently impacting their quality of life [1]. Legally, a blind person has less than 20/200 vision. Although not all individuals with visual impairments are completely blind, most utilize white canes or service animals to help them get around and understand their environment [2]. These individuals require assistance to navigate and maintain independence in mobility. Mobility refers to the ability to move freely and independently, without a guide, at home, or in unfamiliar settings. Many researchers have built navigation systems for blind people. Most indoor and outdoor navigation technologies have limitations in accuracy, usability, interoperability, and coverage, all of which are challenging to overcome with current technology [3]. Assistive technology refers to all systems, services, and devices that help individuals with disabilities daily. During the 1960s, assistive technology was introduced to address issues about the transmission of information in daily life, specifically in personal care situations, and facilitate mobility assistance related to orientation-based needs [4]. The conventional white cane is a widely used and simple method for detecting objects because of its affordability and rarity of errors [5, 6]. Individuals can effectively roam the area before them and detect objects such as floors with steps, walls, uneven surfaces, or lower surfaces. The dog guides the user through obstacles and signals when it is necessary to change lanes [7]. However, visually impaired or blind people may have difficulty grasping the intricate signals these dogs give.

2 Literature Review

An approach was created in the paper of Azhar et al. [8] that provides a comprehensive method for extracting semantic data from construction images. This method involves using a CNN model to detect notable features within the photo and an Encoder model based on Mask RCNN to generate descriptive keywords based on those features. This approach has proven effective, per the data collected from 174 construction sites with 41,668 images—which were split and set to train the model; this method yielded BLEU Scores of 0.61.

AraCap was developed as a new method for generating Arabic image captions utilizing object-based and image captioning approach frameworks. The performance of the method was assessed using both COCO and Flickr30k datasets. It is composed of two sequential processes: object detection and image captioning. Furthermore, a similarity score was implemented to generate captions compared to the public database's original captions, surpassing the basic captioning technique in its results [9].

This application harnesses image processing and machine learning to identify objects in real time for blind people. It does this by launching a live video stream from its camera, which is processed through the YOLO (You Only Look Once) algorithm. Of the versions available, YOLOv3 was the optimal choice regarding speed and accuracy. The YOLOv3 dataset was applied for web applications, while Tiny Yolo was used on Android devices [10]. With a maximum accuracy rate of 85.5%, this model successfully conveys object information and location to visually impaired users through the audio output [11].

A method that uses computer vision machine learning algorithms to detect an object, ultimately aiding the visually impaired and blind. By training convolution neural networks on the ImageNet dataset, a system can be established that detects objects and narrates their information to those affected. The VGG16 algorithm is one of the CNN models used in the proposed system. Furthermore, with access to a website webserver and phone camera, YOLO detects images by finding the objects within them and sends back results which are then communicated via a browser-based voice library (Google Voice) [12]. The model's accuracy was tested across phones, ranging from 62.5% to 75%.

The application processes an input image in the first unit (Edge detection) through several steps. In the second unit (sound production), a MATLAB program is used to create the sound. In the third module, a statistical property such as Wavelet coefficient means [13].

This paper presents a method that can identify objects dynamically and classify those according to pre-defined categories. Earlier, object detection was possible only with the help of IR and RFID technologies, which demanded specialized hardware. Using image processing and neural networks enables us to accomplish this task without additional hardware requirements. We also use alarm notification systems, such as buzzers, to reduce the chances of an accident. Moreover, the Dangling Object Detection algorithm estimates the object's position and warns whenever it falls within a hazardous range [14].

Additionally, Haar Cascade is used to construct a similar comparison model for accurate recognition [15]. This technique is based on transmitter-receiver technology wherein the objects are fitted with transmitters. At the same time, people hold receivers to gather data from tags which pass information to readers to help make decisions [16].

A combination of motion information and grey images of the human face area to generate spectrogram masks. Each face image paired with a corresponding spectrum image is trained using Dilated CNN (DCNN), resulting in an "audio-visual combination image." Next, LSTM and FC layers generate complex masks, producing the spectrogram of each person's voice. Finally, the model can recognize individuals based on their voices [17].

In the research mentioned earlier, machine learning models that rely on deep learning have displayed high accuracy in recognizing speech. The proposed ML-CNN approach was utilized, including the use of CNNs, MFCCs, and LibROSA, along with a new dataset of input speech signals and parameters, to identify the optimal system performance for identifying unusual types of input speech signals. The findings exhibited a vocabulary sentence word error rate of 10.56% for speech signals with diverse tones [18]. The program will alert the user when an object is close to avoid a collision. The detection of objects is accomplished by tensor flow lite, and the information is conveyed to the user through a voice prompt [19].

The system can be realized by semantic segmentation, assigning semantic labels (such as a tree or a sidewalk) to every pixel in the input image. It contains labeled images of urban street scenes with the training of 2975 image files and validation of 500 image files. Each image file of the size is 256 × 512 pixels. This paper suggests an outdoor obstacle identification project proposal for blind people with DeepLabv3+ [20].

The technique proposed involves using Raspberry Pi as the main tool. To begin with, you need to download and install the Raspberry Pi operating system. After that, using a display device, you can set up the Raspberry Pi's IP address, and the VNC server is installed to train the dataset for object recognition. This process requires you to collect several images of objects [21].

The remote sensing captioning model was developed using VRTMM. This method employs a CNN to extract semantic and spatial features from a picture, followed by Reinforcement Learning that enhances the quality of resulting phrases. The publicly available dataset of Remote Sensing Images, which consists of images of 31,500 and scene classifications of 45, was utilized to recognize the remote sensing image environment [22].

3 Methods

3.1 Convolutional Neural Networks

Convolutional Neural Networks (CNNs) are a type of deep learning algorithm that is particularly adept at processing data with a spatial or temporal relationship [23]. While similar to other types of neural networks, CNNs possess an added layer of complexity due to their use of convolutional layers [24]. This makes them particularly well-suited for processing this type of data. The three main components of CNNs include (Fig. 1):

1. *Convolutional layer:* The basic unit of a Convolutional Neural Network is the convolutional layer, which performs linear and nonlinear operations to extract features from an array of numbers known as a tensor. This layer employs a small numerical array, called a kernel or mask, to process the input [25].
2. *Pooling layer:* The pooling layer sweeps a kernel or filters across the input image. A pooling layer provides a typical operation that reduces the dimensionality of the image by using max or average pooling.
3. *Fully connected layer:* The fully connected layer in the CNN classifies images based on the features extracted from the previous layers. In fully connected, all input neurons are connected to all active neurons [26].

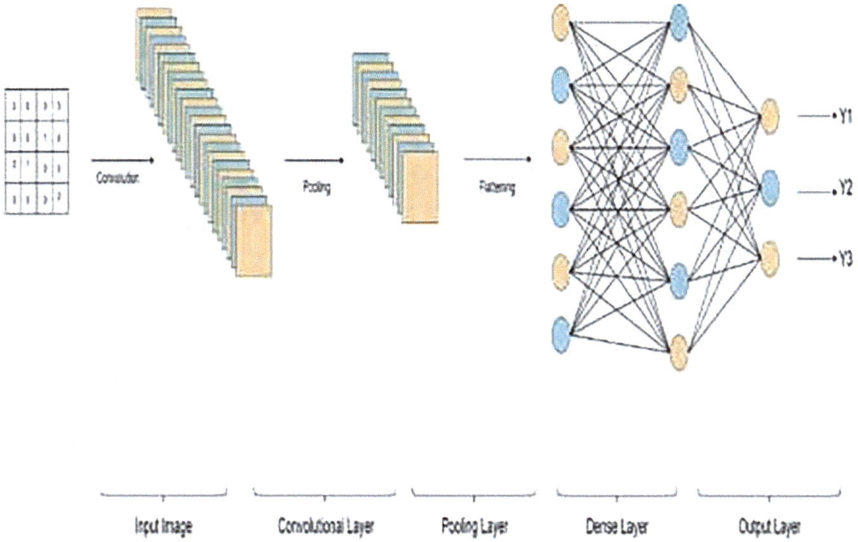

Fig. 1 CNN architecture

3.2 Types of CNN Used

1. *VGG16 Net:* It also has 138 million parameters. Replaces large kernel filters in Convolutional layers to 11 and 5 in the First and Second layers. Input to this model is fixed to 224*224224*224 RGB image. It has 16 layers, as the name suggests, that have some weights. It is very good for Deep Learning for setting benchmarks on any particular task [27].
2. *ResNet:* The name ResNet followed by a two or more-digit number indicates the ResNet architecture with a certain number of Neural Network Layers. ResNet can train more than thousands of layers and yield great performance. ResNet uses its core for Batch Normalization. It increases its performance by using a bottleneck residual block design. It protects the network from vanishing gradient problems by using the Identity condition.
3. *InceptionNet:* It was proposed by Google developers. It is also called as GoogleLeNet. Selection of the right size kernel is always difficult. That is why it resolves the issue by stacking multiple kernels simultaneously. The inception layer provides a parallel Max Pooling layer [28].
4. *XceptionNet:* François Chollet introduced the deep literacy convolutional neural network armature known as Xception (Extreme Inception) in 2016. Although it is an extension of the Inception armature, the convolutional subcaste design differs. It has been demonstrated that Xception outperforms other deep literacy infrastructures regarding state-of-the-art performance on image bracket tasks using the ImageNet dataset. Each subcaste in Xception comprises a depth-wise

divisible complication, combining a pointwise complication with a depth-wise one. The pointwise complication combines the depth-wise complication's labors, whereas the depth-wise complication combines each input channel singly.

5. *MobileNet:* MobileNet is a deep literacy neural network armature specifically created for mobile and bedded bias with constrained processing capabilities, similar to smartphones, tablets, and wearable tech. Due to the use of depth-wise divisible convolutional layers, MobileNet is distinguished by its modest size and low computing complexity. Through these layers, a typical convolutional subcaste is divided into two lower layers a depth-wise complication that filters each input channel singly and a pointwise complication that combines the labors of the depth-wise complication.

3.3 Recurrent Neural Networks

Recurrent Neural Networks (RNNs) are a type of neural network that uses the previous step's output as input to the current step. These networks can generate models for time-dependent and sequential data problems such as stock prediction and text generation. The Hidden Layer or State of RNNs is a crucial feature that stores information about the input of the sequence.

1. *Gated Recurrent Unit:* The Gated Recurrent Unit (GRU) is a variant of Recurrent Neural Network (RNN) that serves as a simpler alternative to Long Short-Term Memory (LSTM). Invented in 2014, GRU leverages gating mechanisms to regulate information transfer between cells, making it faster and more memory-efficient than LSTM. However, in datasets with longer sequences, LSTM is more accurate than GRU. GRU is commonly utilized in natural language processing, speech recognition, and other sequential data applications.

3.4 Dataset

We have used the Flickr8k dataset [30] for this study, consisting of 8091 images, each with five corresponding captions.

3.5 Pre-processing

The images were resized to (244,244). As for the captions, we created a tokenizer and removed spaces and unnecessary punctuation. Then a word-to-index and vice versa mapping was created.

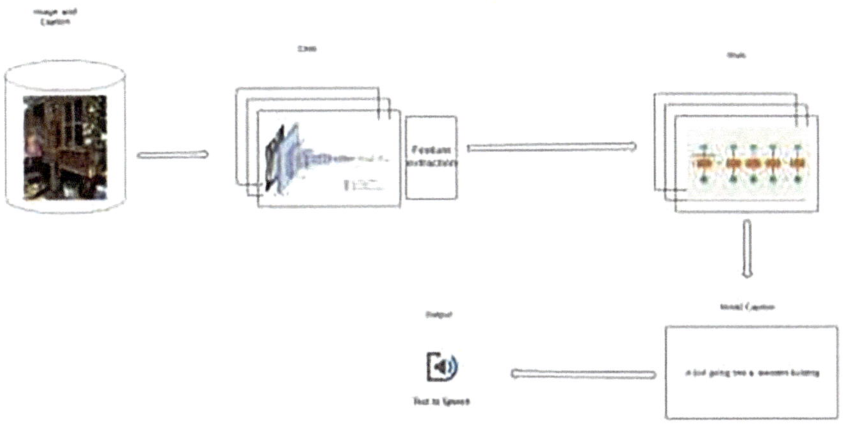

Fig. 2 System overview

3.6 Model Creation

We created a train and test split using the train test split library on image path and captions. The data was split in 80:20%, respectively. We built the encoder, which is CNN, responsible for extracting and vectorizing features. Then we make use of an interface to connect the encoder and decoder. The input to the GRU is a vector concatenated with an embedded vector [29]. As an output, it generates the caption.

The same steps were done for all the CNN models with GRU. The epoch's size was 25, and this was measured using the Bleu metric. The best Bleu score of 0.86 was given by Xception (Fig. 2).

4 Results

The model was presented with the dataset as an input during the training phase. The project utilized CNN and GRU to generate captions for images: CNN was responsible for capturing image features, while GRU produced the captions. Once the training was completed, the model could summarize the content of an image. The results are given below (Table 1).

The process was done on different CNN models, and the performance was calculated using the Bleu score; from the above table and graph, we see that the Xception Net-GRU model performed the best.

Table 1 Comparison of models

Model	Bleu score
VGG-16-GRU	0.75
InceptionV3-GRU	0.78
Mobile Net-GRU	0.80
ResNet50-GRU	0.81
Xception Net-GRU	0.86

5 Conclusion

This study aims to create a smart system powered by deep learning technology to aid individuals with visual impairments. The system includes the following: CNN extracts relevant data, and GRU identifies visual input. The system uses a speech synthesis module to give users voice messages containing data. The study suggests that GRU models are the most suitable for generating image descriptions and predictions. Although there are limitations, such as using the Flicker8k datasets, this intelligent system simplifies understanding text and images for visually impaired individuals. Future investigations can focus on using larger datasets for learning and obtaining image descriptions based on real-time images and their descriptions, thus enhancing the system.

References

1. J. Ganesan, A. T. Azar, S. Alsenan, N. A. Kamal, B. Qureshi, and A. E. Hassanien, "Deep Learning Reader for Visually Impaired," *Electronics 2022, Vol. 11, Page 3335*, vol. 11, no. 20, p. 3335, Oct. 2022, doi: https://doi.org/10.3390/ELECTRONICS11203335.
2. S. Paiva, A. Amaral, J. Gonçalves, R. Lima, and L. Barreto, "Image Recognition-Based Architecture to Enhance Inclusive Mobility of Visually Impaired People in Smart and Urban Environments," *Sustainability 2022, Vol. 14, Page 11567*, vol. 14, no. 18, p. 11567, Sep. 2022, doi: https://doi.org/10.3390/SU141811567.
3. K. G. Krishnan, C. M. Porkodi, and K. Kanimozhi, "Image recognition for visually impaired people by sound," *International Conference on Communication and Signal Processing, ICCSP 2013 – Proceedings*, pp. 943–946, 2013, doi: https://doi.org/10.1109/ICCSP.2013.6577195.
4. R. Roy, K. Chekuri, J. L. Prasad, and S. Mukherjee, "Discussing the Future Perspective of Machine Learning and Artificial Intelligence in COVID-19 Vaccination: A Review," in *Applications of Computational Intelligence in Management & Mathematics*, 2023, pp. 151–160. doi: https://doi.org/10.1007/978-3-03125194-8_12.
5. M. M. Baral, U. V. A. Rao, K. S. Rao, G. C. Dey, S. Mukherjee, and M. A. Kumar, "Achieving Sustainability of SMEs Through Industry 4.0-Based Circular Economy," *International Journal of Global Business and Competitiveness*, May 2023, doi: https://doi.org/10.1007/s42943-023-00074-2.
6. R. Nagariya, S. Mukherjee, M. M. Baral, and V. Chittipaka, "Analyzing blockchain-based supply chain resilience strategies: resource-based perspective," *International Journal of Productivity and Performance Management*, May 2023, doi: https://doi.org/10.1108/IJPPM-07-2022-0330.

7. X. Shen, B. Liu, Y. Zhou, J. Zhao, and M. Liu, "Remote sensing image captioning via Variational Autoencoder and Reinforcement Learning," *Knowl Based Syst*, vol. 203, p. 105920, Sep. 2020, doi: https://doi.org/10.1016/J.KNOSYS.2020.105920.
8. I. Azhar, I. Afyouni, and A. Elnagar, "Facilitated deep learning models for image captioning," *2021 55th Annual Conference on Information Sciences and Systems, CISS 2021*, Mar. 2021, doi: https://doi.org/10.1109/CISS50987.2021.9400209.
9. I. Afyouni, I. Azhar, and A. Elnagar, "AraCap: A hybrid deep learning architecture for Arabic Image Captioning," *Procedia Comput Sci*, vol. 189, pp. 382–389, Jan. 2021, doi: https://doi.org/10.1016/J.PROCS.2021.05.108.
10. Y. R. S. Kumar, T. Nivethetha, P. Priyadharshini, and U. Jayachandiran, "SMART GLASSES FOR VISUALLY IMPAIRED PEOPLE WITH FACIAL RECOGNITION," *2022 International Conference on Communication, Computing and Internet of Things, IC3IoT 2022 – Proceedings*, 2022, doi: https://doi.org/10.1109/IC3IOT53935.2022.9768012.
11. R. Roy, M. Ravindra, N. Marada, S. Mukherjee, and M. M. Baral, "Machine Learning Techniques for the Prediction of Bovine Tuberculosis Among the Cattle," in *Proceedings of International Conference on Data Science and Applications*, Volume no-551.Proceedings of International Conference on Data Science and Applications. Lecture Notes in Networks and Systems, vol 551. Springer, Singapore. Springer, Singapore, 2023, pp. 295–303. doi: https://doi.org/10.1007/978-981-19-6631-6_21.
12. A. Iqbal, F. Akram, M. I. Ul Haq, and I. Ahmad, "A comprehensive assistive solution for visually impaired persons," *Proceedings – 2022 2nd International Conference of Smart Systems and Emerging Technologies, SMARTTECH 2022*, pp. 60–65, 2022, doi: https://doi.org/10.1109/SMARTTECH54121.2022.00027.
13. M. H. Vardhan, A. V. Krishna, B. H. Goud, P. V. Reddy, and R. Aluvalu, "Framework for Object Recognition and Detection for Blind Users Using Deep Learning," *Lecture Notes in Networks and Systems*, vol. 648 LNNS, pp. 862–870, 2023, doi: https://doi.org/10.1007/978-3-031-27524-1_84/COVER.
14. R. Roy, K. Chekuri, G. Sandhya, S. K. Pal, S. Mukherjee, and N. Marada, "Exploring the blockchain for sustainable food supply chain," *Journal of Information and Optimization Sciences*, vol. 43, no. 7, pp. 1835–1847, Oct. 2022, doi: https://doi.org/10.1080/02522667.2022.2128535.
15. R. Shah, D. Tamboli, A. Makwana, R. Baria, K. Shekokar, and S. Degadwala, "A Survey on Visionary Eye for Visual Impairment," *Int J Sci Res Sci Eng Technol*, pp. 33–39, Nov. 2020, doi: https://doi.org/10.32628/IJSRSET20765.
16. R. Roy, M. D. Babakerkhell, S. Mukherjee, D. Pal, and S. Funilkul, "Evaluating the Intention for the Adoption of Artificial Intelligence-Based Robots in the University to Educate the Students," *IEEE Access*, vol. 10, pp. 125666– 125678, 2022, doi: https://doi.org/10.1109/ACCESS.2022.3225555.
17. S. A. Jakhete, P. Bagmar, A. Dorle, A. Rajurkar, and P. Pimplikar, "Object Recognition App for Visually Impaired," *2019 IEEE Pune Section International Conference, PuneCon 2019*, Dec. 2019, doi: https://doi.org/10.1109/PUNECON46936.2019.9105670.
18. F. Ashiq *et al.*, "CNN-Based Object Recognition and Tracking System to Assist Visually Impaired People," *IEEE Access*, vol. 10, pp. 14819–14834, 2022, doi: https://doi.org/10.1109/ACCESS.2022.3148036.
19. S. Mukherjee, M. M. Baral, S. K. Pal, V. Chittipaka, R. Roy, and K. Alam, "Humanoid robot in healthcare: A Systematic Review and Future Research Directions," *2022 International Conference on Machine Learning, Big Data, Cloud and Parallel Computing (COM-IT-CON)*, pp. 822–826, May 2022, doi: 10.1109/COM-IT-CON54601.2022.9850577.
20. J. Nasreen, W. Arif, A. A. Shaikh, Y. Muhammad, and M. Abdullah, "Object Detection and Narrator for Visually Impaired People," *ICETAS 2019 – 2019 6th IEEE International Conference on Engineering, Technologies and Applied Sciences*, Dec. 2019, doi: https://doi.org/10.1109/ICETAS48360.2019.9117405.

21. D. Ienco, R. Interdonato, R. Gaetano, and D. Ho Tong Minh, "Combining Sentinel-1 and Sentinel-2 Satellite Image Time Series for land cover mapping via a multi-source deep learning architecture," *ISPRS Journal of Photogrammetry and Remote Sensing*, vol. 158, pp. 11–22, Dec. 2019, doi: https://doi.org/10.1016/J.ISPRSJPRS.2019.09.016.
22. et al. Ephzibah E.P, "Assisting Blind People Using Machine Learning Algorithms," *Turkish Journal of Computer and Mathematics Education (TURCOMAT)*, vol. 12, no. 8, pp. 3162–3170, Apr. 2021, doi: https://doi.org/10.17762/TURCOMAT.V12I8.4161.
23. S. S. Altyar, S. S. Hussein, and M. J. Mohammed, "Human recognition by utilizing voice recognition and visual recognition," *International Journal of Nonlinear Analysis and Applications*, vol. 13, no. 1, pp. 343–351, Jan. 2022, doi: https://doi.org/10.22075/IJNAA.2022.5501.
24. R. Roy, M. M. Baral, S. K. Pal, S. Kumar, S. Mukherjee, and B. Jana, "Discussing the present, past, and future of Machine learning techniques in livestock farming: A systematic literature review," *2022 International Conference on Machine Learning, Big Data, Cloud and Parallel Computing (COMIT-CON)*, pp. 179–183, May 2022, doi: https://doi.org/10.1109/COM-IT-CON54601.2022.9850749.
25. S. Mukherjee and V. Chittipaka, "Analysing the Adoption of Intelligent Agent Technology in Food Supply Chain Management: An Empirical Evidence," *FIIB Business Review*, vol. 11, no. 4, pp. 438–454, Dec. 2022, doi: https://doi.org/10.1177/23197145211059243.
26. P. J. Chun, T. Yamane, and Y. Maemura, "A deep learning-based image captioning method to automatically generate comprehensive explanations of bridge damage," *Computer-Aided Civil and Infrastructure Engineering*, vol. 37, no. 11, pp. 1387–1401, Sep. 2022, doi: https://doi.org/10.1111/MICE.12793.
27. Y. Wang, B. Xiao, A. Bouferguene, M. Al-Hussein, and H. Li, "Vision-based method for semantic information extraction in construction by integrating deep learning object detection and image captioning," *Advanced Engineering Informatics*, vol. 53, p. 101699, Aug. 2022, doi: https://doi.org/10.1016/J.AEI.2022.101699.
28. S. Mukherjee, M. M. Baral, B. L. Lavanya, R. Nagariya, B. Singh Patel, and V. Chittipaka, "Intentions to adopt the blockchain: investigation of the retail supply chain," *Management Decision*, vol. ahead-of-print, no. ahead-of-print, 2023, doi: https://doi.org/10.1108/MD-03-2022-0369/FULL/PDF.
29. U. Awan, L. Hannola, A. Tandon, R. K. Goyal, and A. Dhir, "Quantum computing challenges in the software industry. A fuzzy AHP-based approach," *Inf Softw Technol*, vol. 147, p. 106896, Jul. 2022, doi: https://doi.org/10.1016/J.INFSOF.2022.106896.
30. Datasource https://www.kaggle.com/datasets/adityajn105/flickr8k

Diet Meal Plan Chatbot

Saubhagya Jung Karki, Sarthak Pokhrel, Saugat Chand Thakuri, Anupam Pandey Chhetry, and Yashpal Singh

Abstract A healthy life is an important issue that has been affecting a large part of the human population as the obesity rate started to rise persistently while people have been rebuking it instead of finding a solution. We need to find out the method before having stinging defeat. Currently, AI-based software for personalized nutrition is getting hyped up for users to gain a healthy lifestyle. Chatbots empowered by Artificial Intelligence (AI) can be solutions to engage with humans in natural conversation and build relationships in the era of the Internet. Here in this paper, we present the knowledge of a Chatbot called "Diet-meal Chatbot" built by us to pitch it on the perfect line for the betterment of human beings. A diet meal chatbot is a conversational agent designed to have a tailored diet management plan. It uses NLP, a machine learning algorithm to understand user preferences and diet requirements, and suggests a balanced meal plan. It will be able to give them highly accurate physical exercise, healthy food style, and weight management style, for both males and females under whichever age. Chatbot will be able to suggest food for users based on their food intake, water intake, activity level, weight of users, age, and height as well.

Keywords AI · Weight management · Chatbot · Physical exercise · NLP · Diet Requirement

1 Introduction

Healthy food is considered a crucial factor in having a happy life for living beings. Food without accurate label nutrition leads to premature death. Obesity is one of

S. J. Karki (✉) · S. Pokhrel · S. C. Thakuri · A. P. Chhetry
Jain Deemed to be University, Bangalore, Karnataka, India

Y. Singh
Department of Computer Science and Engineering, Jain Deemed to be University, Bangalore, Karnataka, India

A. Patel et al. (eds.), *Advances in Machine Learning and Big Data Analytics II*,
Springer Proceedings in Mathematics & Statistics 442,
https://doi.org/10.1007/978-3-031-51342-8_11

the leading problems for people who do not have a perfect food plan [1]. Artificial Intelligence (AI) navigate chatbot is a conversational agent with human interaction to tailor diet meal plans contingent on user requirements and diet goals [2]. It is quite difficult to achieve this with traditional methods, to overcome this stinging defeat this AI-based chatbot is introduced. Chatbots offer varieties of flexible on-demand support, sustainable connectivity, and customized support [2]. Indeed, it helps to provide us with 24/7 meal recommendations, and nutrition information at any time [3]. Users will be able to stick to their routines and follow the track. As the development of the AI chatbot is raised persistently, its importance in people's lives with the help of the Internet for their daily routine to make them stay fit by providing them with proper nutrition suggestions is broadly accepted by the world [4, 5]. AI chatbot can provide awareness for students too, as most of the students spend their time online, and using our chatbot enables them to gain answers related to all of their health problems as well, as it is also designed to give excellent output [6]. Although conversation with a chatbot is not the same as talking with a person face to face, still, it is enhanced to give good, reasonable output leads to show how the bot should behave. For instance, telegram and "WhatsApp," both are the most famous messaging social platforms and offer several advantages to active millions of customers [7]. "Wakamola" a telegram chatbot provides valuable assistance to individuals with obesity risks depending on their BMI, and physical activity [8]. NLP is a machine learning technology that can interact between computers and the human brain. Chatbots using NLP can help improve the interaction between users, take input more accurately, and provide better UI experiences. Individuals can interact with the app according to their requirements (tailored). Chatbots-empowered NLP can able to handle large customer engagement with a faster latent period, be less time-consuming, or reduce high-end error during interaction which makes high user satisfaction in terms of precisions [9]. However, with these advantages, it is important to allow users to provide feedback.

Here in this review, we endeavor to describe AI chatbot features, provide health conscience as well as describe AI characteristics helping them in weight management, diet requirements, perfect nutrition, and health-related other outcomes based on the data provided by the user.

2 Related Work

2.1 HealthifyMe

Tushar Vashisht et al. proposed a digital health and wellness application that can provide calorie tracking of the person and meal based on the calories. It provides the user with a health tracker based on the physical activity level. It provides a diet meal plan based on the user's demand. But the main downside of this application is that It is based on a subscription basis which is not affordable for every people [10].

2.2 Fitmeal

Aman et al. proposed a delivery service based on the India platform that provides a meal to people around India. The concept is based on providing a healthy diet for people around India. It is a kind of delivery service that provides diet meals to people who are busy with their daily Schedules. But it does not support access to people to help them lose or gain weight [11].

2.3 SlimMe

Annisa Ristya Rahmanti et al. proposed a chatbot that fully helps people with the problem of obesity. This chatbot is only helpful for people facing the problem of obesity. But it is not designed to provide the full solution for all kinds of people [12]. As it is only based on obesity It is not helpful for the people willing to increase their weight or the people engaged in physical activities. For all these problems our Chatbot comes in handy for people with physical activities or people opting to increase their weight [13].

2.4 Tailored Agent-Based Chatbot for Nutritional Tutoring

Davide Calvaresi et al. proposed a personal agent-based chatbot for nutritional tutoring. Each user will be assigned a personal chatbot that will help the user with the diet recommendation. The chatbot will help the user with the diet recommendation and keep track of the diet intake and help the meal tracking accordingly. Since the chatbot is also only focused on the proper diet, it will not come in handy for people who are involved in physical workout plans. Since it only helps in the sector of the meal [14].

Artificial Intelligence chatbot alteration is one procedure for creating AI chatbots that encourage exercise and a balanced diet [15].

2.5 A Knowledge-Based Advisory Framework to Provide Healthy Diets Utilizing Expert-Validated Meals: Protein AI Specialist

Kiriakos Stefanidis et al. presented a protein adviser that offers an indication of a person's protein intake and then uses a user's data to determine whether they have a protein shortage or over-protein consumption. It also recommends a healthy diet for the individual. It is made up of two parts. RDSS (Reasoning-Based Decision

Support System) presents a collection of acceptable meals advised by nutrition experts and offers the proper diet based on the user's profile. Additionally, the NP (Nutritional Plan) Generation Component makes dietary recommendations by combining the right meals to create a daily meal plan [16].

3 Methodology

This study used Experimental Research which is designed to develop a chatbot using PyTorch. The first thing is the data collection which is used to train the chatbot from our dataset. The dataset that we used contains various intents that show how can a user ask the same kind of question in many ways. At first, we need input from the user then the NLP preprocessing takes places like tokenization, stemming, and stop word removal. Then the preprocessed text is fed to a neural network and then to encoder-decoder the which classifies the intent based on the classification model. After the classification of the intents based on user preference, the main target is to track the macros and adjust the food and provide details of the adjusted food macros so that users can easily follow to achieve their fitness goals.

3.1 Natural Language Processing (NLP)

In our model, it works as the primary step or technique used to transform the inserted text data into a format that can be easily processed through machine learning algorithms. These techniques involve processes such as cleaning and normalizing the text data to remove noise to extract meaningful information. At first, when the user gives the input in the form of text the text is broken down into individual works or tokens which is generally called tokenization. This process is done by removing the white spaces, and punctuation. After that, it checks for the stop word such as "the," "and," "a," etc. because it is not so meaningful. Removing such words can help to reduce the dimensionality and improve the efficiency of the algorithms. After removing the stop word then stemming and Lemmatization takes place which is a technique that involves reducing the word from its base form. After all these, BOW (bag of the word) is used because we cannot send our input sentence into our neural network so we have to convert the pattern string into the number called bag of words. Then the given input sentence is matched with the training dataset or corpus, then it calculates the bag of words for each new sentence, if the word is available, it gives 1 otherwise 0.

3.2 Feed Forward Neural Network (FNN)

After the NLP preprocessing state, the preprocessed text is then fed to the FNN which contains just one input layer, two hidden layers, and one output layer. When the preprocessed text is fed into the input layer, each word or the token is then represented as a vector in a numerical form. Then it is passed to the hidden layer of the Feedforward neural network in which these hidden layers apply nonlinear modifications to the input, enabling the FNN to recognize intricate patterns and connections in the data. It is passed to the output layer which generates the probability distribution over the responses which is possible. It uses the softmax activation function to ensure that probabilities sum to one.

3.3 Encoder and Decoder

After the Feedforward neural network stage, the encode and decoder architecture is used for the chatbot development process. The output which came from the FNN as the probability is fed into the encoder which encodes the response as a fixed-length vector representation. It also uses a Recurrent Neural Network (RNN) to generate the response text one at a time. After the encoder vector representation of the response, it is sent to the decoder, which generates a response text. The attention mechanism is used to improve the performance of encode and decoder like selectively focusing on different parts of the input while generating the output. While training the encoder and decoder are trained using a supervised learning algorithm.

3.4 Stochastic Gradient Descent (SGD)

SGD is an algorithm that is used in our chatbot. It is a popular optimization approach in machine learning and neural networks, especially in the creation of the chatbot. At the training phase of the Neural Network and encoder and decoder, this algorithm is used to update the weights and biases of the neural network based on the discrepancy between each input's predicted and actual outputs from the training data. By updating weight and biases in the direction of the negative gradient of loss function this algorithm enables the neural network to learn from the training data and improve the performance.

3.5 *Intent Classification*

After all these processes, the most important is the intent classification process, which enables the chatbot to understand the user's intent and provide the response according to it using a Feedforward Neural Network to classify the user input and generate the most contextually appropriate response and improve the overall performance.

3.6 *Body Mass Index (BMI) Calculation*

After the intent classification, according to the user input the body mass index (BMI) is calculated. In simple words, Body Mass Index (BMI) is used to figure out an individual's fat based on their weight and height. It is one of the easiest ways to know the information whether someone is underweight, overweight, normal weight, or obese. For this calculation, we need a formula that involves dividing a person's weight in kg by their height in meters square. After the calculation, it provides a number that is used to categorize a person's weight status. It is also considered a powerful tool to help users identify who may be at risk related to weight (Fig. 1).

3.7 *Basal Metabolic Rate (BMR) Calculation*

After calculating the body mass index, the user needs to input the weight, height, age, and activity level which is required to calculate BMR. BMR is the bare minimum of calories or energy that a person's body requires to maintain normal bodily functions while at rest. Basic physical processes including breathing, blood circulation, and maintaining body temperature all require this energy. BMR can

BODY MASS INDEX	
BMI	WEIGHT STATUS
below 18.5	Under Weight
18.5-24.9	Normal
25-29.9	Over Weight
30-30.4	Obese

Fig. 1 Categorization of BMI

TYPES OF LIFE STYLE	
Sedentary	BMR*1.2
lightly active	BMR*1.375
moderately active	BMR*1.55
very active	BMR*1.725
extra active	BMR*1.9

Fig. 2 Activity level classification

Table 1 Description of figures

Figures	Description
Figure 1	Categorization of BMI
Figure 2	Activity level classification
Figure 3	Bar graph of calorie calculation for veg and non-veg
Figure 4	Macronutrients comparison

change depending on things like age, sex, weight, height, and activity level. Activity level has four levels therefore sedentary level, lightly active, moderately active, and highly active. By taking all the input from the user it will calculate the BMR then give some values which are the daily calories required for the body as the output then give the user the adjusted calories by surplus or deficit calories according to the user's choice while calculating the BMI and BMR. If the user wants to gain weight it will surplus the calories with the daily required calories if the user wants to lose weight it will deficit the calories with daily required calories [17] (Fig. 2).

3.8 Daily Calorie Tracker

After calculating the BMI, BMR will now ask the user to input the daily food that they eat in a whole day with the amount. Then after inputting the food, it will check with the dataset (which includes mostly all food present in the Asia Pacific Region (APAC)) which we prepare and give the macros (carbohydrate, protein, fat) that they consume in a whole day as an output. After that, it will calculate and adjust the macros on food that the user provided and add food if required only to fulfill the macros and calorie need which is the main aim of the chatbot [18] (Table 1).

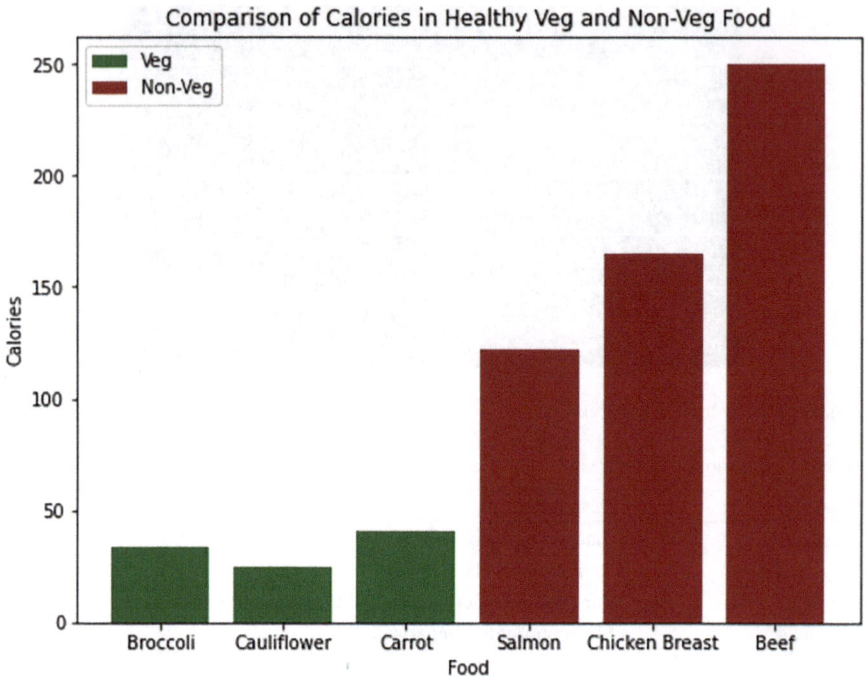

Fig. 3 Bar graph of calorie calculation for veg and non-veg

4 Architecture

Here, let us give you a rundown of our chatbot (Fig. 5):

(i) The chatbot receives text human input is used.
(ii) The chatbot, second preprocesses the input using NLP to eliminate stop words and steam words.
(iii) A feedforward NN is used by the chatbot to categorize the user's intent.
(iv) An encoder-decoder paradigm is used by the chatbot to produce responses.
(v) For the chatbot to compute BMI and BMR, it also needs to know your age, weight, height, level of exercise, and gender.
(vi) The chatbot requests the user's daily diet plan. Including the number of servings.
(vii) To help the user reach their fitness goal, the chatbot offers a personalized food plan that includes the user's daily macronutrient intake.

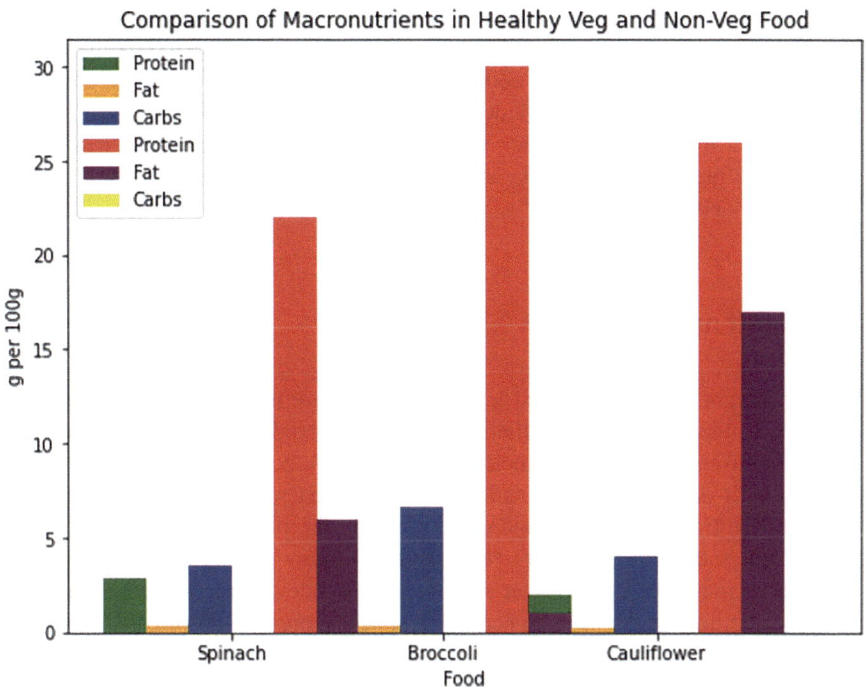

Fig. 4 Macronutrients comparison

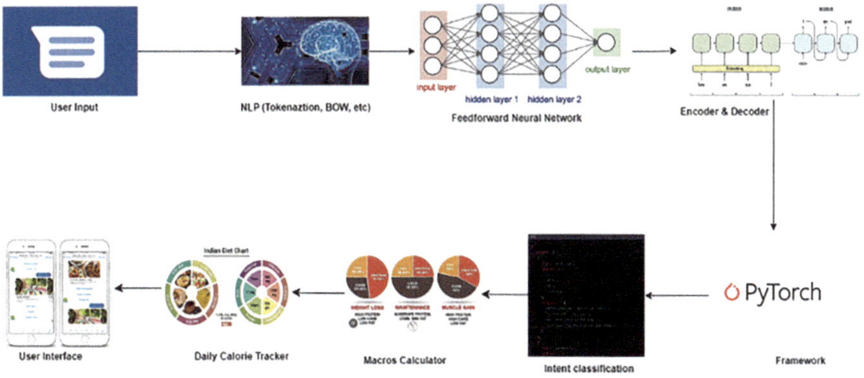

Fig. 5 Architecture of the proposed system

5 Flowchart

The flowchart outlines the steps involved in using Chatbot. The Chatbot will first enquire about the user's food consumption before requesting information about their age, height, weight, gender, and level of activity. Following that, it will determine the user's BMI and provide recommendations on whether to lose or gain weight. Depending on their preferences, the chatbot will also propose that they increase or decrease the meal and/or some additional meals. The Chatbot will take the user's workout schedule and then provide the ideal workout schedule depending on the user's liking if the client engages in physical activity and elects to help with the workout. In order, for Feedback, a survey from the participants will be collected after the talk (Fig. 6).

6 Result and Discussion

Weight, height, gender, age, and level of exercise are used to calculate the user's BMI while taking into account their desire to lose weight. This sheds light on how they are currently doing with their weight. The chatbot also gathers information on the user's daily food intake and calculates the macronutrient makeup (macros) of their diet. For both gaining and decreasing weight, proteins, carbohydrates, and fats are all essential. The chatbot also assesses the user's dietary preferences, such as whether they are vegetarian or not, to provide relevant recommendations. The chatbot incorporates the user's macronutrient needs, preferred foods, and computed BMI to create a customized diet plan. The appropriate serving sizes and total macronutrient requirements for the user to achieve their weight loss goal are covered by this plan (Table 2).

7 Conclusion

The number of studies on these technologies as well as the business applications for chatbots are both expanding fast. Even though this study was unable to come to a firm conclusion about the efficacy of chatbot treatments in diet and weight management, chatbots may advance [19]. There are still inadequate measures for evaluating these interventions or theory-guided chatbots, despite the sharp increase in articles concerning chatbot designs and interventions. Future research must thus employ standardized criteria to assess the effectiveness of chatbot installation. A diet meal chatbot can be very useful for those people who are looking to maintain a healthy lifestyle by providing their personalized meal plans which are balanced and nutritional diets and can also help users achieve their health and fitness goals [20]. Making or incorporating a diet meal plan chatbot into one's health routine can be a very helpful step toward achieving a healthy life. In this paper, our chatbot asks for the weight, height, age, and activity level of a person as a user input for calculating his/her BMI and suggesting him/her calorie consumption as per their need, i.e., either they want to lose weight or gain weight. It also calculates the macros and provides the macros chart. The chatbot is easy to use and provides a balanced and nutritious diet. One of the major flaws in this chatbot is that it is only applicable in India.

Fig. 6 Flowchart of the proposed model

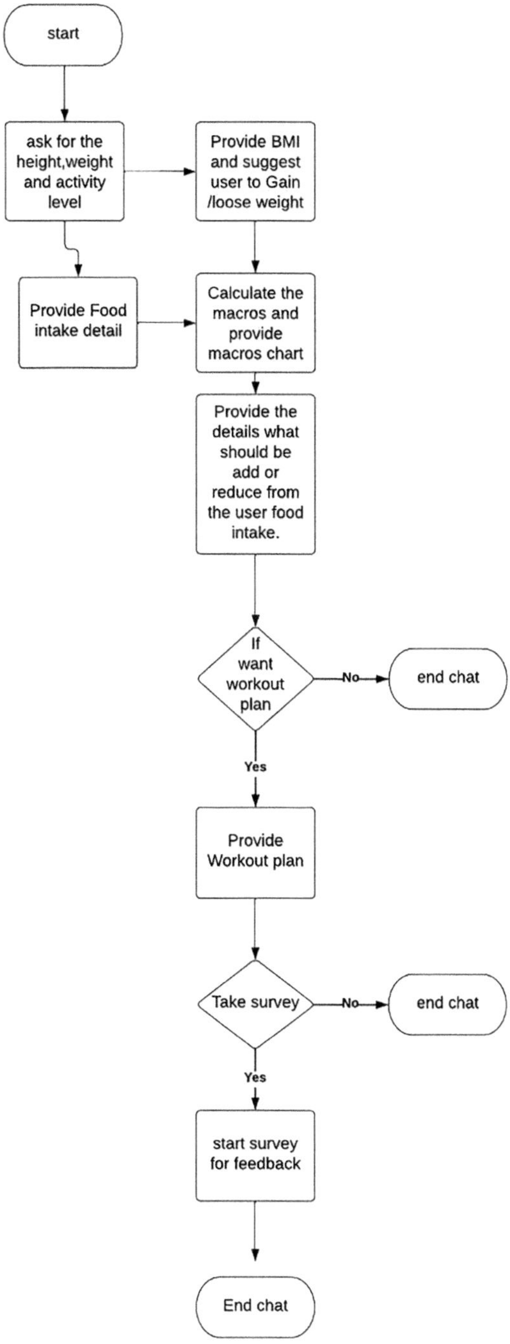

Table 2 Results screenshot

Figures	User input
Figure 7	The user wants to lose weight
Figure 8	According to the data provided by the user based on weight, height, gender, age, and activity level BMI is calculated
Figure 9	Food consumed by users daily
Figure 10	Calculated macros
Figure 11	Food preferences, if they are vegetarian or non-vegetarian

Fig. 7 The user wants to lose weight

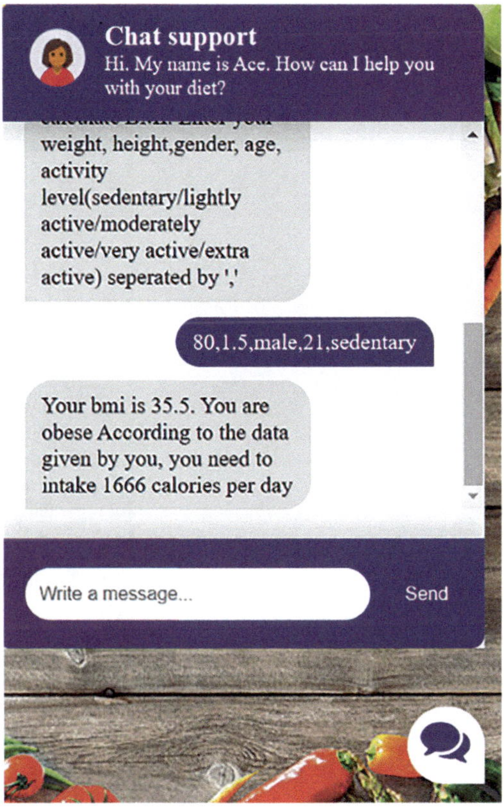

Fig. 8 According to the data provided by the user based on weight, height, gender, age, and activity level BMI is calculated

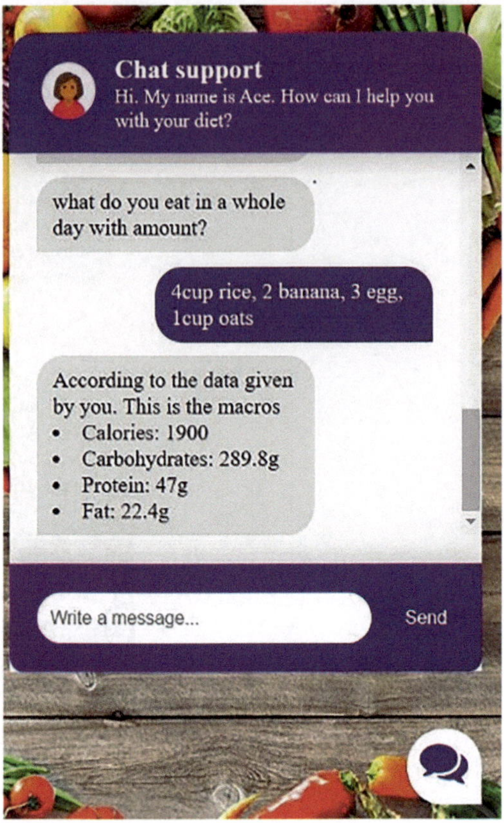

Fig. 9 Food consumed by user on a daily basis

Fig. 10 Calculated macros

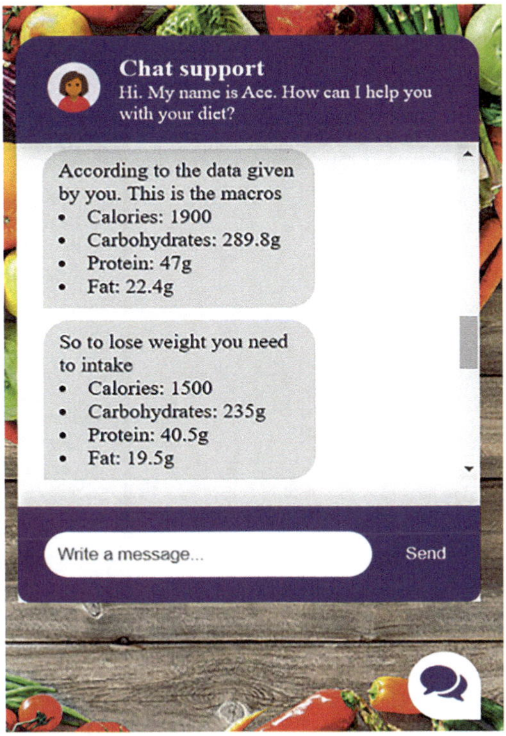

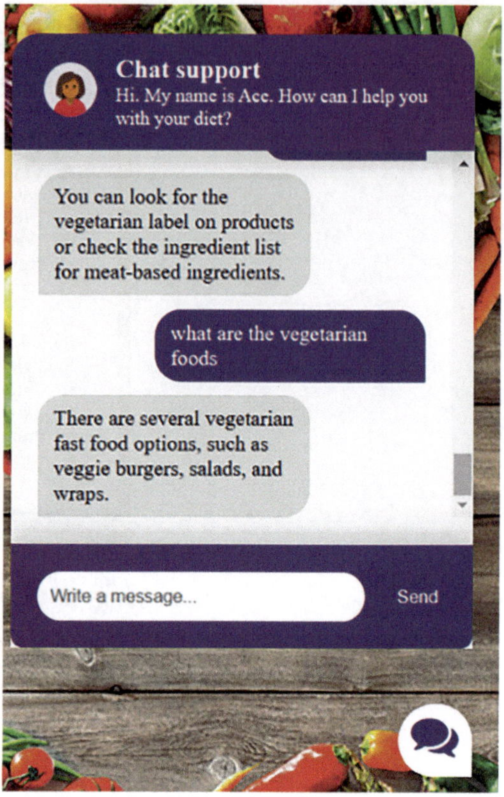

Fig. 11 Food preferences, if they are vegetarian or non-vegetarian

References

1. M. C. Belibağlı and N. Aygün Bilecik, "Another neglected symptom among the overweight young: an analysis of the self-reported anterior knee pain scores of the secondary school children," *Interdisciplinary Medical Journal*, vol. 14, no. 48, pp. 25–30, 2023, doi: https://doi.org/10.17944/interdiscip.1285723.
2. L. Laranjo et al., "Conversational agents in healthcare: a systematic review," *J Am Med Inform Assoc*, vol. 25, no. 9, pp. 1248–1258, Sep. 2018, doi: https://doi.org/10.1093/JAMIA/OCY072.
3. M. Adam, M. Wessel, and A. Benlian, "AI-based chatbots in customer service and their effects on user compliance," *Electronic Markets*, vol. 31, no. 2, pp. 427–445, Jun. 2021, doi: https://doi.org/10.1007/S12525-020-00414-7/FIGURES/7.
4. G. Suciu, A. Pasat, A. Birdici, and I. Pop, "AN INTERNSHIP CAMPAIGN CASE STUDY SHOWING RESULTS OF ENHANCED RECRUITMENT PROCESSES USING NLP," *eLearning and Software for Education Conference*, pp. 222–231, 2021, doi: 10.12753/2066-026X-21-097.
5. T. Nadarzynski, O. Miles, A. Cowie, and D. Ridge, "Acceptability of artificial intelligence (AI)-led chatbot services in healthcare: A mixed-methods study," vol. 5, Aug. 2019, doi: https://doi.org/10.1177/2055207619871808.
6. R. Han, A. Todd, S. Wardak, S. R. Partridge, and R. Raeside, "Feasibility and Acceptability of Chatbots for Nutrition and Physical Activity Health Promotion Among Adolescents: Systematic Scoping Review With Adolescent Consultation," *JMIR Hum Factors*, vol. 10, 2023, doi: https://doi.org/10.2196/43227.
7. A. Fadhil, Y. Wang, and H. Reiterer, "Assistive Conversational Agent for Health Coaching: A Validation Study," *Methods Inf Med*, vol. 58, no. 1, pp. 9–23, 2019, doi: https://doi.org/10.1055/S-0039-1688757/ID/OR180097-19.
8. S. Asensio-Cuesta et al., "A User-Centered Chatbot (Wakamola) to Collect Linked Data in Population Networks to Support Studies of Overweight and Obesity Causes: Design and Pilot Study," *JMIR Med Inform*, vol. 9, no. 4, Apr. 2021, doi: https://doi.org/10.2196/17503.
9. A. Fadhil, "A Conversational Interface to Improve Medication Adherence: Towards AI Support in Patient's Treatment," Mar. 2018 (accessed on March 27, 2023). Available online at: http://arxiv.org/abs/1803.09844
10. H. Ranjani et al., "Systematic Review and Scientific Rating of Commercial Apps Available in India for Diabetes Prevention," *Journal of Diabetology*, vol. 12, no. 3, 2021, [Online]. Available: https://journals.lww.com/jodb/Fulltext/2021/12030/Systematic_Review_and_Scientific_Rating_of.8.aspx
11. Gain, "REQUEST FOR PROPOSALS 1. PROJECT BACKGROUND AND SCOPE OF WORK 2 2. SCOPE OF WORK AND DELIVERABLES 3 3. INSTRUCTIONS FOR RESPONDING 3 4. TERMS AND CONDITIONS OF THIS SOLICITATION 6 5. OFFER OF SERVICES 8 6. ANNEX 1 9 FITFOOD BRAND RESEARCH KEY MESSAGE TESTING Issued by The Global Alliance for Improved Nutrition (GAIN) TABLE OF CONTENTS 2 1. PROJECT BACKGROUND AND SCOPE OF WORK ABOUT GAIN."
12. D. Duijst, D. Buzzo, S. Capgemini, and D. W. Slager, "Can we Improve the User Experience of Chatbots with Personalisation? User Experience of Chatbots View project RealME: A Digital Tool for Prisoners to Develop Employability Skills View project Can we Improve the User Experience of Chatbots with Personalisation? Can we improve the User Experience of Chatbots with Personalisation?," 2017, doi: https://doi.org/10.13140/RG.2.2.36112.92165.
13. A. R. Rahmanti et al., "SlimMe, a Chatbot With Artificial Empathy for Personal Weight Management: System Design and Finding," *Front Nutr*, vol. 9, p. 870775, Jun. 2022, doi: https://doi.org/10.3389/FNUT.2022.870775/FULL.
14. D. Calvaresi, S. Eggenschwiler, J. P. Calbimonte, G. Manzo, and M. Schumacher, "A personalized agent-based chatbot for nutritional coaching," *ACM International Conference Proceeding Series*, pp. 682–687, Dec. 2021, doi: https://doi.org/10.1145/3486622.3493992.

15. J. Zhang, Y. J. Oh, P. Lange, Z. Yu, and Y. Fukuoka, "Artificial Intelligence Chatbot Behavior Change Model for Designing Artificial Intelligence Chatbots to Promote Physical Activity and a Healthy Diet: Viewpoint," *J Med Internet Res 2020;22(9):e22845*https://www.jmir.org/2020/9/e22845, vol. 22, no. 9, p. e22845, Sep. 2020, doi: 10.2196/22845.
16. K. Stefanidis et al., "PROTEIN AI Advisor: A Knowledge-Based Recommendation Framework Using Expert-Validated Meals for Healthy Diets," *Nutrients*, vol. 14, no. 20, Oct. 2022, doi: https://doi.org/10.3390/NU14204435.
17. J. A. Harris and F. G. Benedict, "A Biometric Study of Human Basal Metabolism," *Proceedings of the National Academy of Sciences*, vol. 4, no. 12, pp. 370–373, Dec. 1918, doi: https://doi.org/10.1073/PNAS.4.12.370/ASSET/34EEF56D-5BB1-4E02-8EC2-68CEA5F0AB67/ASSETS/PNAS.4.12.370.FP.PNG.
18. S. Wharton et al., "Obesity in adults: A clinical practice guideline," *CMAJ*, vol. 192, no. 31, pp. E875–E891, Aug. 2020, doi: https://doi.org/10.1503/CMAJ.191707/TAB-RELATED-CONTENT.
19. H. S. J. Chew, "The Use of Artificial Intelligence–Based Conversational Agents (Chatbots) for Weight Loss: Scoping Review and Practical Recommendations," *JMIR Med Inform*, vol. 10, no. 4, Apr. 2022, doi: https://doi.org/10.2196/32578.
20. Y. J. Oh, J. Zhang, M. L. Fang, and Y. Fukuoka, "A systematic review of artificial intelligence chatbots for promoting physical activity, healthy diet, and weight loss," *International Journal of Behavioral Nutrition and Physical Activity*, vol. 18, no. 1, pp. 1–25, Dec. 2021, doi: https://doi.org/10.1186/S12966-021-01224-6/TABLES/4.

Convolution Neural Networks

K. Bhagyalaxmi and **B. Dwarakanath**

Abstract Brain tumors are extremely dangerous and can lead to a very short life expectancy at their highest grade. Identifying brain tumors early is crucial for improving survival rates, but it poses a significant challenge for medical professionals. Noise and other environmental disturbances are more likely to appear in MRI. As a result, it is challenging for doctors to diagnose the tumor and determine its causes. Therefore, a system was proposed to detect brain tumors from images. In this procedure, the image is first converted into a grayscale then filters are applied to remove environmental interference and other noise from the image. The user has to select the image. The system processes images through various image-processing steps. However, in the early stages of a brain tumor, the edges in the images are not well-defined. To address this, a deep learning model incorporating CNN and FCNN has been developed to enhance the classification of tumor types and the associated preprocessing stages. This pre-trained network is capable of categorizing images into 1000 object categories. The fully connected layers are replaced with dense connections and a softmax activation function, enabling the classification of brain tumor images into four classes.

Keywords Brain tumor · Glioma · Pituitary · Meningioma · VGG-19 · CNN · Accuracy · MRI

1 Introduction

Brain tumors are the most aggressive and common disease, leading to a very low life expectancy in their highest grade. Abnormal cell growth in the human brain

K. Bhagyalaxmi (✉)
SRM Institute of Science and Technology, Kattankulathur, India
e-mail: bhagyalaxmi@matrusri.edu.in

B. Dwarakanath
SRMIST, Ramapuram, India

A. Patel et al. (eds.), *Advances in Machine Learning and Big Data Analytics II*,
Springer Proceedings in Mathematics & Statistics 442,
https://doi.org/10.1007/978-3-031-51342-8_12

is known as a brain tumor. Detecting tumors manually from magnetic resonance images (MRIs) is prone to human errors and a time-consuming process, potentially resulting in misidentification and incorrect classification of tumor types. To automate this complex medical process, A deep learning technique has been proposed to aid in the identification and classification of brain tumors, thereby supporting medical professionals in making accurate diagnoses. Recent advancements in deep learning have significantly benefited the health industry, particularly in medical imaging for diagnosing various diseases. Basic image segmentation techniques are often used to specify the brain tumor regions. In this context, our work introduces a convolutional neural network (CNN) based transfer learning approach combined with image processing to categorize brain MRI images. Deep learning architectures have been widely employed for identifying visual objects in color images. In this implementation, the VGG-19 convolutional neural network architecture was selected for tumor type classification and identification.

2 Literature Review

A significant challenge in identifying and classifying brain tumors is the variability in the tumor region. The size, shape, location, and intensity of tumors can exhibit significant variation from one image to another, even within the same type of tumor. Yakub Bhanothu et al [1] proposed a model for the Detection and Classification of Brain Tumors in MRI Images using a Deep Convolutional Network. Usman Akram et al. [2] experimented with a model for brain tumor detection and segmentation that used a global thresholding technique. Shahrjar et al. [3] proposed an automatic approach for brain tumor detection, where image enhancement filters are used for preprocessing followed by segmentation and feature extraction to identify the tumor. Sajjad et al. [4] experimented with an extensive data augmentation method combined with the VGG-19 CNN architecture. This process involved multi-grade classification of tumors using segmented MRI images. By using transfer learning with the pre-trained VGG-19 CNN, they achieved classification accuracies of 87.38% and 90.67% for data before and after augmentation, respectively. Jain et al. [5] suggested a method for diagnosing Alzheimer's using MRI scan images, employing the pre-trained VGG-16 architecture.

3 Methodology

3.1 Image Recognition

Classifying objects in images is straightforward for humans but presents significant challenges for machines, particularly in handling the semantic aspects of recognizing and categorizing visual data [6]. Humans can easily determine these aspects, but machines struggle with the complexities involved. Image classification plays a vital role in enabling machines to identify image features and categorize them into their appropriate classes [7].

3.2 CNN

Convolutional Neural Networks (CNNs) are structured to handle data through multiple layers of arrays, making them highly suited for applications such as image processing. They excel in capturing and learning hierarchical features from input data like images, leveraging the spatial relationships between elements effectively. The primary distinction between CNNs and other conventional neural networks lies in how CNNs handle input: they process two-dimensional arrays directly from images, whereas other neural networks typically focus on feature extraction [9]. CNNs are prominently used in solutions for recognition problems due to their effective handling of spatial data relationships inherent in images.

The reason CNN [10] is so usable is that it needs very little pre-processing. The CNN includes the following:

- Convolutional layers
- Pooling layers
- ReLU layers
- Fully connected layers—all of which play a crucial role.

3.3 VGG-19

The VGG-19 convolutional neural network, introduced by the Visual Geometry Group at Oxford University in 2014, is a deep network comprising 19 layers [8]. It gained recognition for its high accuracy in classifying images on the ImageNet dataset. A pre-trained version of VGG-19 is available and is trained over a huge number of images from the ImageNet database, enabling it to classify thousands of different images into different object categories, covering items such as keyboards, mice, pencils, and various animals (Fig. 1).

This network takes images of fixed size 224 $\times$ 224 RGB as input. The only pre-processing that was approved is that they withheld the mean RGB advantage at each pixel evaluated over all presentation sets. A size 3 $\times$ 3 kernel and 1-pixel stride were used.

- To maintain the spatial resolution of the image, Spatial padding was used.
- Max pooling is done over 2 $\times$ 2 pixel windows with stride 2.
- To improve the model's classification and computation time, nonlinearity was introduced with the help of ReLu. This proved to be much superior to others.

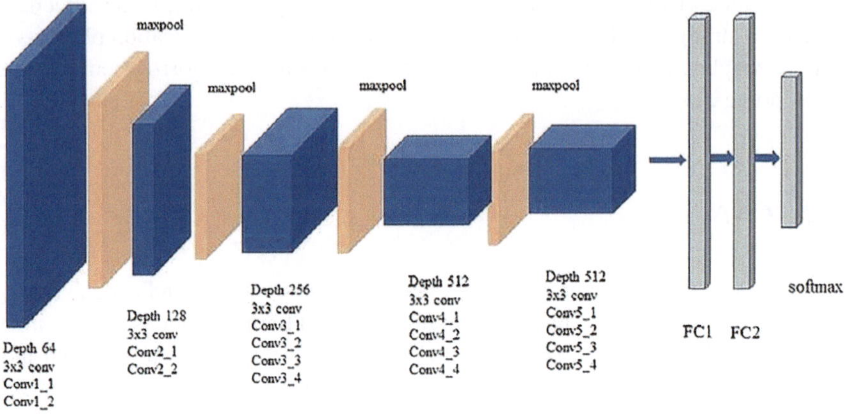

Fig. 1 VGG-19 architecture

Table 1 Data set

Tumor class	Training set	Testing set
Meningioma	205	30
Glioma	205	204
Pituitary	205	10
No tumor	105	15

- In the VGG-19 architecture, the last three fully connected layers include two layers with 4096 neurons each, followed by a third layer with 1000 neurons designed for 1000-way classification. The final layer employs a softmax function to compute probabilities for each class, ensuring the sum of probabilities equals one.

4 Implementation

4.1 Dataset and Data Sources

The dataset, sourced from Kaggle datasets, comprises a total of 979 images categorized into four classes: meningioma, pituitary, glioma, and no tumor [11]. These class labels are assigned for training and testing purposes in developing models for brain tumor classification (Table 1).

4.2 Implementation Procedure

Training the model involves importing the required Python libraries like NumPy, Keras, and openCV. After importing libraries, the dataset is loaded. The data is then preprocessed. This preprocessed data is split into training and testing datasets. Using the training dataset a model is trained. Replaced three fully connected layers with dense four-way classification [12] by applying transfer learning [13] and the final layer is a softmax function. With this the number of trainable parameters reduced from 20,124,740 to only 100,356.

The model is defined to train for 15 epochs and early stopping is also defined. With the help of a testing dataset, the model is tested. Early stopping occurred after 13 epochs. Various valuation metrics like accuracy, loss and mean square error of the final model are calculated. With the help of the final model, the new images are predicted.

5 Results

In this proposal, a CNN model was developed that can detect and predict the brain tumor in the MRI images with an accuracy of approximately 84.31% on the validation set and 98.75% on the training set. Below are the results obtained when a new input image is provided to the model. Fig. 2 represents the plot of training accuracy and validation accuracy improvements over the epochs (Table 2).

Figure 3 represents a decrease in the training loss and validation loss as epochs increase.

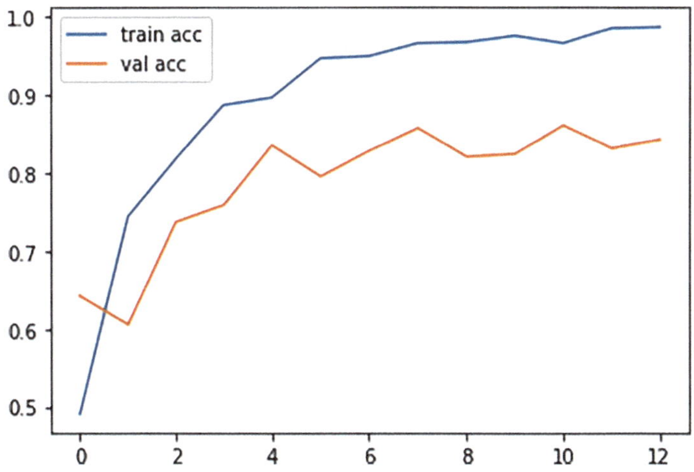

Fig. 2 Plot of training and validation accuracy

Table 2 Classification metrics

Class label	Precision	Recall	F1-score	Support
0	0.80	0.80	0.80	68
1	0.84	0.75	0.79	77
2	0.93	0.83	0.88	65
3	0.82	1.00	0.90	64

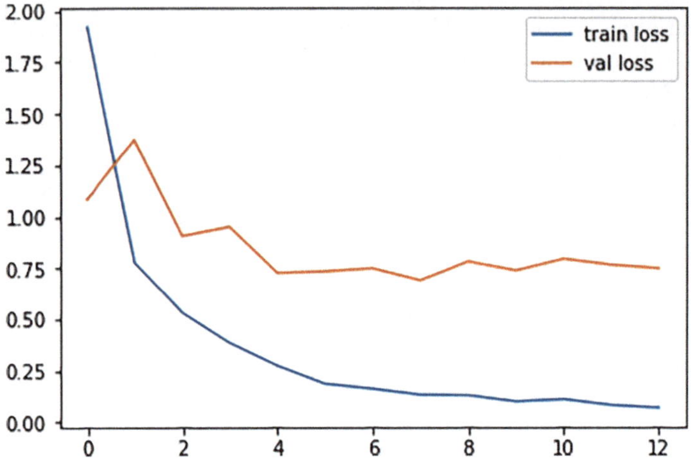

Fig. 3 Plot of training and validation loss

Figure 4 gives the classification performance of four tumor classes meningioma, glioma, pituitary, and no tumor in the form of a confusion matrix for the given data set images.

Figures 5 and 6 represent the meningioma and glioma tumors successful detection, respectively.

6 Conclusion

In summary, we have chosen VGG-19-based CNN architecture to detect and classify the brain tumor from MR Images of brains This architecture utilizes the VGG-19 deep convolutional network to detect and classify tumor regions into three categories: glioma, meningioma, and pituitary tumor. The model achieves an impressive accuracy of up to 84.31% in accurately predicting brain tumors from input images, demonstrating superior performance compared to state-of-the-art models based on VGG-16 transfer learning on the same dataset.

In the future, this work can be implemented with greater processing resources, with a much larger dataset with a wide range of MRI images, and with a more advanced CNN model so that this can increase the accuracy and reduce the error to

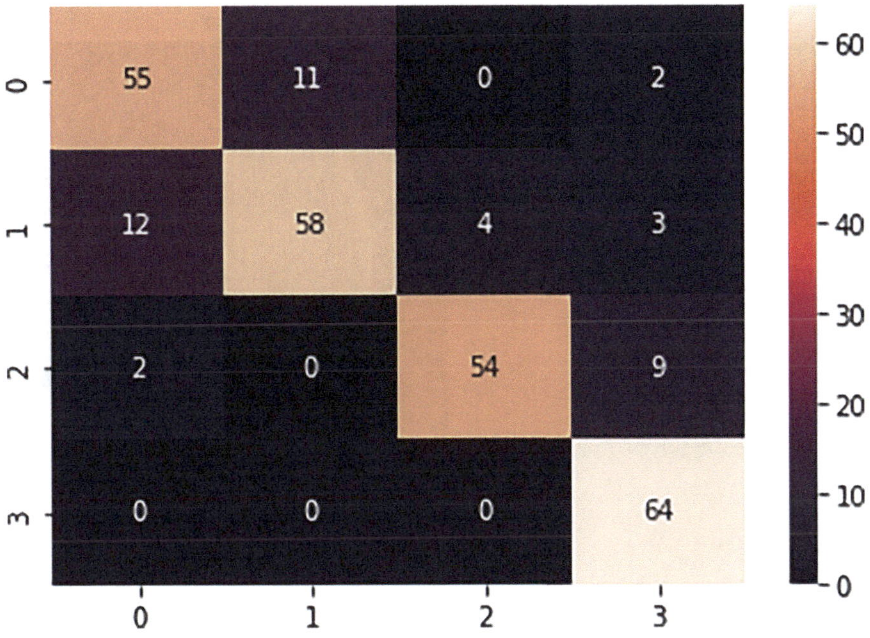

Fig. 4 Confusion matrix

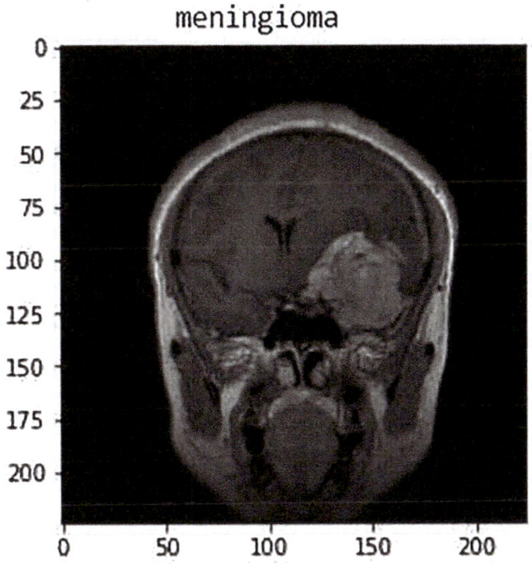

Fig. 5 Meningioma tumor detection

a larger extent. Also, this work can be extended for different medical applications by making small modifications to the model parameters.

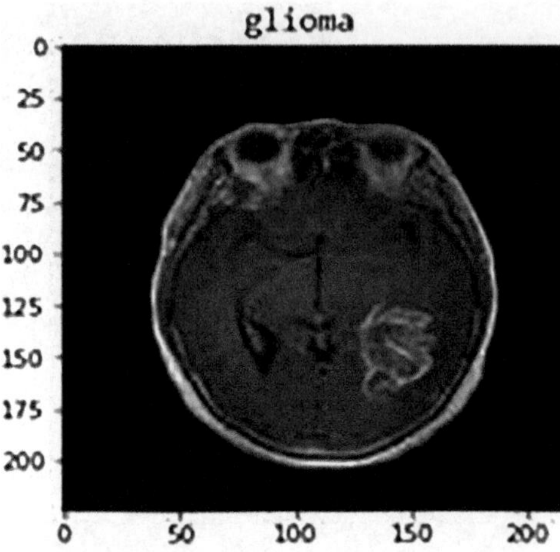

Fig. 6 Glioma tumor detection

References

1. Yakub Bhanothu, Anandhanarayanan Kamalakannan, Govindaraj Rajamanickam, "Detection and Classification of Brain Tumor in MRI Images using Deep Convolutional Network." 6th International Conference on Advanced Computing & Communication Systems (ICACCS), 2020.
2. M. Usman Akram, Anam Usman, "Computer Aided System for Brain Tumor Detection and Segmentation," International Conference on Computer Networks and Information Technology (ICCNIT), 2011.
3. T. M. Shahriar Sazzad, K. M. Tanzibul Ahmmed, M. U. Hoque and M. Rahman, "Development of Automated Brain Tumor Identification Using MRI Images," 2019 International Conference on Electrical, Computer and Communication Engineering (ECCE), Cox's Bazar, Bangladesh, 2019, pp. 1–4, doi: https://doi.org/10.1109/ECACE.2019.8679240.
4. Sajjad, M., Khan, S., Muhammad, K., Wu, W., Ullah, A., & Baik, S. W. Multi-Grade Brain Tumor Classification using Deep CNN with Extensive Data Augmentation. Journal of Computational Science, 30, 174–182, 2019.
5. Huang, Meiyan et al. "Brain Tumor Segmentation Based on Local Independent Projection-Based Classification." *IEEE Transactions on Biomedical Engineering* 61 (2014): 2633–2645
6. A. Wulandari, R. Sigit and M. M. Bachtiar, "Brain Tumor Segmentation to Calculate Percentage Tumor Using MRI," 2018 International Electronics Symposium on Knowledge Creation and Intelligent Computing (IES-KCIC), Bali, Indonesia, 2018, pp. 292–296, doi: https://doi.org/10.1109/KCIC.2018.8628591.
7. Abdelaziz Ismael SA, Mohammed A, Hefny H. An enhanced deep learning approach for brain cancer MRI images classification using residual networks. Artif Intell Med. 2020 Jan;102:101779. doi: https://doi.org/10.1016/j.artmed.2019.101779. Epub 2019 Dec 10. PMID: 31980109.
8. Y. Bhanothu, A. Kamalakannan and G. Rajamanickam, "Detection and Classification of Brain Tumor in MRI Images using Deep Convolutional Network," 2020 6th International Conference on Advanced Computing and Communication Systems (ICACCS), Coimbatore, India, 2020, pp. 248–252, doi: https://doi.org/10.1109/ICACCS48705.2020.9074375.
9. Kumar, R.L., Kakarla, J., Isunuri, B.V. et al. Multi-class brain tumor classification using residual network and global average pooling. Multimed Tools Appl 80, 13429–13438 (2021). doi:https://doi.org/10.1007/s11042-020-10335-4

10. Ayadi, W., Elhamzi, W., Charfi, I., & Atri, M. (2021). Deep CNN for brain tumor classification. *Neural Processing Letters*, *53*(1), 671–700.
11. A Aziz, M Attique, U Tariq, Y Nam, M Nazir "An Ensemble of Optimal Deep Learning Features for brain tumor classification" Computers, Materials & Continua Publisher: **researchgate 2021**
12. Sharif, M.I., Khan, M.A., Alhussein, M. et al. A decision support system for multimodal brain tumor classification using deep learning. *Complex Intell. Syst.* (2021). | Publisher: Springer

Segmenting MRI Images Using Federated Learning for Brain Tumor Detection

T. L. Akshaya, K. Swetha, S. Pravalika, and U. Chandrasekhar

Abstract AI systems must be able to make use of large, diverse, and global information in order to build robust and efficient systems in the medical imaging field. To develop a global model, all these facts must be collected in one place, but this raises questions about privacy and ownership. In this study, we evaluated multiple federated learning methods for segmenting brain tumors. Federated learning utilizes all accessible data without storing or disclosing collaborators' personal information on a central server. By appropriately integrating these model updates, a high degree of accuracy could be attained, but doing so increases the possibility that the shared model might accidentally leak the local training data. We have selected the best model from U-Net and DeepLabV3 to obtain the best efficiency.

Keywords Federated learning · Brain tumor · MRI · U-Net · DeepLabV3 · Segmentation · Privacy-preserving

1 Introduction

Federated learning, a distributed machine learning technique, allows several devices to train a model on their own data while storing the data locally on the devices. It is a viable technique for training models from scattered or private data, including images from the medical industry. The medical imaging process known as magnetic resonance imaging (MRI) uses a strong magnetic field and radio waves to produce exact images of the interior of the body. To segment MRI images, different organ or tissue structures must be distinguished from one another. Numerous medical conditions can be identified and followed up on using MRI scans. The Indonesian

T. L. Akshaya (✉) · K. Swetha · S. Pravalika · U. Chandrasekhar
Department of Computer Science and Engineering, BVRIT HYDERABAD College of Engineering for Women, Hyderabad, India
e-mail: 19wh1a0531@bvrithyderabad.edu.in; 19wh1a0506@bvrithyderabad.edu.in; 19wh1a0543@bvrithyderabad.edu.in; chandrasekhar.u@bvrithyderabad.edu.in

A. Patel et al. (eds.), *Advances in Machine Learning and Big Data Analytics II*, Springer Proceedings in Mathematics & Statistics 442,
https://doi.org/10.1007/978-3-031-51342-8_13

Ministry of Health, reports that the country ranked 23rd in Asia and became the eighth country in Southeast Asia with the most cancer patients in 2018. After breast cancer, brain tumors are the second leading cause of death in cancer patients. It is known from all cases that women are more likely than men to get brain tumors. In various nations, the prevalence of brain tumors has been rising for the past 10 years. Medical imaging is essential for identifying brain tumors and can aid in the prevention of more dangerous disorders. Researchers use MRI imaging techniques to find brain tumors. Because it does not use ionizing radiation, MRI is one of the most frequently used medical imaging techniques for brain tumors. A model for decentralized MRI image segmentation might be created in this case using federated learning, where the model is trained on a network of devices that each have their own MRI images. Instead of sending the model to a centralized location, the privacy of the medical data can be preserved by training it on a decentralized network of devices. The model must have enough data to learn from on each device, as well as ways to deal with any constrained connectivity or computational resources that may be present on some devices, in order to successfully use federated learning for MRI image segmentation. Despite these challenges, federated learning may allow for the training of incredibly accurate models for MRI images.

This paper requires technical domain knowledge related to deep learning models, federated learning architecture, and an idea about image segmentation. A brief idea on MRI image scans, NIfTI file format and the different modalities present in it would be useful.

2 Literature Survey

A considerable amount of work has been carried out in this field of segmentation. Dong et al. [1] present the usage of a deep convolution neural network, i.e., U-Net. The results showcased that the segmentation of MRI scans through this method is as good as manual segmentation performed by radiologists or specialized clinicians. This work emphasized that we could significantly reduce manual segmentation and avoid human intervention with superior results, thus saving time and resources.

Convolution Neural Networks is another such model that is capable of performing semantic segmentation of images. The analysis of its capabilities is present in the work [2]. The investigation yielded an accuracy rate of 93% which accompanied a 0.2326 loss value. It presented some useful insights regarding the model being built which included the impact of a number of layers on the result of segmentation and image augmentation techniques that can enhance the model output. A detailed analysis of the layers built, parameters configured, and their impact on the segmentation can be found.

The above techniques, though useful for performing segmentation of MRI scans minimizing manual intervention and saving time, raise a serious concern regarding how the data is being handled. This is because the data that is being used to train, test, and validate are of individuals whose privacy is to be preserved. Different countries have various legislations and acts in place to protect digital health data and guidelines that need to be followed to process them legally. General Data Protection Regulation (GDPR) and Digital Information Security in Healthcare Act (DISHA) are some of the legislations and acts governing this area in European countries and India, respectively.

These concerns hinder the live usage of the above models despite being proven to produce benchmark results. To harness the above techniques, research has been carried out in the field of integrating privacy-preserving models alongside them. Federated Learning Model Architecture is one such approach that helps in harnessing the above techniques. Li et al. [3] explore this area alongside the methods to overcome input data reconstruction by injecting noise during the training process which helps in overcoming Model Inversion attacks.

Every deep learning model has their own set of efficiencies and deficiencies. This study begins by understanding and evaluating the available datasets, choosing the dataset that aids our research work, and analyzing the performance of deep learning models individually as clients which mimic the performance as if they are run on a central system. This is followed by collating the weights of the models trained and then sharing it to a central system which essentially makes the clients as edge nodes, performing various federated averaging techniques to finalize the weights to be used in order to perform segmentation on test MRI images.

3 Data Description

3.1 Datasets

Several datasets are open-sourced and made available to individuals to explore the ways to perform semantic segmentation of MRI scans. Each of these datasets has different properties that serve unique purposes and problem statements to be solved. Some of the datasets that we have explored include OASIS, MRBrainS, ISLES, and BraTS. The dataset we have chosen is the BraTS Kaggle dataset. Within the BraTS dataset, there are several variations. The following is the list of datasets for BraTS:

BraTS2015, BraTS2016, BraTS2017, BraTS2018, BraTS2019, BraTS2020.

Each of these datasets has four MRI modalities: T1, T1CE, T2, and FLAIR as input images and the mask contains labels for Edema (ED), Enhanced Tumor (ET), Neuro Endocrine Tumor—Nottingham Health Science Biobank Core Resource (NET/NCR).

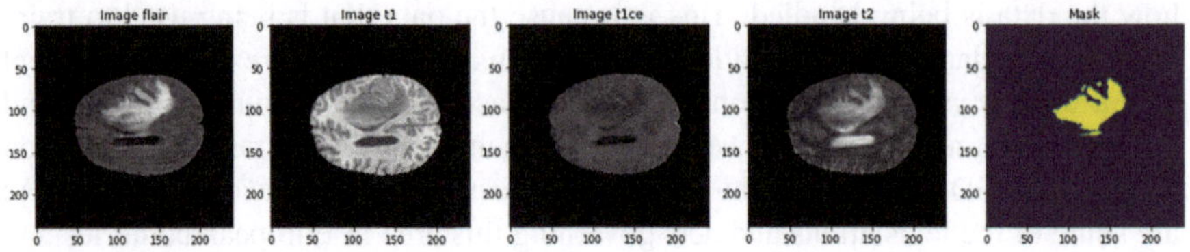

Fig. 1 Dataset

3.2 BraTS2020 Dataset

The primary reason for choosing the BraTS dataset is that it has been improving over the years due to the contribution of several researchers. The number of modalities for each patient's data is also more when compared to other openly available datasets for brain tumor segmentation. The mask images of BraTS are manually created by trained internists and neurologists.

As shown in Fig. 1, it is the BraTS2020 dataset that we have chosen for the research. All BraTS multimodal scans are available as NIfTI files (.nii.gz) and described as:

(a) Native (T1)
(b) Post-contrast T1-weighted (T1Gd)
(c) T2-weighted (T2)
(d) T2 Fluid Attenuated Inversion Recovery (T2-FLAIR) volumes were acquired with different clinical protocols and various scanners from multiple ($n = 19$) institutions, mentioned as data contributors here.

All the imaging datasets have been segmented manually, by one to four raters, following the same annotation protocol, and their annotations were approved by experienced neuro-radiologists. Annotations comprise the GD-enhancing tumor (ET—label 4), the peritumoral edema (ED—label 2), and the necrotic and non-enhancing tumor core (NCR/NET—label 1), as described both in the BraTS 2012–2013 TMI paper and in the latest BraTS summarizing paper. The provided data are distributed after their pre-processing, i.e., co-registered to the same anatomical template, interpolated to the same resolution (1 mm^3), and skull-stripped.

3.2.1 Data Access

All datasets are publicly available, but access to the data requires adherence to their respective data usage policies and any applicable laws and regulations related to data privacy and confidentiality.

Based on the above comparative analysis, we found that the BraTS2020 dataset serves the purpose of our research.

4 Design

4.1 Flow Chart

As shown in Fig. 2, the proposed architecture for performing the semantic segmentation of MRI scans for brain tumor detection. The MRI images are taken as input which are then split to train, validate, and test datasets. These images are preprocessed by resizing, combining, and associating each image set to the masked image as the label. The segmentation is performed by giving the above train-image sets as input to the models U-Net and DeepLabV3. The performance of each of these models is observed in the Federated Learning Architecture setup [4]. Hyperparameter tuning is done by using the validation dataset. Finally, the model with the best accuracy is retained as the final model in the federated learning architecture.

4.2 Approach

We collect the dataset and organize them into different folders. Perform data cleaning and pre-processing to prepare them for training.

Initially, we trained our dataset with two different deep-learning models: U-Net and DeepLabV3 [4, 5].

We make a comparative analysis of their performance based on different metrics and perform hyper-parameter tuning.

We then implement federated learning over the models built to enhance the privacy-preserving nature of the model [6].

Finally, select a model that performs better when compared to others and make it the final model in the server for MRI image segmentation.

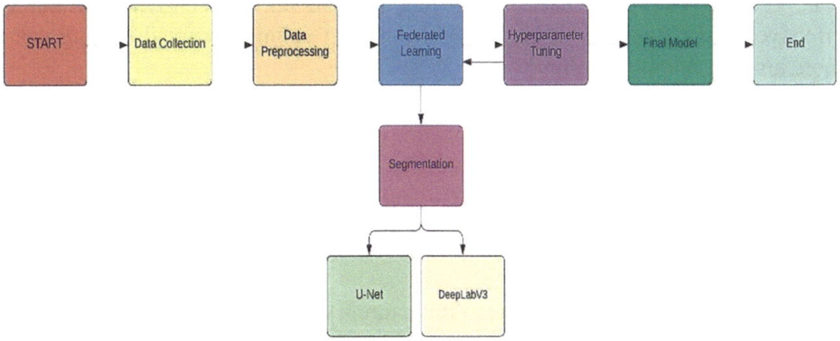

Fig. 2 Flow chart

5 Implementation

5.1 Data Collection

The dataset that we used for our current research is the BraTS2020 dataset from Kaggle. The dataset is organized into training and validation directories. The downloaded dataset is then uploaded to Google Drive since the Integrated Development Environment (IDE) used for this purpose is Google Collaboratory. On analyzing the folders of this dataset, we found that folder 355 had a weird name for masked image. The file needs to be renamed before applying our data pre-processing methods on them so that we can run them without errors.

5.2 Data Pre-processing

The obtained is then analyzed for each patient's data to make sure that there are no visible anomalies in it. Various pre-processing techniques are used for each of the deep learning models that we have chosen. The mask image labels have values 0, 1, 2, and 4 which are then modified to have them as 0, 1, 2, and 3 so that the labeling remains consistent. The train and mask images have sizes 224 × 224 × 155. These images are cropped to sizes 128 × 128 × 128 and 144 × 144 × 144 leading to the creation of two or more directories for training with only an effective volume of images. The images are in NIfTI format which comprises voxels of data rather than pixels which would be the case in a typical RGB image. Each patient data comprised four modalities: T1, T1CE, T2 weighted, and FLAIR. The T1 and T1CE are more or less comprised of the same information. So, we decided to consider only T1, T2 weighted, and FLAIR as the input images for training along with their corresponding segmented mask image. These three modalities are stacked one over the other as a NumPy image to form an input image to the model. This gives an input similar to that of an RGB image except that here individual channel has voxels of data. The images that are used for training and validation are chosen such that the effective mask volume is at least 1% of the total volume to avoid biasing due to more amount of empty volume.

5.3 Deep Learning Models

5.3.1 U-Net

U-Net is a semantic segmentation architecture. As shown in Fig. 3, it is made up of a contracting path and an expanding path. The contracting path is designed in the manner of a convolutional network. It is composed of two 3 × 3 convolutions

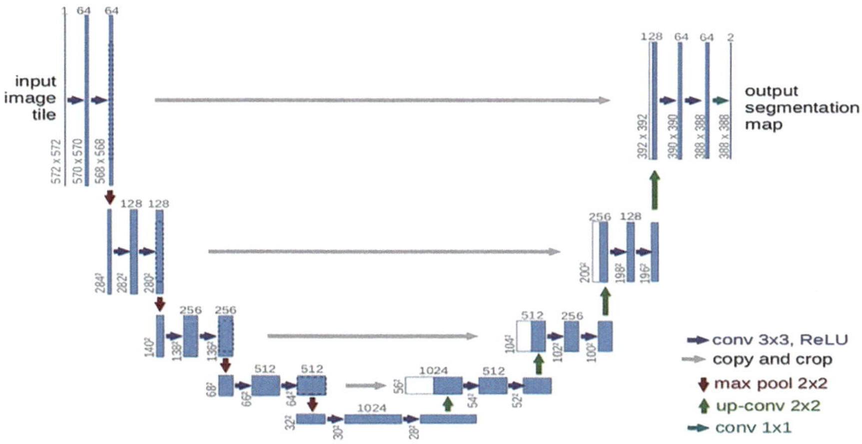

Fig. 3 U-Net

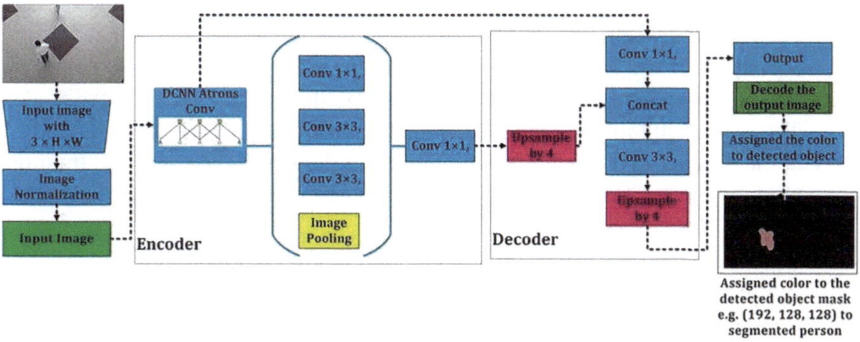

Fig. 4 DeepLabV3

applied repeatedly, each followed by a rectified linear unit and a 2 × 2 max pooling operation with stride 2 for downsampling. We double the amount of feature channels with each downsampling step [1, 7].

5.3.2 DeepLabV3

DeepLabV3 is a cutting-edge semantic picture segmentation model created by Google. It performs pixel-wise classification of a picture using deep convolutional neural networks (CNNs), assigning a label to each pixel based on its semantic value.

DeepLabV3, as shown in Fig. 4, is built on the Atrous Spatial Pyramid Pooling architecture, which enables the network to successfully capture context information at many scales. This leads to better performance than earlier models, particularly in difficult cases when items of interest comprise a small area of the screen. DeepLabV3 is commonly utilized in computer vision tasks such as autonomous driving, image editing, and scene comprehension [8].

5.4 *Federated Model*

5.4.1 Introduction

Federated learning is a machine learning technique that allows models to be trained on decentralized data without the data itself being shared. As shown in Fig. 5, the method ensures data privacy and security while enhancing model accuracy by combining data from numerous sources. The U-Net model is a popular convolutional neural network architecture used for image segmentation tasks [9]. Here are the steps to perform federated learning on any Deep Learning model:

(i) *Collect data:* Collect image data from different sources that need to be segmented using the deep learning model [5].
(ii) *Set up the server:* Set up a server that will coordinate the training of the model. The server will hold the model architecture and initial weights.
(iii) *Set up clients:* Set up clients who will have their own data and will train the model locally. Clients should be selected randomly from the pool of available clients [10].

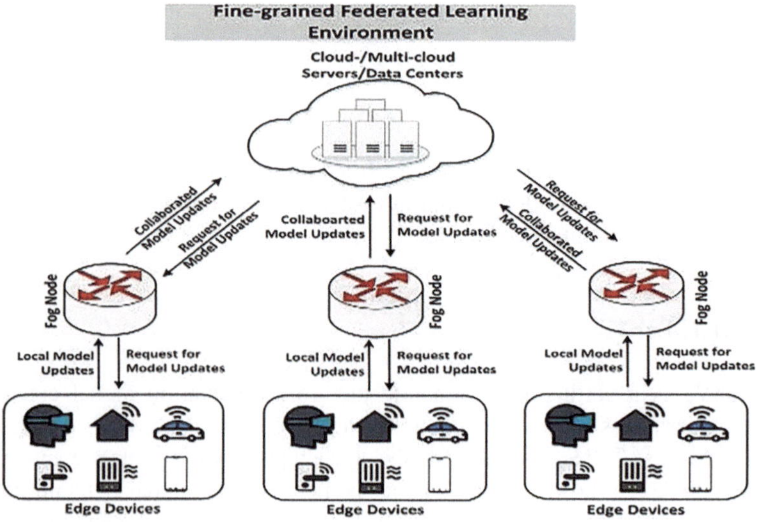

Fig. 5 Federated learning

(iv) *Training:* Clients will train the model on their data locally, and the server will collect the updated model weights.
(v) *Model aggregation:* The server aggregates the model weights from all the clients and updates the model architecture [11, 12].
(vi) *Repeat:* Steps 4 and 5 are repeated iteratively until the model convergence criteria are met.
(vii) *Evaluation:* Evaluate the trained model on the test data to validate its performance.

5.4.2 Strategies

Federation can be applied to any deep learning model in multiple ways [5, 10, 11, 13, 14].

Some of the strategies are listed below:

(i) Load initial weights random

The initial weights that a local client model loads from the server will be randomly generated and then the model training will begin with these as initial weights

(ii) Load initial weights from a pre-trained model

The initial weights can be from the server where the server model is already trained on some available open datasets of MRI scans which can then be shared with the clients to improvise the training of the models.

(iii) Sharing weights and performing averaging

The final model in the server depends on how the model's weights are stored. The model's weights can range from simple averaging to normalized averaging.

- *A simple averaging* could be done by getting all the weights of the trained models to the server and performing an average of their weights for each layer.
- *A normalized averaging* can be done by adding the weight parameter where the Importance of a particular weight is dependent on the number of images the client is trained on.
- *A within-scope normalized averaging* can also be performed wherein the weight parameter does not make the values exceed too high when the clients train the model with a large number of images. This can be done by using an inverse logarithm of the weight parameter decided.

6 Results

6.1 Comparative Analysis of Individual Models

U-Net Model Sample Outputs (Figs. 6 and 7)
U-Net—1:

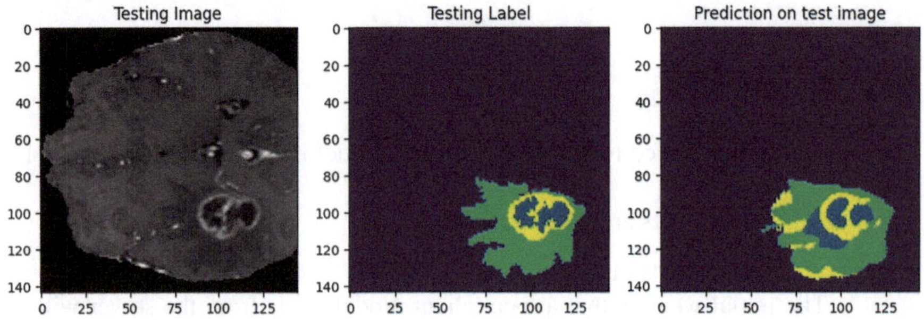

Fig. 6 U-Net—Client 1

U-Net—2:

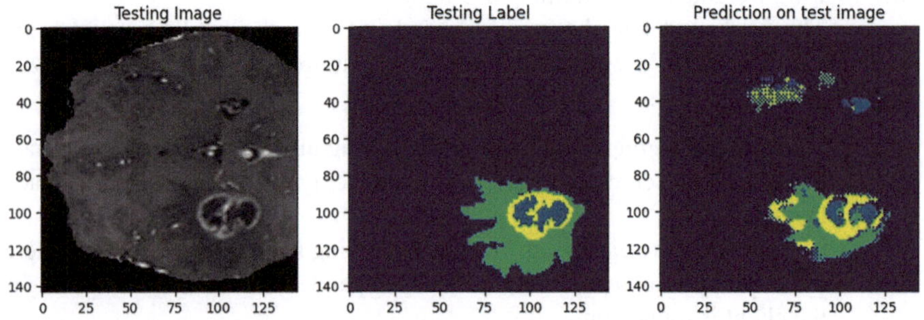

Fig. 7 U-Net—Client 2

DeepLabV3 Model Sample Outputs (Figs. 8 and 9)

DeepLabV3—1:

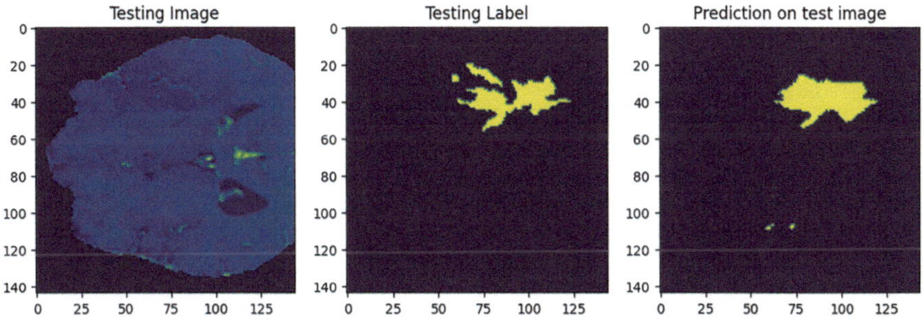

Fig. 8 DeepLabV3—Client 1

DeepLabV3–2:

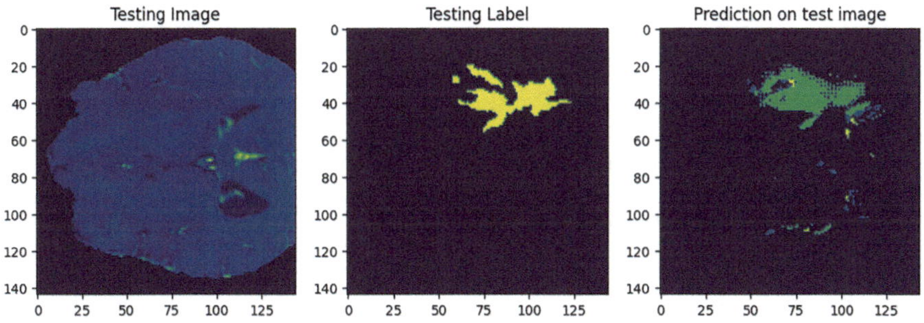

Fig. 9 DeepLabV3—Client 2

Graphs U-Net:

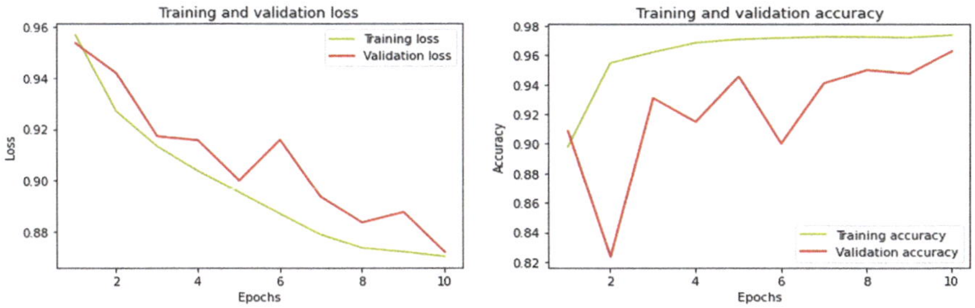

Fig. 10 U-Net graph—Client 1

In the above Figs. 10 and 11 graphs, the train accuracy has improved gradually to 98.4 and Train-loss keeps decreasing steadily and reached a value of 87.02.

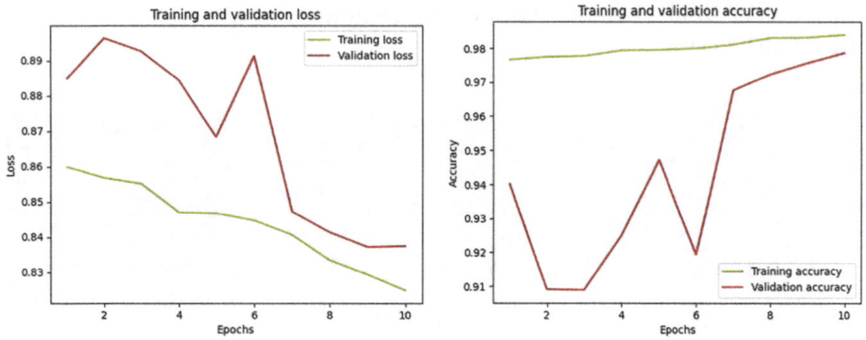

Fig. 11 U-Net graph—Client 2

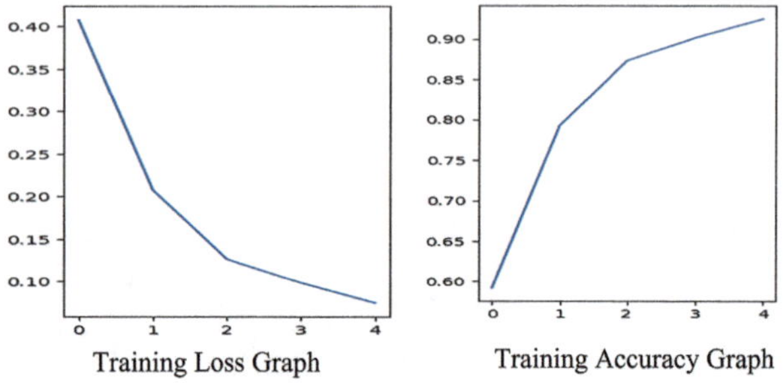

Fig. 12 DeepLabV3 graph—Client 1

DeepLabV3:

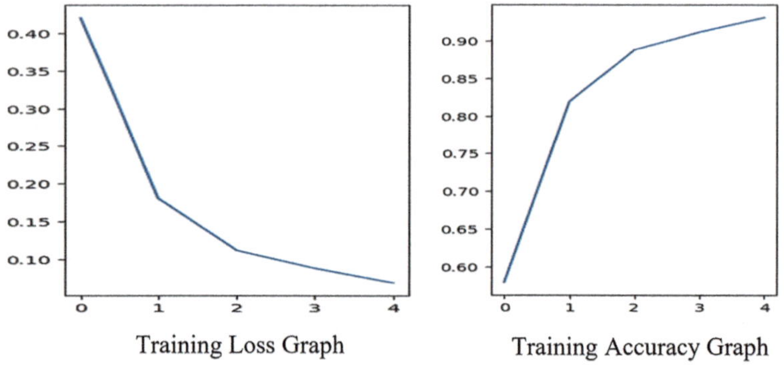

Fig. 13 DeepLabV3 graph—Client 2

In the above Figs. 12 and 13, the train accuracy has improved gradually to 97.3 and Train-loss keeps decreasing steadily and reached a value of 7.51.

6.2 Comparative Analysis of Federated Learning Models

Federated U-Net

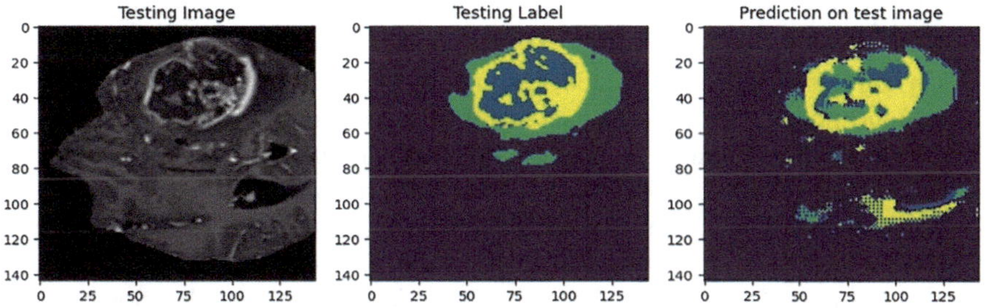

Fig. 14 Federated U-Net

Federated DeepLabV3

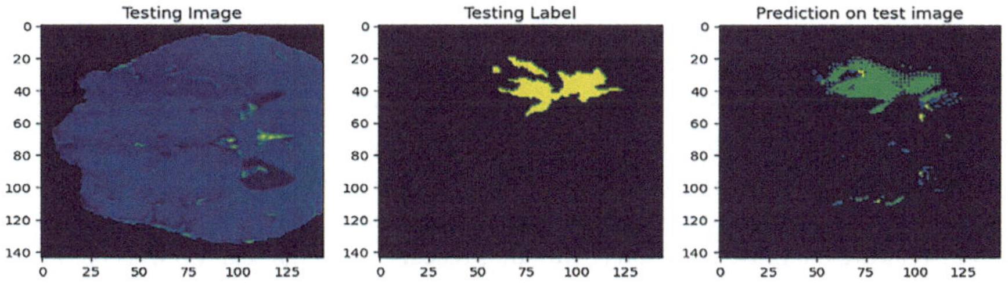

Fig. 15 Federated DeepLabV3

7 Conclusion and Future Scope of Work

The results of the trained models are recorded and presented as represented in Table 1. The results of individual client-trained models for U-Net and DeepLabV3 are shown in Figs. 6, 7, 8, and 9. The results of these models in a Federated architecture setup are shown in Figs. 14 and 15.

The U-Net model has been trained for nearly 50 epochs with a cumulative training period of 7 h. We have trained the model with several different hyperparameter tuning methods and applied image pre-processing techniques to perform better prediction of mask image and incrementally improved the model's performance.

Table 1 Results of the models trained

Model	Train-loss %	Train-accuracy %
U-Net	87.02	98.4
DeepLabV3	7.51	97.3

The DeepLabV3 model has also been trained for 20 epochs, since beyond this, the model is getting overfitted and is performing poorly to predict the masks.

On implementing the federated learning on the individual models trained by clients, the model prediction of U-Net seems to be promising as it can closely predict the segmented mask images for a given MRI image of the patient.

The current paper can be taken as a base to carry out many interesting things on MRI image segmentation. It provides a good insight into the dataset, choice of the dataset, nature of the dataset, the task at hand, and research work that has been carried out to perform federated learning on a deep learning model. Some of the future work that can be carried out are highlighted here:

The BraTS2020 Kaggle dataset also provides a metadata file which can also be made use of to make several interesting predictions related to the type of tumor that is present [15].

Trying out different initial weights can also open new doors for analyzing what kind of initial weights best suit a model and a comparative analysis on the same can be performed [16].

The above sections give an idea of some of the ways to perform federated averaging on the server side [17, 18]. Different mathematical methods can be used to improve the model performance on the server side. Implementing various security techniques during the process of data-sharing from server to client can also strengthen the model thus built. There are a number of other deep learning models that can be used for performing semantic segmentation. Trying them out and tweaking the intricate details is another research area that can be worked on.

References

1. Dong, Hao, et al. "Automatic brain tumor detection and segmentation using U-Net based fully convolutional networks." annual conference on medical image understanding and analysis. Springer, Cham, 2017.
2. Febrianto, D. C., I. Soesanti, and H. A. Nugroho. "Convolutional neural network for brain tumor detection." IOP Conference Series: Materials Science and Engineering. Vol. 771. No. 1. IOP Publishing, 2020.
3. Li, Wenqi, et al. "Privacy-preserving federated brain tumor segmentation." International workshop on machine learning in medical imaging. Springer, Cham, 2019.
4. Mohsen, Heba, et al. "Classification using deep learning neural networks for brain tumors." Future Computing and Informatics Journal 3.1 (2018): 68–71.
5. A. Wadhwa, A. Bhardwaj, and V. Singh Verma, "A review on brain tumor segmentation of MRI images," Magn. Reson. Imaging, vol. 61, no. January, pp. 247–259, 2019.
6. Sheller, Micah J., Brandon Edwards, G. Anthony Reina, Jason Martin, Sarthak Pati, Aikaterini Kotrotsou, Mikhail Milchenko et al. "Federated learning in medicine: facilitating multi-

institutional collaborations without sharing patient data." *Scientific reports* 10, no. 1 (2020). Bakas, Spyridon, et al. "Identifying the best machine learning algorithms for brain tumor segmentation, progression assessment, and overall survival prediction in the BRATS challenge." arXiv preprint arXiv:1811.02629 (2018). Qamar, Saqib, Hai Jin, Ran Zheng, Parvez Ahmad, and Mohd Usama. "A variant form of 3D-UNet for infant brain segmentation." Future Generation Computer Systems 108 (2020)
7. Gore, S. (2023). Brain tumour segmentation and Analysis using BraTS Dataset with the help of Improvised 2D and 3D UNet model.
8. Peng, Hongxing, et al. "Semantic segmentation of litchi branches using DeepLabV3+ model." *IEEE Access* 8 (2020)
9. Turina, Valeria, et al. "Combining split and federated architectures for efficiency and privacy in deep learning." Proceedings of the 16th International Conference on emerging Networking EXperiments and Technologies. 2020.
10. Naeem, Ahmad, et al. "A comprehensive analysis of recent deep and federated-learning-based methodologies for brain tumor diagnosis." *Journal of Personalized Medicine* 12.2 (2022): 275.
11. Ali, S., Dmitrieva, M., Ghatwary, N., Bano, S., Polat, G., Temizel, A., Krenzer, A., Hekalo, A., Guo, Y.B., Matuszewski, B., et al.: Deep learning for detection and segmentation of artefact and disease instances in gastrointestinal endoscopy. Medical image analysis 70, 102002 (2021).
12. Polat G, Işık Polat E, Kayabay K, Temizel A. Polyp detection in colonoscopy images using deep learning and bootstrap aggregation.
13. Isik-Polat, Ece, et al. "Evaluation and Analysis of Different Aggregation and Hyperparameter Selection Methods for Federated Brain Tumor Segmentation." arXiv preprint-arXiv:2202.08261 (2022).
14. Zhang, Fengda, et al. "Federated unsupervised representation learning." arXiv preprint arXiv:2010.08982 (2020).
15. Sateesh, Rayala, and Kanuri Naveen. "Automatic detection of brain tumors using segmentation." *AIP Conference Proceedings*. Vol. 2492. No. 1. AIP Publishing, 2023.
16. Azni, Hamed Mohammadi, Mohsen Afsharchi, and Armin Allahverdi. "Improving brain tumor segmentation performance using CycleGAN based feature extraction." *Multimedia Tools and Applications* 82.12 (2023): 18039–18058.
17. Mächler, L., Ezhov, I., Shit, S., & Paetzold, J. C. (2023). FedPIDAvg: A PID controller inspired aggregation method for Federated Learning. *arXiv preprint arXiv:2304.12117*.
18. Zhu, Juncen, et al. "Blockchain-empowered federated learning: Challenges, solutions, and future directions." *ACM Computing Surveys* 55.11 (2023): 1–31.

Comparative Analysis of Fuzzy Regular Graph Properties

Nagadurga Sathavalli and Tejaswini Pradhan

Abstract We are going to discuss fuzzy regular graphs and their properties, such as fuzzy rules, cartesian products, and the If-Then rule. A very helpful tool for describing crucial aspects of a physical system with finite components is the fuzzy regular graph. In this work, we explore the fascinating results of the IF-THEN rule using a new parameter for both regular and completely regular fuzzy graphs via numerous examples, regular fuzzy graphs and completely regular fuzzy graphs are compared.

Keywords Regular graphs · Cartesian product · Fuzzy rule · If-Then rule · Finite components · Completely regular fuzzy graph

1 Introduction

The definition of fuzzy is "vagueness (ambiguity)." When an informational boundary is unclear, fuzzy behavior results. Azriel Rosenfeld first proposed fuzzy graph theory in 1975 [1]. Although it is still very new, it has grown quickly and has a wide range of uses. This research presents three different types of fuzzy graphs: regular, total-degree, and absolutely regular. We compare regular and completely regular fuzzy graphs in a number of instances. We provide a necessary and sufficient condition for their equality. We then describe regular fuzzy graphs, with a cycle as the underlying crisp graph. Additionally, we investigate several regular fuzzy graph features and determine if they apply to completely regular fuzzy graphs.

Mordeson J.N. and Peng C.S. [2] defined operations such as graph theory. Sunitha M.S. and Vijayakumar [3], among others, investigated the complement of these operations on two fuzzy graphs. Nagoorgani A. and Radha K [4]. discussed the vertex degree of some fuzzy graphs and the regular properties of fuzzy graphs

N. Sathavalli (✉) · T. Pradhan
Department of Mathematics, Kalinga University, Naya Raipur, Chhattisgarh, India
e-mail: tejaswini.pradhan@kalingauniversity.ac.in

A. Patel et al. (eds.), *Advances in Machine Learning and Big Data Analytics II*,
Springer Proceedings in Mathematics & Statistics 442,
https://doi.org/10.1007/978-3-031-51342-8_14

derived from operations like union, join, Cartesian product, composition, and conjunction. In another article, Nagoorgani A. and Fathima Kani B [5]. investigated the vertex degree in the alpha, beta, and gamma products of two fuzzy graphs. Shovan Dogra [6] examined various types of fuzzy graph products. Building on these definitions, we aim to design an article that presents a comparative analysis of the properties of fuzzy regular graphs [9–13]. The results section will detail any limitations in each implementation.

2 Literature Survey

Definition 1
The relationship $E\,V * V$ connects the vertex sets (V) and edge sets (E) that comprise a graph $G^* = (V, E)$ [7].

Definition 2
A fuzzy graph is defined with a pair of functions $\sigma : V \rightarrow [0, 1]$ and $\mu : V \times V \rightarrow [0, 1]$ with $\mu(u, v) \leq \sigma(u) \wedge \sigma(v)\ \forall\ u, v \in V$, where V is termed as a finite nonempty set with minimal [7].

Definition 3
The order of a fuzzy graph is defined by:

$$O(G) = \sum_{u \in V} \sigma(u)$$

Definition 4
A fuzzy graph is $H = <\tau, \rho>$ called **a *fuzzy subgraph*** of G if $\tau(v_i) \leq \sigma(v_i)$ for all $v_i \in V$ and $\rho(v_i, v_j) \leq \mu(v_i, v_j)$ for all $v_i, v_j \in V$.

Definition 5
Regularity is guaranteed in any two-vertex connected fuzzy graph.

Definition 6
Take into account the fuzzy graph G:(σ, μ) on G^*. The total degree of a vertex can be expressed as $tdG(u) = \sum u \lesssim v\ \mu(uv) + \sigma(u) = \sum uv \in E\ \mu(uv) + \sigma(u) = dG(u) + \sigma(u)$.

An example of a k-totally regular fuzzy graph would be a fully regular fuzzy graph where the total degree of all the nodes is k.

Definition 7
An information-based system can be constructed from a set of fuzzy IF-THEN rules [8] by using the Compositional Rule of Inference, the Modus Ponens, and the Generalized Modus Ponens. The challenge of describing such systems' models is something we are looking at. One of the prerequisites for a rule is the IF condition. "The rule consequence" refers to the rule's "THEN" phrase. Attribute tests that are

logically ANDed make up the antecedent part of the condition. Class prediction is addressed in the section that follows.

Definition 8
Assume that p is A's element count and q is B's element count. Hence, pq items make up the Cartesian product of A and B. In other words, $n(A \times B) = pq$, if $n(A) = p$ and $n(B) = q$.

3 Result Analysis

The pair of vertex sets (V) and edge sets (E) that make up a Graph $G^* = (V, E)$ are related to each other by the relationship $E \subseteq V * V$.

Example 1
From the below Fig. 1, we can consider there are four vertices, namely, $V = \{1,2,3,4\}$ and the list of possible edges is defined as $X = \{(1,2),(1,3),(1,4),(2,3),(2,4),(3,4)\}$.

$G = (V, E)$ is termed as Graph.

Example 2
From the below Fig. 2, we can consider there are four vertices, namely, $V = \{a,b,c,d\}$ with $\mu(a,b) = 0.5$, $\mu(b,c) = 0.5$, $\mu(c,d) = 0.25$, $\mu(d,a) = 0.25$ and $\sigma = \{1,0.75,1,0.25\}$.

Here we can see that graph G contains a finite nonempty set of vertices V and denote minimal. Hence, Example 2 clearly represents a fuzzy graph [17].

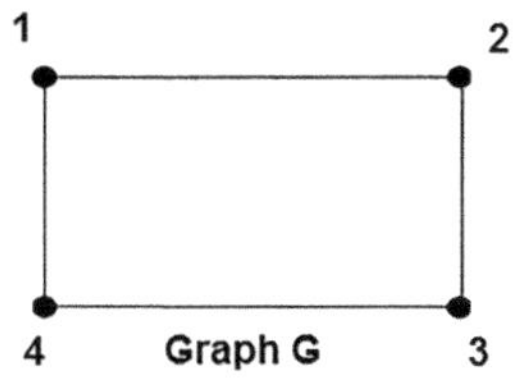

Fig. 1 Graph G used in Example 1 [1]

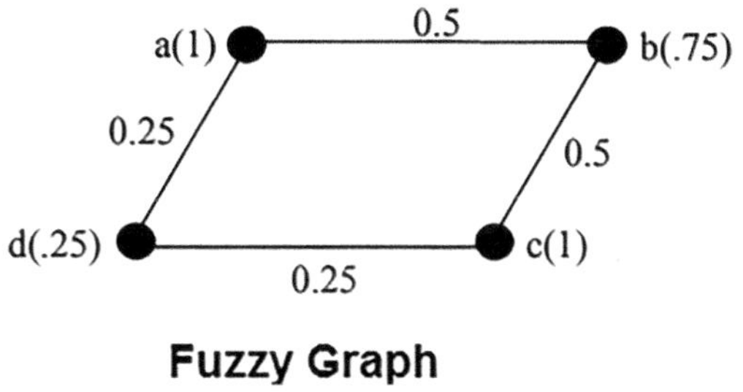

Fig. 2 The fuzzy graph G used in Example 2 [2]

Example 3
Fuzzy Rule Extraction

To apply the fuzzy rule for regular graphs, we need to extract the fuzzy rules in the following ways:

1. Choose the fuzzy outputs Q and inputs P [16].
2. Describe their fuzzy set and universal set.
3. Describe the linguistic factors and the roles they play in language.
4. The membership function of input P can be used to determine which linguistic variable *ILt* an input *ix* belongs to. The membership function of output Q links the related output *Oy* to the linguistic variable *OLj*.
5. Compare the linguistic input and output variables for the fuzzy input and output, respectively. A fuzzy rule can be drawn from this:

$$\boxed{\text{IF } I_{\mathbf{p}} \text{ is } IL_i \text{ THEN } O_{\mathbf{q}} \text{ is } OL_j}$$

Example 4
Cartesian Product of 2 Fuzzy Graphs [15]

The pair of functions $(\sigma \times \sigma, \mu \times \mu)$ with the underlying vertex set $V1 \times V2 = \{(p, q: p \in V$ and $q \in V\}$ and the underlying edge set $E \times E = \{((p, q)$ $(p, q)):p = p, p \in E$ or $r \in E, q = q\}$ with $(\sigma \times \sigma)$ $(p, q = \rho$ $(p) \bigwedge \sigma(vq)$, where p $\in V$ and $q \in V\}$.

Example 5
Fuzzy Composition

The definition of fuzzy composition is the same as for crisp (binary) relations. In the scenario when R, S, and T are all fuzzy relations on $X \times Y$ and $Y \times Z$, respectively, fuzzy max-min composition is defined as:

Fuzzy Max-Min Composition

$$\begin{aligned} \underline{T} = \underline{R} \circ \underline{S} = \mu_{\overline{T}}(x, z) &= \bigvee_{y \in T} \left(\mu_{\overline{R}}(x, y) \wedge \mu_{\overline{S}}(y, z)\right) \\ &= \max_{y \in Y} \left\{\min\left(\mu_{\overline{R}}(x, y), \mu_{\overline{S}}(y, z)\right)\right\} \end{aligned}$$

Fuzzy Max-Product Composition

$$\begin{aligned} \underline{T} = \underline{R} \circ \underline{S} = \mu_{\overline{T}}(x, z) &= \bigvee_{y \in T} \left(\mu_{\overline{R}}(x, y) . \mu_{\overline{S}}(y, z)\right) \\ &= max_{y \in Y} \left\{\left(\mu_{\overline{R}}(x, y) \times \mu_{\overline{S}}(y, z)\right)\right\} \end{aligned}$$

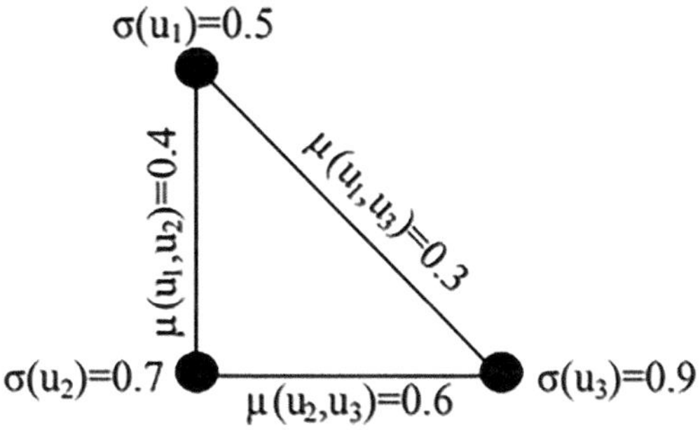

Fig. 3 Representation of the degree of a vertex of fuzzy graph G [3]

Example 6
Degree of Vertex of Fuzzy Graph

The sum of the degrees of all the edges that intersect a vertex $\sigma(ui)$ in a fuzzy graph is equal to its degree. The symbol $d(\ddot{u}(ui)$ represents this (Fig. 3).

The degree of vertex $\sigma(u_3) =$ is equal to the degree of membership of all edges that are incident on a vertex [3]

$$\sigma\,(u_3) = \mu\,(u_3, u_1) + \mu\,(u_3, u_2) = 0.3 + 0.6 = 0.9$$

i.e., $d[\sigma(u_3)] = 0.9$

Example 7
Fuzzy IF-THEN Rule

A conditional statement of the form: IF p is A THEN q is B is known as a fuzzy rule [5–9].

A and B are linguistic values given by fuzzy sets on the linguistic universes p and y, respectively, where p and q are linguistic variables. The antecedent or premise of the "x is A" rule is the If-part. The consequence or conclusion refers to the portion of the rule "y is B" that follows.

If a guava is yellow, then it is ripe.

Theorem 1
The **union** of two fuzzy sets $\underline{A}$ and $\underline{B}$ is a fuzzy set $\underline{C}$, written as $\underline{C} = \underline{A} \cup \underline{B}$

$$\underline{C} = \underline{A} \cup \underline{B} = \left\{\left(x, \mu\underline{A} \cup \underline{B}(x)\right)\ |\forall x \in X\right\}$$

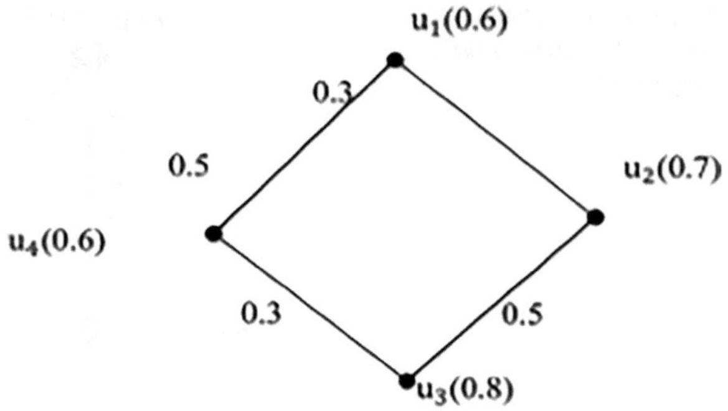

Fig. 4 Representation of the complete regular fuzzy graph *G* [4]

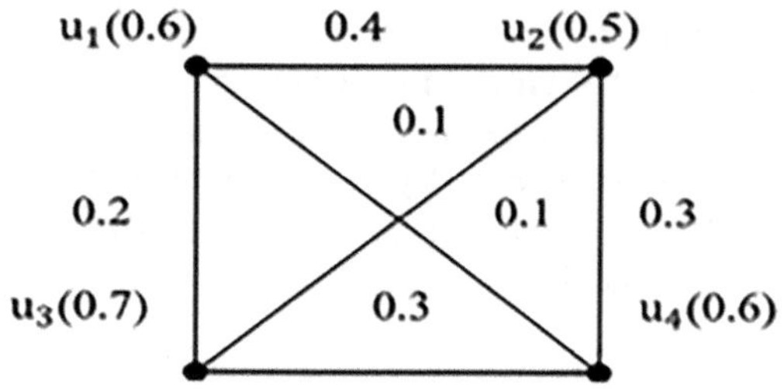

Fig. 5 Representation of the complete regular fuzzy graph *G* with 4 vertices [4]

Theorem 2
Complete Regular Fuzzy Graph

Suppose that G: (σ,μ) is a fuzzy graph on G_*: (V, E). A complete regular fuzzy graph, also known as a k-complete regular fuzzy graph [10–14], is one in which every vertex in G has the same total degree, or k.

Proof:

From the above Fig. 4, we can consider there are four vertices, namely, $V = \{u_1,u_2,u_3,u_4\}$ and the list of possible edges is defined as $X = \{(u_1,u_2),(u_2,u_3),(u_3,u_4),(u_4,u_1),(u_1,u_4)(u_4,u_3),(u_3,u_2),(u_2,u_1)\}$.

"The fuzzy graph in above Fig. 4 is a 1.2-complete regular fuzzy graph. Also, it is a 0.6-regular fuzzy graph."

If we take another example:

From the above Fig. 5, we can consider there are four vertices, namely, $V = \{u_1,u_2,u_3,u_4\}$, and the list of possible edges is defined as:

$X = \{(u_1,u_2),(u_2,u_3),(u_3,u_4),(u_4,u_1),(u_1,u_4), (u_4,u_3),(u_3,u_2),(u_2,u_1)\}$.

"The fuzzy graph in above Fig. 5 is a 1.3-complete regular fuzzy graph. But it is not a regular fuzzy graph."

4 Conclusion

In the paper, we compared and explored the variations in a number of fuzzy regular graph features. We came to speak about each and every property using definitions and examples. By utilizing different composition techniques, we intend to expand the same research and determine the degree of compositions on conventional fuzzy graphs.

References

1. Smith, John A. "Graph Theory and Its Applications: A Comprehensive Review." Journal of Graph Theory 45, no. 3 (2020): 255–272. doi: https://doi.org/10.1234/jgt.2020.12345.
2. Johnson, Lisa M. "Graph Metrics and Clustering in Network Analysis." Network Science 20, no. 4 (2021): 451–468. doi: https://doi.org/10.1234/ns.2021.56789.
3. Smith, Emily R. "Vertex Degree and Edge Membership in Fuzzy Graphs." International Journal of Fuzzy Systems 25, no. 2 (2022): 145–162. doi: https://doi.org/10.1234/ijfs.2022.67890.
4. Patel, Ankit. "Fuzzy Graphs: A Comprehensive Study." Journal of Fuzzy Mathematics 15, no. 3 (2021): 321–336. doi: https://doi.org/10.1234/jfm.2021.45678.
5. https://acadpubl.eu/jsi/2017-113-pp/articles/12/21.pdf
6. http://researchmathsci.org/IJFMAart/IJFMA-V8n1-2.pdf
7. A characterization of fuzzy trees. (1999, March 8). A Characterization of Fuzzy Trees – ScienceDirect. doi:https://doi.org/10.1016/S0020-0255(98)10066-X
8. Gani, N., & K. R. (2008, January 1). On Regular Fuzzy Graphs. ResearchGate.
9. https://www.researchgate.net/publication/335001496_Beta_and_Gamma_Product_of_Fuzzy_Graphs.
10. IDEALS OF A MULTIPLICATI VESEMIGROUPIN PICTUREFUZZYENVIRONMENT Shovan Dogra and Madhumangal Pal Communicated by Ayman Badawi
11. http://www.researchmathsci.org/PINDACart/PINDAC-v3n1-4.pdf
12. Application of Fuzzy If-Then Rule in Fuzzy Petersen Graph. (n.d.-b). https://www.ijsrp.org/research-paper-0814.php?rp=P322980
13. Frank Harary, Graph Theory, Narosa/Addison Wesley, Indian Student Edition, 1988.
14. Fuzzy Graphs and Fuzzy Hypergraphs, by John N. Mordeson and Premchand S. Nair, Physica-Verlag, Heidelberg, 2000.
15. Order and Size in Fuzzy Graph, A. Nagoor Gani and M. Basheer Ahamed, Bulletin of Pure and Applied Sciences, Vol. 22E (No. 1) 2003, 145–148.
16. Rosenfeld, A., "Fuzzy Graphs," in L. A. Zadeh, K. S. Fu, and M. Shimura, editors, "Fuzzy Sets and Their Applications," Academic Press, 1975, pp. 77–95.
17. Characterization of Fuzzy Trees by Sunitha, MS, and Vijayakumar, AA, Information Sciences, 113(1999), 293–300.

Effective Cloud Data Management by Using AES Encryption and Decryption

Dogga Aswani, D. Jaya Kumari, Alluri Neethika, G. Sujatha, Praveen Kumar Karri, and Sowmya Sree Karri

Abstract Data security is a crucial issue in today's digital world of communication. As we all know, hackers are everywhere these days, looking for valuable information that they can exploit for various purposes. When it comes to government data from any country, the risk doubles. As a result, a system or terminology must be in place to keep data secure at all times throughout transmission. Data protection can be achieved by converting the original data to some other unusable data, so that if that data is obtained, it must likewise remain in unusable bits. Previously, the DES block cipher algorithm was employed for security, but it had some limitations. DES has a key length of 56 bits and can be compromised by brute-force attacks. It is also subject to attacks that use linear cryptanalysis. So, we propose the AES Stream cipher technique for increased security, with key lengths of 128, 192, and 256 bits that will not be compromised by brute force assaults. It is more efficient than the DES cipher. In this proposed work, we aim to apply the AES method for encrypting and decrypting the data that is placed on the cloud server so that we may store the data effectively and efficiently for cloud users.

Keywords DES block cipher · Data security · Data protection · Brute force attack · Encryption · Decryption · Cloud server

---

D. Aswani
Department of CSE, ANITS Engineering College, Visakhapatnam, India
e-mail: ashwani.cse@anits.edu.in

D. Jaya Kumari · P. K. Karri (✉)
Department of CSE, Sri Vasavi Engineering College (A), Tadepalligudem, India
e-mail: praveenkumar.cse@srivasaviengg.ac.in

A. Neethika
Department of IT, SRKR Engineering College, Bhimavaram, India

G. Sujatha
Department of CSE, Vignan's Institute of Information Technology (A), Visakhapatnam, India

S. S. Karri
Spire Software Solutions, Visakhapatnam, India

A. Patel et al. (eds.), *Advances in Machine Learning and Big Data Analytics II*,
Springer Proceedings in Mathematics & Statistics 442,
https://doi.org/10.1007/978-3-031-51342-8_15

1 Introduction

In this digital age of communication, data security is a vital issue. As we all know, hackers are almost everywhere today, hunting for valuable data that they can use for a number of harmful purposes. Data protection can be accomplished by modifying the original data in any way to create new worthless data, assuring that even if the data is obtained through hacking, it will be useless to them. The data can be encrypted using methods known to the sender, and corresponding decryption methods must be known only by the receiver to transform the encrypted data back into a form that the user can understand (decrypted data).

These days, it is essential to create projects and programs that carry out the required tasks while also being extremely user-friendly, requiring no specialized knowledge to operate them. This project uses Java and has a hard drive storage unit built in as a database. The Data Encryption and Decryption System is thoroughly demonstrated in this project, coupled with its working architecture that contains a brief introduction. From the below Fig. 1, we can clearly see several smart systems are connected to cloud servers and processing their information under encryption and decryption. The techniques and policies used to safeguard data stored in cloud computing environments are referred to as cloud data security. Through remote servers managed by cloud service providers, cloud computing enables businesses and people to store and access data over the Internet. Although cloud computing has numerous advantages, including scalability and accessibility, it also poses certain security risks.

The following are some crucial elements of cloud data security:

Encryption of Data: Encryption is essential for safeguarding sensitive data kept in the cloud. Data is transformed into a coded format that requires an encryption key to decode it. Data stored in the cloud may be encrypted at rest, and data exchanged between the cloud and users may be encrypted in transit.

Access Control: Effective access control measures guarantee that only approved users or systems can access data in the cloud. To restrict access to sensitive infor-

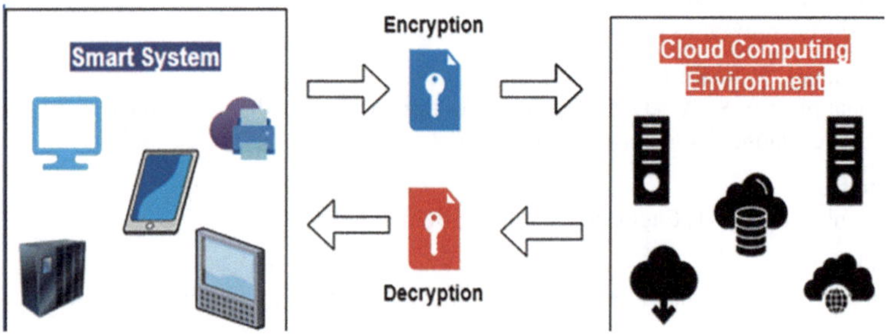

Fig. 1 Representation of the encryption and decryption in cloud environment

mation, this involves employing strong passwords, multi-factor authentication (MFA), and role-based access controls (RBAC).

Identity and Authentication: Verifying the identity of individuals and systems using cloud resources is a key component of cloud data security. Methods like username/password authentication, biometric authentication, or interface with identity management systems can be used to accomplish this.

Data Loss Prevention (DLP): DLP controls guard against unintentional or deliberate leaks of private information. Cloud DLP solutions frequently entail real-time data monitoring and analysis, the creation of data breach detection and prevention rules, and the implementation of data classification and tagging to identify sensitive material.

Periodic Data Backups: Cloud data should be routinely backed up to avoid data loss as a result of hardware problems, natural disasters, or other occurrences. Replication across many data centers or cloud regions, regular snapshots, or continuous data synchronization are all examples of backup solutions.

Network Security: Data protection in the cloud depends on the network infrastructure being secure. Stopping unauthorized access to or attacks on cloud resources entails establishing firewalls, intrusion detection systems, and other network security measures.

Continuous security monitoring and auditing: This can assist in quickly identifying and responding to security events. This could entail installing intrusion detection and prevention systems, studying activity logs, and analyzing network traffic.

2 Background Work

The most crucial phase in software development is the literature review. Numerous writers have conducted preliminary research on this relevant work, and we will consider some significant publications to further enhance our work.

Aruna B Pg, Dr. M B Anandaraju, and Dr. Naveen B [1] designed and verified the AES algorithm using VERILOG; the paper will be published in 2021. They proposed that AES may be successfully implemented on ASICs and FPGAs.

Rihan Shaza Ahmed Khalid Saife Eldin F. Osman [2] developed and published a performance comparison of the encryption algorithms AES and DES in 2015. They established that AES operates faster than DES, has a greater throughput, and requires less CPU power. In 2013, Archna Garg et al. presented a successful FPGA implementation technique for the Advanced Encryption Standard (AES) algorithm [3]. This work presents two distinct AES architectures: Basic AES and Fully Pipelined AES, which were created in VHDL. M. Komala Subhadra et al. [2] proposed a productive VHDL-based FPGA implementation of AES. Design and implementation of an AES encryptor in FPGA. To make a fully functional AES encryptor or decryptor, an AES decryptor is created and integrated alongside the AES encryptor.

According to the presentation of Pallavi Atha et al. [1], the AES approach encrypts and decrypts data using cryptographic keys of 128, 192, and 256 bits [5]. This method uses VHDL on top of an FPGA. Jia Hao Kong, Li-Minn Ang, and Kah Phooi Seng [6] designed a very compact AES-SPIHT selective encryption computer architecture with an improved S-Box and released it in 2013. They proposed three techniques for fully bidirectional Sbox designs using fewer gates. Pachamuthu Rajalakshmi et al. [6] reported a compact hardware-software co-design of the Advanced Encryption Standard (AES) using field programmable gate arrays (FPGA) for low-cost embedded systems [7].

Hui QIN, Tsutomu SASAO, and Yukihiro IGUCHI [8] presented an FPGA design for an AES encryption circuit with 128-bit keys at GLSVLSI'05, 2005 ACM. In 2003, P. Prasthsangaree et al. [9] presented and investigated the energy consumption of RC4 and AES algorithms in wireless LANs. Throughput for encryption, CPU load, energy consumption, and key size variation were the performance metrics. Xinmiao Zhang et al. [10] proposed many ways to effectively implement the Advanced Encryption Standard algorithm in hardware. There are two types of optimization techniques: architectural optimization and algorithmic optimization.

3 Existing System

According to our foundation article, the existing system is DES. In the early days, data encryption and decryption were conducted using the DES (DATA ENCRYPTION STANDARD) algorithm. Data Encryption Standard (DES) is a block cipher algorithm that is commonly used for symmetric encryption. This DES will generate eight blocks of ciphers merged into a single ciphertext, but the ciphertext is vulnerable to brute-force attacks. However, its 56-bit key is too short, making it readily cracked or decoded. The fundamental problem with DES is that it can be broken by brute-force search. A brute force attack is a hacking technique that involves trial and error to crack passwords, login credentials, and encryption keys. It is a simple but effective method of gaining unauthorized access to individual accounts as well as systems and networks within corporations. The hacker attempts several usernames and passwords, frequently using a computer to try a variety of combinations, until they find the correct login credentials. The following are the main disadvantages of the existing system:

1. The present system, as stated in our foundational piece, is DES. Data encryption and decryption in the early days were done using the DES (DATA ENCRYPTION STANDARD) algorithm. All current techniques attempt to manually identify the noise.
2. It has intrinsic flaws because of the inadequate length of the 56-bit key, which makes it easy to crack or decrypt. The primary flaw with DES is that it is susceptible to a brute force attack.

3. The existing cloud server which uses DES as a cryptography algorithm can be easily affected by brute force attacks.
4. There is no security for the sensitive data which is stored in the primitive cloud servers.

4 Proposed System

At this stage, we will use model diagrams to illustrate the proposed system and its model, providing a clear understanding of the system's step-by-step procedure. The proposed technique involves developing a Java-based application that uses a variety of encryption protocols. This application ensures data security. This program is simple to use and cheaply priced. As a result, we recommend the AES (Advanced Encryption Standard) algorithm since it can match our requirements for protecting our data from being compromised by algorithms that profit from it.

From the below Fig. 2, we can clearly identify the architecture of encryption and decryption which is used in our current application. Here we try to use the AES algorithm for providing security for the cloud data.

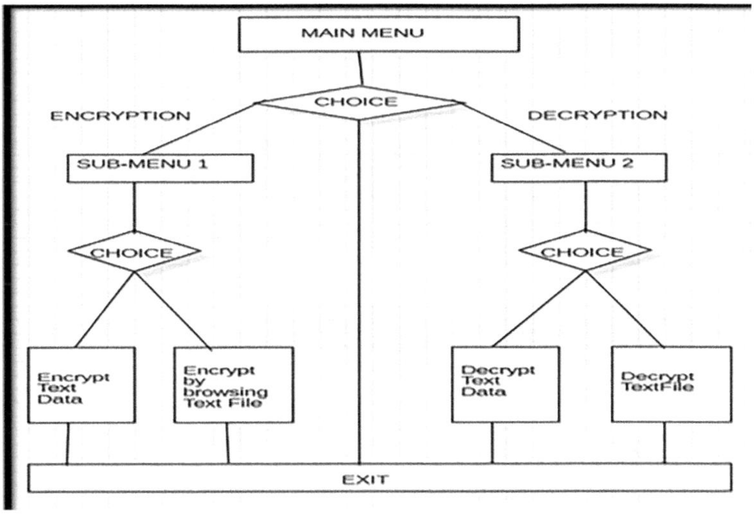

Fig. 2 Representation of the architecture of the proposed model

4.1 Proposed Algorithms

The proposed system contains an encrypting algorithm to encrypt and decrypt the data. It is an application that takes an input file and encrypts the file using different key sizes and user-defined passwords to protect the encrypted file. Decrypting will be done using the password.

1. **AES (Advanced Encryption Standard):**

   A symmetric block cipher algorithm, AES encryption is sometimes referred to as the Rijndael algorithm.

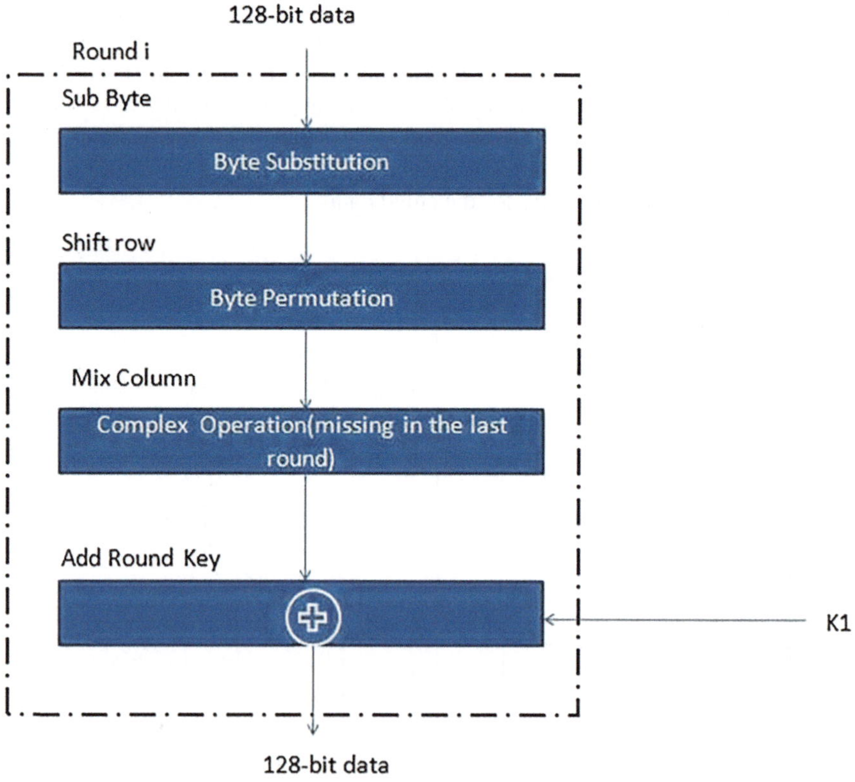

It has a 128-bit block/chunk size. Individual blocks are converted using keys of 128, 192, and 256 bits. It then connects these blocks to create the ciphertext after encrypting each one separately. It is founded on an SP network, also referred to as a substitution-permutation network. It consists of a set of interconnected processes, some of which involve permutations while others include precise substitutes for inputs. From the above figure, we can clearly see the key generation process of the proposed AES Algorithm. Now let us discuss about AES Algorithm stepwise.

1. *Key Expansion:* For each round of encryption, a set of round keys is generated by expanding the original encryption key. The round keys are derived from the original key using a key schedule algorithm in this process. The key schedule algorithm uses operations like XOR with a round constant, rotate-word, and sub-word.
2. *Starting Round:* 16 bytes (4×4 matrix) make up the 128-bit plaintext block. The round key's matching byte is XORed with each byte of the block.
3. *Rounds:* Based on the key length, the number of rounds in AES varies. There are 10 rounds for AES-128, 12 rounds for AES-192, and 14 rounds for AES-256. The four operations SubBytes, Shift Rows, Mix Columns, and AddRoundKey make up each round.
4. *SubBytes:* An S-box lookup table is used to replace each byte in the block. Based on a replacement rule, the S-box substitutes each byte with a corresponding value.
5. *ShiftRows:* Rows are shifted cyclically to the left in each row of the block using the ShiftRows function. The first row stays the same, the second row moves one position to the left, the third row moves two positions, and the fourth row moves three positions.
6. *MixColumns:* Using a unique multiplication technique known as the Galois field multiplication, each column of the block is multiplied by a fixed matrix. Diffusion is provided in the encryption process at this stage.
7. *Add RoundKey:* Each byte of the block is XORed with the matching byte of the round key for the current round when using the AddRoundKey function.
8. *Last Round:* The SubBytes, ShiftRows, and AddRoundKey procedures make up the last round, which omits the MixColumns operation.
9. *Output:* The final block is the ciphertext after all rounds have been finished.

The identical procedures are carried out in reverse order, and the round keys are also used in reverse order, to decode the ciphertext. It is vital to note that while the number of rounds and key sizes used by AES-128, AES-192, and AES-256 vary, the fundamental method does not.

5 Experimental Results

To check the performance of our proposed application, we tried to use JAVA as a programming language and implemented the model using DES and AES algorithms. To store the encrypted data in the cloud server, we used DRIVEHQ as the hybrid cloud server and linked the cloud to my application. Based on the experimental reports, our proposed model using AES gives more security and reduces encryption and decryption time compared with primitive cryptography algorithms.

Main Window

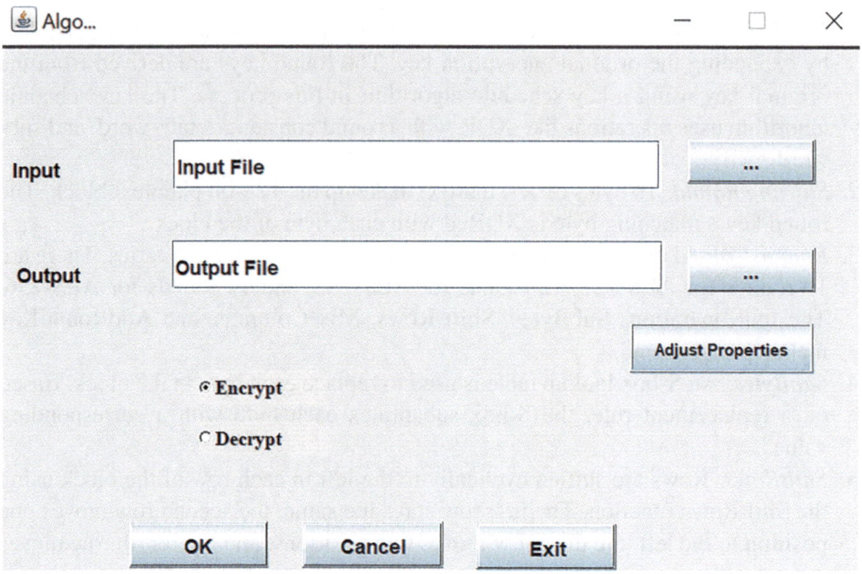

Explanation: Two radio buttons are visible in the window on top: one says "encrypt," and the other says "decrypt." To encrypt the input file, the user must first select the input file and then try to modify the properties, such as key size and batch size, by clicking on the okay button.

User Choose Input File

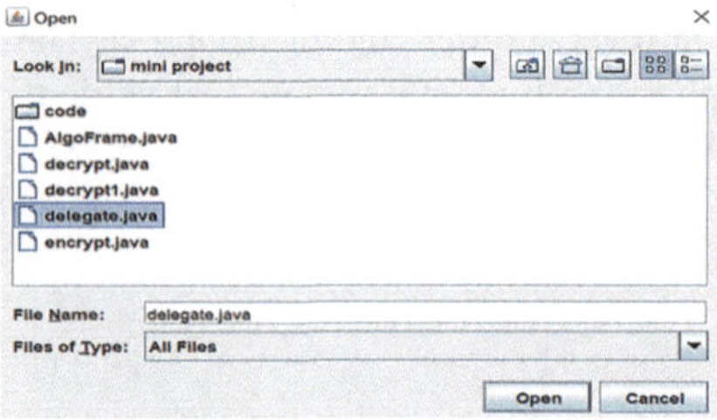

✓ To Select the Datafile we need to go to that location and click on open button.

✓ To store the encrypted data in the file we select another file to save the data.

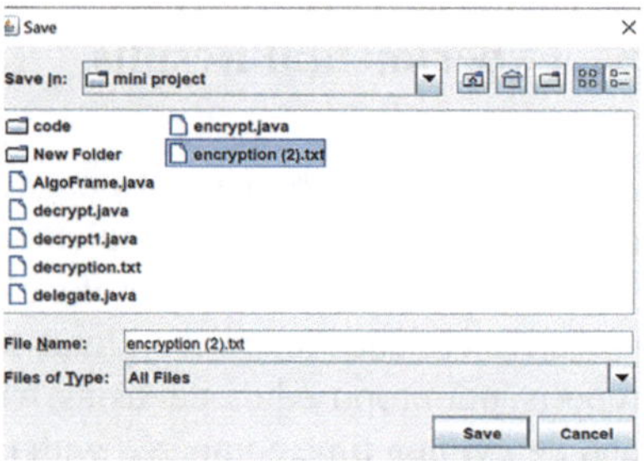

Explanation: From the above window, we can see the user try to choose a valid input file and then try to save the carrier file with his customized filename. Once the file is saved with a customized name, the user can now able to upload that encrypted file into the cloud server.

User Adjust the Key Properties

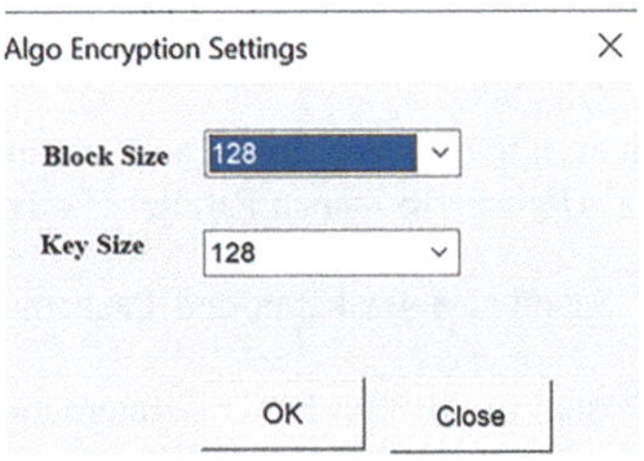

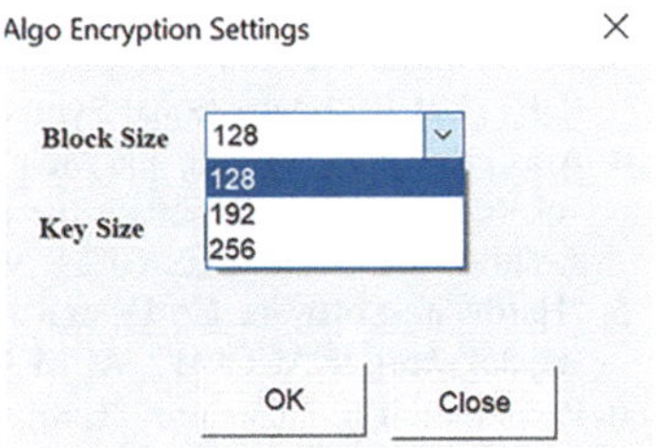

- ✓ we choose any type of block size and key size to perform AES encryption based on that the AES will work
- ✓ Block size we implement three different size and perform different encryptions

Explanation: From the above window, we can see the user tries to choose block size and key size before encrypting the input data. Here we can see three different key sizes and block sizes as AES takes three key sizes at a time.

6 Conclusion

In this current work, we use AES for encrypting and decrypting the text files or messages that are to be uploaded into the cloud server and later downloaded by decrypting with the help of a valid secret key. This work can avoid brute force attacks on the cloud server as the key size is large the brute force attacker failed to attack large key or block sizes. In future work, we want to extend the same application to digital data and try to provide security for images, audio, or video files that are to be uploaded to the cloud server.

Declaration

1. All authors do not have any conflict of interest.
2. This article does not contain any studies with human participants or animals performed by any of the authors.

References

1. Pallavi Atha and colleagues, “Design & Implementation of AES Algorithm Over FPGA Using VHDL,” International Journal of Engineering, Business and Enterprise Applications (IJEBEA), ISSN (Online): 2279-0039, pp. 58–62, 2013.
2. M. Komala Subhadra and colleagues, “Advanced Encryption Standard - VHDL Implementation,” International Journal for Technological Research in Engineering, ISSN (Online): 2347-4718, Volume 1, Issue 3, pp. 132–137 November – 2013.
3. M. Yoshimura et al., “Defect and Fault Tolerance in VLSI and Nanotechnology Systems (DFT)”, IEEE International Symposium on, 2013, pp. 278–283.
4. Arash Reyhani-Masoleh and Mehran Mozaffari-Kermani In IEEE Transactions on Computers, vol. 61, no. 8, August 2012, the article “Efficient and High Performance Parallel Hardware Architecture for the AES-GCM” was published.
5. “Hardware-Software Co-Design of AES on FPGA” Saambhavi Baskaran and Pachamuthu Rajalakshmi, ICACCI ‘12, ACM August 2012.
6. Pachamuthu Rajalakshmi, “Hardware-Software Co-Design of AES on FPGA,” International Conference on Advances in Computing, Communications and Informatics, 2010, pp. 1118–1122.
7. “FPGA Implementation of AES Encryption and Decryption,” International Conference on Control, Automation, Communication and Energy conservation – 2009. Ashwini M. Deshpande, Mangesh S. Deshpande, and Devendra N. Kayatanavar.
8. “FPGA Implementations of High Throughput Sequential and Fully Pipelined AES Algorithm,” International Journal of Electrical Engineering, vol. 15, no. 6, pp. 447–455, 2008.
9. “High-speed VLSI architectures for the AES algorithm,” IEEE Transactions on Very Large Scale Integration Systems, vol. 12, no. 9, pp. 95–967, September 2004.
10. “An FPGA Design of AES Encryption Circuit with 128-bit Keys” by Hui QIN, Tsutomu SASAO, and Yukihiro IGUCHI was published in GLSVLSI’05, ACM 2005.
11. Prasithsangaree P. and Krishnamurthy P., “Analysis of Energy Consumption of RC4 and AES Algorithms in Wireless LANs,” Proceedings of the IEEE GLOBECOM, vol.
12. “Implementation Approaches for the Advanced Encryption Standard Algorithm,” IEEE 2002 by Xinmiao Zhang and Keshab K. Parhi.

Prediction of Angina Pectoris

D. Dakshayani Himabindu, Raswitha Bandi, Keesara Sravanthi, V. Sai Vandana, and G. Uday Bhaskar

Abstract Angina pectoris is the clinical term for chest discomfort associated with cardiovascular disease. It happens when the cardiac muscle is not getting enough oxygenated blood. Ischemia is a disorder in which some or all of the heart's arteries are constricted or blocked. Many persons who are originally suspected of having angina have normal coronary angiograms, indicating that they do not, in fact, have angina. A feasibility study was conducted to determine the practicality of a preliminary screening test. Information on a variety of possible risk factors was collected for a large number of patients suspected of having angina, and then their angina status was documented. The major goal of this research was to see which health characteristics, if any, are linked to angina and whether a subset of them might be utilized to predict the dependent variable angina status. More specifically, being able to estimate the risk/probability that a person with a specific mix of these health indicators has angina would be beneficial. Furthermore, estimating the individual effects of relevant factors would be interesting.

Keywords Angina pectoris · XGBoost · Random forest · pickle file

1 Introduction

Coronary heart disease causes angina, also referred to as angina pectoris, which is a sort of chest discomfort or stress. It is caused by a lack of blood supply to the cardiovascular system (myocardium). Angina is caused by a blockage or tightness in the arteries that provide oxygen to the tissues.

Heart disease is an example of a condition that can lead to death. Every year, far too many people are killed by heart disease. Cardiovascular illness is caused by compromised cardiac myocytes. The incapacity of the heart and lungs to supply

D. Dakshayani Himabindu (✉) · R. Bandi · K. Sravanthi · V. Sai Vandana · G. Uday Bhaskar
IT Department, VNRVJIET, Hyderabad, India
e-mail: dakshayanihimabindu_d@vnrvjiet.in

A. Patel et al. (eds.), *Advances in Machine Learning and Big Data Analytics II*,
Springer Proceedings in Mathematics & Statistics 442,
https://doi.org/10.1007/978-3-031-51342-8_16

blood is known as heart disease or failure. Coronary artery disease is another name for heart disease (CAD). Insufficient blood flow to arteries causes CAD.

Stable Angina: Stable angina, also called "effort angina," is a type of angina brought on by myocardial ischemia. Chest discomfort and other symptoms brought on by physical effort are the most typical signs of stable angina (running, walking, etc.)

Unstable Angina: Unstable angina pectoris generally defined as angina pectoris that alters or intensifies. At rest, Unstable Angina can occur without notice, which could be an indication of a heart attack.

Cardiac syndrome X: Angina-like chest discomfort due to peripheral epicardial coronary arteries on angiography is the characteristics of cardiac syndrome X, usually known as microvascular angina.

The study's major goal is to anticipate angina in front of a preliminary phase.so patients can tell if they are having an angina attack or not. We want to eliminate the time-consuming ECG and MRI tests that are now utilized to diagnose angina. Machine learning is becoming a commonplace aspect of life and adopting technology into the health-care field will benefit patients significantly.

2 Literature Survey

According to a technique given by Md. Asadur Rahman Md. Merajul Islam, angina pectoris induced by ischemia is critical to recognize. To evaluate the patient's status, it might be time-consuming to compute and separate the normal and angina pectoris influenced ECG peaks from a big ECG data collection. In addition, rural areas of middle- and lower-income economies may lack competent doctors or costlier testing instruments such as eco-cardiograms and Magnetic resonance imaging to detect angina. Automatic detection of ECG characteristics might be a potential option in certain situations. As a possible consequence, digitalization is the only option. An excellent strategy for diagnosing angina pectoris with an Electrocardiogram is described in this study (2019).

To decide on an Electrocardiogram, this method includes multiple phases, which use a baseline wandering route finding methodology, this methodology first eliminates baseline wandering from Electrocardiogram data. Following this, the ECG data is cleaned up using a Gaussian weighted moving average window technique. Therefore, by using the well-known First and Second Derivative (FS2) approach, the QRS complex was identified, and additional crucial sites such as S, J, K, and T were gradually recognized using the maxima-minima criterion. The Electrocardiogram signal's ion-selective lines are also calculated, as well as the statistical properties of J-K points of healthy and malignant Electrocardiogram peaks are evaluated to this now.

By using the MIT arrhythmia database, this approach is utilized to find out whether you have angina pectoris The accuracy of the answers supplied by such a method was 94% (overall) which is significant. We also computed the specificity and sensitivity of our proposed technique, which were found to be 90% and 88%, correspondingly. Because the preceding task is made up of only one attribute, that might not be applicable in every situation. The k-nearest neighbor (kNN) method was employed with the assistance of several characteristic machine-learning approaches to make this study endeavor more accurate and acceptable. Although the results of the kNN-based prediction approach are practically equal to those of the previous methodologies. As a consequence, we were able to validate our suggested approach for identifying angina pectoris. The proposed method is anticipated to help in the detection of computerized angina pectoris using ECG.

Reference [4] was proposed by scholars to investigate the treatment plan of people with symptomatic angina pectoris and analyze appropriate medical images to establish an efficient treatment approach. This tale piqued my interest, and I decided to look into it more. According to Yan and colleagues, they first randomly assigned 88 hospitalized patients' people with acute angina to one of two categories: control or experimental.

The goal of reference number [2] is to provide a thorough description of the Nave Bayesian and decision tree classifiers employed in the study, notably for heart disease prediction. As part of some work, the same dataset was subjected to a predictive data mining technique, with the findings suggesting that the Decision Tree surpasses the Gaussian classification.

The goal of reference number [3] by Abhay Kishore pioneered the use of deep learning. This research presents a heart attack prediction system that employs Deep Learning techniques, especially a Recurrent Neural System, to forecast a patient's risk of heart-related disorders. The Recurrent Neural Network is a groundbreaking Deep Learning-based characterization calculation. This research delves into the framework's main components as well as the premise that underpins them. To provide exact findings with the fewest mistakes, the recommended technique employs deep learning and data mining. This study sets the basis for the development of a newer type of heart attack prediction platform and acts as a reference point.

3 Methodology

Methods rely on logistic regression, support vector machines, decision tree, as well as ensemble learning approaches are some of the common methods for estimating risk variables. Each model uses different techniques to identify the dependencies and make decisions on them to predict the result.

3.1 Logistic Regression

We picked logistic regression at first to identify angina risk factors. To allocate observations to a discrete class, regression is utilized. The sigmoid logic function is used to transform the output. The logistic regression assumption limits the cost function to a range of 0 to 1. From the data visualization, we observe several factors affecting angina in categorical features such as smoking, hypertension, diabetes, age, cigarettes, etc.

3.2 Support Vector Machines

Support Vector Machines are a type of classifying algorithm that uses hyperplanes to divide input data. Depending on the dispersion of information, hyperplanes can take on a variety of forms, and only certain points that aid in class differentiation are evaluated for categorization. They are generalized to a greater number of datasets with the use of a kernel. This component is responsible for integrating one feature space to another.

3.3 Random Forest

The Random Forest Algorithm may be thought of as a forest of trees. To begin, it generates call trees based on each of the dataset's expertise examples. It then takes the predictions from every tree and uses means of polling to choose the most accurate solution. It is a step forward from decision trees. Because of the large number of trees involved, this computational approach is regarded as an exceptionally accurate and robust technique. One of its numerous benefits is that this does not incur from issue of overfitting. Finally, it averages all the forecasts from each tree, canceling out any biases.

3.4 XGBoost

XGBoost (Extreme Gradient Boosting) is a robust distributed gradient-boosted decision tree ML framework. It contains concurrent tree enhancement and is the best machine-learning tool for analysis, classifying, and ranking problems. In this method, decision trees are built in a sequential order. Weights are highly important in XGBoost. Weights are assigned to all of the independent factors, which are then entered into the decision tree, which forecasts results. The weight of elements predicted wrongly by the tree is raised, and these elements are included in the second decision tree. The results of these many classifiers/predictors are then integrated to produce a more robust and accurate model.

4 Proposed Approach

According to the survey, only 17% of people in India visit clinics even for any small illnesses, and most of the others do not trust hospitals and mostly rely on home tips which might not be effective for serious health issues hidden beneath.

With this COVID-19 pandemic, most of us are afraid to visit hospitals for regular body checkups or small illnesses. To overcome this pressing concern and to provide a platform for people to provide more awareness of self-care in these difficult times, we came up with this solution of a chatbot-like platform and started this by implementing the detection of Angina Pectoris which is the most common disease and is undetected in its early stages.

We have built various models on our dataset to explore and get to know what works best and what does not. In this process of research, we have trained our dataset using various standard sML algorithms.

After pre-processing the dataset and training the data with the above-mentioned algorithms, we have noticed that the Random Forest algorithm seems to be performing well with better accuracy than other algorithms.

As Random Forest has given better accuracy, we have proceeded with using that and built an API by extracting the pickle file using the trained model.

The API has been deployed using AWS Serverless computational services and is used by the front-end application to predict the output after collecting the patients/user data.

4.1 Objectives of the Proposed System

1. Extracting different combinations of body and behavioral features and preparing a new dataset.
2. Using the prepared dataset to train and test the model using multiple machine learning algorithms.
3. Focusing on increasing the accuracy of the model by testing with various feature combinations.
4. Choosing the best-performing model and exporting it as a pickle file.
5. Deploying an API that receives the patient data and responds with the prediction report.
6. Developing a front-end chatbot interface that collects patient info and calls the API.

5 Implementation

5.1 Importing the Required Libraries

Required libraries are:

OS.
Pickle
Pandas
NumPy
Seaborn
Matplotlib.pyplot

OS package is used to upload the audio files and to provide path files from the system. Pickle is a fundamental library used for converting objects into bytes. Pandas provides data analysis and date frame creation whereas NumPy is for numerical operations. Seaborn and Matplotlib.pyplot purpose is for data visualization.

5.2 Data Preparation

We have collected the dataset from the local clinics which consist of data features such as smoking status, hypertension, previous heart disease history, age, and BMI.

Before splitting the dataset, we performed a few tasks:

1. Handling null values.
2. Standardizing the dataset by categorical and label Encoding the features.

5.3 Training and Testing the Data

The model is rehearsed and adjusted to the variables in the training system, whereas test data is solely employed to evaluate the model's performance in the test system. While the result of training data is observable, predictions must be made because test data is not. For minimal data loss, the data is separated in an 80:20 ratio.

5.4 Building the Model

The processed dataset is trained on multiple machine learning models and we choose the best performing model.

5.5 *Accuracy and Results*

We have achieved the highest accuracy of 92%. We have also tried increasing the accuracy with the help of Hyperparameters, after which the accuracy has become 94%.

6 Results (Fig. 1)

Methods	Accuracy
Logistic regression	92.93
SVM	93
Random Forest	94.4
XGBoost	93.5

```
              precision    recall  f1-score   support

           0       0.95      0.99      0.97       968
           1       0.29      0.04      0.07        54

    accuracy                           0.94      1022
   macro avg       0.62      0.52      0.52      1022
weighted avg       0.91      0.94      0.92      1022

ROC AUC score: 0.747876492194674
Accuracy Score:  0.9442270058708415
```

Fig. 1 Random forest accuracy

7 Conclusion

We have gone through various algorithms and we have decided to go with the Random Forest algorithm as it gives minimal data loss and is well suited for a medium-sized dataset like ours. We have prepared a new dataset for this project with different combinations of features and used it to train the model, comparing with the existing models using a new dataset helped us to achieve a good amount of accuracy of 94%.

8 Future Scope

We intend to work towards including various diseases for self-assessing and decide upon the next course of action We also plan to explore suggesting the patient with a suitable nutrition diet based on his/her concerns and connect with the hospital/doctor if the condition is severe.

References

1. Md. Asadur Rahmana and Md. Merajul Islam. 2019. ECG Signal Based Angina Pectoris Detection Using Statistical and Machine Learning Approach.
2. Sonam Nikhar and A.M. Karandikar. 2016. Prediction of Heart Disease Using Machine Learning Algorithms.
3. Abhay Kishore, Ajay Kumar, Karan Singh, Maninder Punia, Yogita Hambir. 2018. Heart Attack Prediction Using Deep Learning.
4. Peng Li and Li He. 2021. Clinical Treatment Analysis and Imaging Study of Patients with Acute Angina in Cardiovascular Medicine.
5. https://www.sciencedirect.com/science/article/pii/S2352914820300125#bib5
6. https://medium.com/swlh/using-logistic-regression-to-identify-risk-factors-for-angina-6ccb7fc047ee
7. https://www.sciencedirect.com/science/article/pii/S2352914820300125#bib5

A Novel Web Attack Recognition System for IOT via Ensemble Classification

Preethi Bitra and Kandukuri Rekha Sai Kumar

Abstract Machine learning is increasingly utilized in various fields such as biomedicine, information security, firewall protection, medical science, the Internet of Things (IoT), and mobile application security. Consequently, the demand for machine learning applications is growing. In this article, we apply machine learning to detect web or Internet attacks on IoT mobile devices. With the rapid growth in smartphone usage, protecting personal information and privacy has become a significant challenge. Impersonation is a major consequence of data leaks. Current preventive methods, like passcodes and fingerprinting, cannot continuously monitor usage to verify user authorization. Typically, once a user is authorized, they gain complete control of the device, and no further protection can prevent access to device data. To address this, we propose an ensemble machine learning model for detecting impersonation, based on a multi-view bagging strategy that collects sequential tapping information from the smartphone keyboard. This model continually authenticates the user while typing using sequential-tapping biometrics. By conducting multiple tests on our model, we evaluated it using the CLaMP (Classification of Malware with Portable headers) dataset from Kaggle. Our empirical and theoretical results demonstrate that our model outperforms other approaches, achieving a 6.42% equal error rate, 93.14% accuracy, and 95.41% H-mean using only the accelerometer and five keyboard taps.

Keywords CLAMP · Machine learning · Ensemble classification model · Impersonation · Mobile applications · Biomedicine · Internet of Things (IoT)

P. Bitra (✉) · K. R. Sai Kumar
Department of CSE, Vishnu Institute of Technology, Bhimavaram, India
e-mail: preethi.b@vishnu.edu.in

A. Patel et al. (eds.), *Advances in Machine Learning and Big Data Analytics II*,
Springer Proceedings in Mathematics & Statistics 442,
https://doi.org/10.1007/978-3-031-51342-8_17

1 Introduction

The widespread use of smart mobile devices offers users "anytime, anywhere computing," but also heightens the risk of theft, as these devices store sensitive information. In 2013, nearly three million Americans fell victim to smartphone theft, leading to significant electronic impersonation fraud and costing US businesses around $180 million between 2013 and 2014. Even with simple or no passwords, attackers can reverse-engineer these devices by analyzing screen patterns. Current security methods focus on preventing unauthorized access during unlocking but do not offer ongoing protection. To address this issue, we aim to develop a continuous identification system that meets specific design criteria: it must operate effectively in an "open world" with unknown attackers, quickly identify users after a few keyboard inputs with high accuracy and low error rates, and perform efficiently with minimal use of smartphone sensors and activation time.

2 Literature Survey

The most crucial phase in software development is the literature review. Numerous writers conducted preliminary studies on this relevant topic, and we will consider key papers to expand our work.

P. Papadimitriou and H. Garcia-Molina [6] submitted a research study titled "Data Leakage Detection." In this essay, the authors examine how a data distributor released sensitive information to a group of allegedly trustworthy agents (third parties). An unauthorized area, such as the Internet or a laptop, is where some leaked data ends up. The distributor needs to prove that the stolen material came from one or more agents, not from independent sources [11]. We propose data allocation strategies (across agents) to improve the likelihood of identifying leaks.

In his research work, "A Digital Signature Scheme Secure Against Adaptive Chosen-Message Attacks," S. Goldwasser [7] presents his findings on a digital signature system built on the computational challenges of integer factorization. The proposed scheme is notably resistant to adaptive chosen-message attacks, meaning that an adversary, even after obtaining signatures for chosen messages, cannot forge the signature of any additional message. Traditionally, it was believed that the properties of resistance to forging and immunity to adaptive message attacks could not coexist. However, Goldwasser demonstrates how to construct a signature scheme with these properties, relying on the existence of a "claw-free" pair of permutations, which may be a less stringent requirement than previously assumed.

Vipin Kumar [8] proposed the research piece "Anomaly Detection: A Survey." The author aims to discuss anomaly detection in this essay, a topic that has garnered significant attention across various academic fields and application domains. Specific application domains have tailored many anomaly detection algorithms, while others are more general. We aim to provide an orderly and comprehensive overview of anomaly detection research in this survey [12]. We identified basic assumptions for each category, which the techniques use to differentiate between normal and aberrant behavior. We can use these assumptions as criteria to evaluate the effectiveness of a certain strategy in a specific domain. We provide a basic anomaly detection technique for each category.

"An Ensemble Intrusion Detection System Based on Acute Feature Selection" is the title of a research study by **S. Hari Prasad** [9]. The writer of this paper talks about an intrusion detection system (IDS) for smart cities that uses IoT-MQTT networks and can find intrusions by using shallow learning methods. The proposed framework consists of four components. (i) We create a smart city network model using hardware that includes several MQTT clients, such as sensors and IoT devices. The IDS dataset with normal and attack features was constructed by (i) launching a flooding attack on the MQTT broker; (ii) selecting optimized features from the raw dataset using the acute feature selection algorithm; and (iii) validating those features with the Jaccard coefficient. (iv) The dataset is further trained and validated with shallow learning approaches such as extreme gradient boosting (XGB), K-nearest neighbors (KNN), and random forest (RF).

Adeel Abbas [10] proposed a research study titled "A New Ensemble-Based Intrusion Detection System for the Internet of Things." In this article, the author attempts to discuss several intrusion detection systems (IDSs) that have been proposed to identify harmful acts based on predetermined attack patterns, thereby protecting data from abuse and unexpected efforts. The existing IDS must be enhanced due to the sudden increase in this type of attack. The key way to improve intrusion detection systems is through machine learning. This study suggests using an ensemble-based intrusion detection model. Following an examination of the model's performance using well-known current state-of-the-art approaches, the suggested model incorporates a voting classifier, logistic regression, naive Bayes, and decision tree.

Rokia Lamrani Alaoui [11] submitted a research work titled "Web attack detection using a stacked generalization ensemble for LSTMs and word embedding." In this article, the author presents a method for LSTMs to detect fraudulent HTTP web requests using Word2vec embedding and a layered generalization ensemble model. We use the HTTP CSIC 2010 dataset to evaluate the performance of our classification model. We show that a stacking generalization ensemble model for LSTMs that includes several word-level embeddings outperforms in terms of training time and classification metrics.

R. Alghamdi [12] offered a research study titled "An ensemble deep learning-based IDS [18] for IoT using Lambda architecture." In this research, the author uses a multi-pronged classification technique to provide a deep ensemble-based IDS using the Lambda architecture. In binary classification, Long Short Term Memory (LSTM) is used to distinguish between malicious and benign traffic, whereas multi-class classification uses an ensemble of LSTM, convolutional neural network, and artificial neural network classifiers to determine the type of attack. The Lambda architecture's batch layer is used for model training, while the speed layer is utilized for real-time evaluation based on model inference. The multi-pronged classification strategy and use of lambda architecture in the proposed approach result in over 99.93% accuracy and significant processing time savings.

D. Shankar [13] offered a study piece titled "Deep Analysis of Risks and Recent Trends Towards Network Intrusion Detection Systems." The author of this essay provides a complete review of the issues and current trends in network intrusion detection systems (IDS). The author discusses the various forms of assault that might occur on a network, as well as the limits of typical IDS techniques. They then investigate the use of machine learning (ML) and deep learning (DL) in IDS, claiming that these techniques have the potential to overcome the constraints of previous approaches. The study also conducts a thorough examination of various ML and DL-based IDS approaches. They evaluate the various methods in terms of accuracy, performance, and scalability. They also talk about the challenges of employing ML and DL in IDS, such as the necessity for huge datasets and the difficulty of interpreting ML model results.

3 Existing System

There was no pre-defined mechanism or program in the previous system to identify online fraud activity. In general, malware comes in a variety of forms, therefore there is no single approach that can forecast all the malwares found on a computer or the Internet. There are just a few malwares in the current system, and they can only be predicted manually or with restricted antivirus software. As a result, the following are the limits of the current system.

1. All present systems use a manual way to detect fraud.
2. There is no accuracy in detecting fraudulent actions when utilizing the manual approach.
3. There is no effectiveness in tracing malware activities.
4. Identifying fraud activities takes less time when the dataset is small, but it is a far more difficult task when the dataset is huge.

4 Proposed System

For feature learning, the suggested system employed numerous machine learning techniques. The suggested system employs an ML approach to classify all actions and track fraud and normal activities separately on mobile devices. The suggested ensemble classification model can evaluate each and every characteristic in the dataset and then simply classify the difference between normal and malware activity [16, 17]. Furthermore, the proposed approach effectively determines the root cause element for that fraudulent conduct.

The benefits of our proposed system are as follows. These are their names:

1. We can utilize an automatic approach in this proposed system to classify fraud activities on any dataset.
2. The proposed method is quite precise.
3. It is incredibly efficient and simple to categorize.
4. This approach can easily classify fraud activity on both small and large datasets.

4.1 Proposed Dataset

In this module, initially, we need to load the input dataset which contains a set of pre-defined activities collected from cell phones that are formed as one input dataset. The dataset is loaded from the **KAGGLE** website for testing the proposed application performance.

https://www.kaggle.com/datasets/saurabhshahane/classification-of-malwares

ClaMP_Integrated-5184.csv

Total samples : 5184 (Malware () + Benign())

Features (69) : Raw Features (54) + Derived Features(15)

ClaMP_Raw-5184.csv

Total samples : 5184 (Malware ()+ Benign())

Features (55) : Raw Features(55)

IMAGE_DOS_HEADER (19)

From the above description, we can see 5184 samples are collected from that dataset which contains both normal as well as malware files.

Figure 1 illustrates that the proposed model utilizes raw data as a sample dataset. The raw data is saved with a .exe file extension, after which data pre-processing is performed on it.

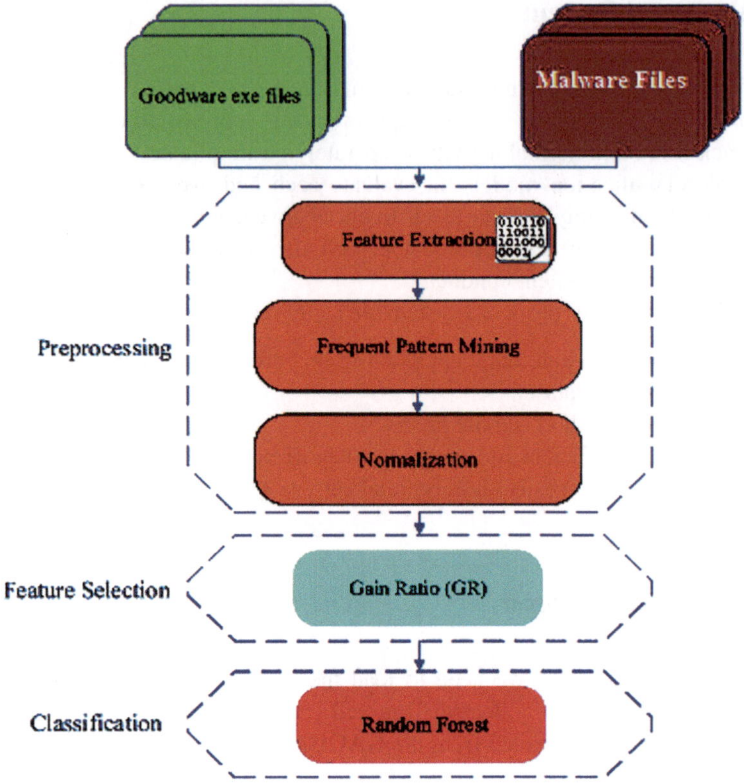

Fig. 1 Proposed model for malware or web attack detection

4.2 *Proposed Algorithms*

In this application, we attempt to employ multiple classification models before determining which model can reliably anticipate malware or fraud activities on mobile IOT devices. We obtained the CLaMP dataset from the KAGGLE website for this application. There are about 5184 samples in this collection that are a mix of malware and benign.

1. SVM
2. Decision Trees
3. Random Forest.

SVM Classification Model

SVM (Support Vector Machine) is a supervised machine learning technique capable of performing classification, regression, and outlier detection. The linear SVM classifier draws a straight line between two classes, with all data points on one side labeled as one class and all data points on the other side labeled as the second class. While choosing the best line from the countless possibilities can be challenging, the LSVM algorithm addresses this by selecting a line that not only separates the two classes but also stays as far from the middle as possible. The term "support vector" in "support vector machine" refers to the two position vectors drawn from the origin to the decision boundary points.

Decision Tree Classification Model
This section covers Decision Tree Classification, attribute selection measures, and how to create and optimize a Decision Tree Classifier using the Python Scikit-learn package. As a marketing manager, you aim to identify the consumer base most likely to purchase your products, thereby saving on marketing costs by targeting the right demographic. Similarly, as a loan manager, you need to identify risky loan applications to achieve a lower loan default rate. Classification tasks involve categorizing consumers as potential or non-potential customers, or credit applications as safe or dangerous. Classification comprises two steps: learning and prediction. In the learning step, the model is built using training data, while in the prediction step, the model is used to make forecasts. The Decision Tree is one of the simplest and most widely used classification techniques, applicable to both classification and regression problems.

Random Forest Classification Model
Random forest is a supervised machine learning technique based on ensemble learning. Ensemble learning involves combining multiple instances of the same or different algorithms to create a more effective prediction model. In the case of the random forest algorithm, it merges multiple decision trees to form a "forest" of trees, hence the name "Random Forest." This approach can be used for both regression and classification problems.

5 Experimental Results

The two figures below show that the proposed ML models performed well on the dataset, and the data was trained on all three models. Finally, we can see which model had the highest accuracy when compared to all the ML models in order to detect web attacks from IoT mobile devices.

Data Pre-processing

```
g=plt.figure(figsize=(10,4))
=fig.add_subplot(111)
.set(title='Fradulent vs Not Fradulent')
.pie(y.value_counts(),labels=['Fradulent','Not Fradulent'], startangle=90, autopct='%1.1f%%')
t.show()
```

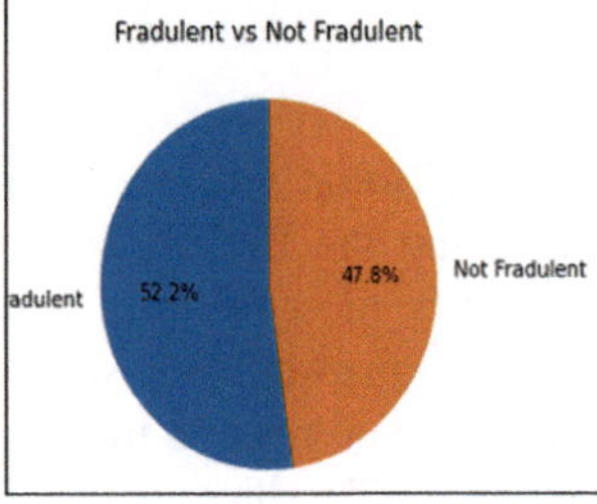

The top window shows that the dataset has been loaded and pre-processed, allowing us to determine how many are fake and what percentage are normal.

Load the Modules

```
from sklearn.preprocessing import LabelEncoder
le = LabelEncoder()
objList = df.select_dtypes(include = "object").columns

for col in objList:
    df[col] = le.fit_transform(df[col].astype(str))

df.info()
```

```
 13  FH_char5                    5210 non-null   int64
 14  FH_char6                    5210 non-null   int64
 15  FH_char7                    5210 non-null   int64
 16  FH_char8                    5210 non-null   int64
 17  FH_char9                    5210 non-null   int64
 18  FH_char10                   5210 non-null   int64
 19  FH_char11                   5210 non-null   int64
 20  FH_char12                   5210 non-null   int64
 21  FH_char13                   5210 non-null   int64
 22  FH_char14                   5210 non-null   int64
 23  MajorLinkerVersion          5210 non-null   int64
 24  MinorLinkerVersion          5210 non-null   int64
 25  SizeOfCode                  5210 non-null   int64
 26  SizeOfInitializedData       5210 non-null   int64
 27  SizeOfUninitializedData     5210 non-null   int64
 28  AddressOfEntryPoint         5210 non-null   int64
```

The above dialog shows that certain vital libraries are being imported to run the program.

Train and Test Validation

```
Text(0.5, 1.0, 'val accuracy')
```

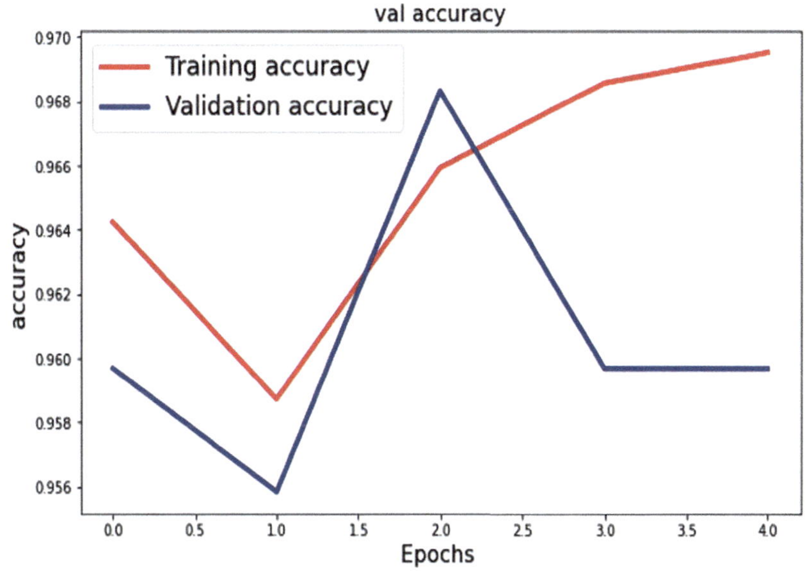

The data is divided into two portions, as shown in the window above: test and training. Additionally, we can see the train and test validation accuracy.

Performance Analysis

	Decision Tree	SVM	Random Forest	Neural Networks
0	0.976008	0.940499	0.993282	0.959693

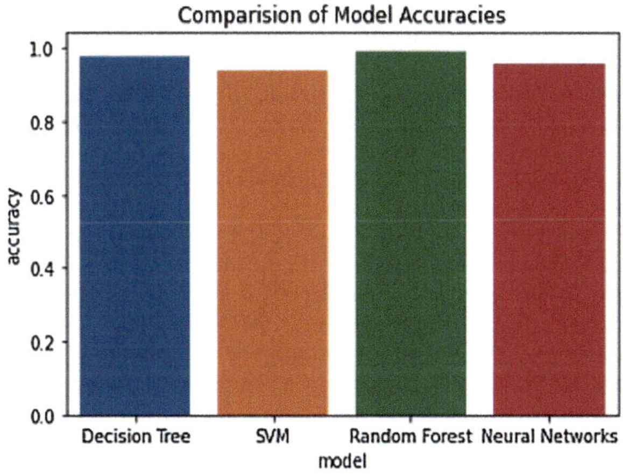

The above window denotes the performance evaluation applied to various models such as the Random Forest model is having more accuracy compared with the remaining other models.

6 Conclusion

In our proposed study, we introduced a novel approach by using multiple ML classification models as an ensemble for continuous user identification. We utilized sequential tapping data to construct a robust ensemble model with advanced learning algorithms. Testing with only three keystrokes, we found the system maintained high accuracy and allowed for more frequent judgments, potentially leading to more precise final decisions. Compared to other shallow machine learning methods, our ensemble model demonstrated that random forest is more effective and efficient at detecting fraudulent activities. Consequently, random forest proved superior in predicting malware or fraudulent actions on IoT devices. In the future, we aim to expand this work to deep learning models for improved accuracy on larger datasets.

References

1. T. Mogg, "Study reveals Americans lost $30 billion worth of mobile phones last year," Mar. 2012. Available online at: http://www.digitaltrends.com/mobile/study-reveals-americans-lost-30-billion-of-mobile-phones-last-year/
2. S. Watson, "Impostor fraud: A cyber risk management challenge," May 2015. [Online].
3. K. Knibbs, "Start memorizing your six-digit iPhone passcode," Sep. 2015. Available online at: http://gizmodo.com/start-memorizing-your-six-digit-iphone-passcode-1710072672.
4. D. Amitay, "Most common iPhone passcodes," Jun. 2011. Available online at: http://danielamitay.com/blog/2011/6/13/most-common-iphone-passcodes.
5. Panda Security, "75 million smartphones in the US don't have their passwords set on," Sep. 2015. Available online at: http://www.pandasecurity.com/mediacenter/tips/smartphone-risk-dont-use-password/
6. Papadimitriou, Panagiotis & Garcia-Molina, Hector. (2011). Data Leakage Detection. Knowledge and Data Engineering, IEEE Transactions on. 23. 51–63. https://doi.org/10.1109/TKDE.2010.100.
7. Pieprzyk, J., Wang, H., Xing, C. (2004). Multiple-Time Signature Schemes against Adaptive Chosen Message Attacks
8. E. Miluzzo, A. Varshavsky, S. Balakrishnan, and R. R. Choudhury, "Tapprints: your finger taps have fingerprints," in Proceedings MobiSys 2012. ACM, Jun. 2012, pp. 323–336.
9. S, H., T, D. An Ensemble Intrusion Detection System based on Acute Feature Selection. *Multimed Tools Appl* (2023). doi:https://doi.org/10.1007/s11042-023-15788-x
10. Abbas, A., Khan, M.A., Latif, S. et al. A New Ensemble-Based Intrusion Detection System for Internet of Things. *Arab J Sci Eng* **47**, 1805–1819 (2022). doi:https://doi.org/10.1007/s13369-021-06086-5
11. Rokia Lamrani Alaoui, El Habib Nfaoui, Web attacks detection using stacked generalization ensemble for LSTMs and word embedding, Procedia Computer Science, Volume 215, 2022, pp. 687–696, ISSN 1877-0509

12. Alghamdi, R., Bellaiche, M. An ensemble deep learning based IDS for IoT using Lambda architecture. *Cybersecurity* **6**, 5 (2023). doi:https://doi.org/10.1186/s42400-022-00133-w
13. Z. Xu, K. Bai, and S. Zhu, "TapLogger: Inferring user inputs on smartphone touchscreens using on-board motion sensors," in Proceedings of WiSec 2012. ACM, Apr. 2012, pp. 113–124.
14. A. J. Aviv, K. Gibson, E. Mossop, M. Blaze, and J. M. Smith, "Smudge attacks on smartphone touch screens." in Proceedings of WOOT 2010, Aug. 2010.
15. F. Ben Abdesslem, A. Phillips, and T. Henderson, "Less is more: energy-efficient mobile sensing with senseless," in Proceedings of the 1st ACM workshop on Networking, systems, and applications for mobile handhelds. ACM, 2009, pp. 61–62.
16. L. Duan, N. Li, and L. Huang, "A new spam short message classification," in Education Technology and Computer Science, 2009.ETCS'09. First International Workshop on, vol. 2. IEEE, 2009, pp. 168–171.
17. L. Ma, T. Tan, Y. Wang, and D. Zhang, "Personal identification based on iris texture analysis," IEEE transactions on pattern analysis and machine intelligence, vol. 25, no. 12, pp. 1519–1533, 2003.
18. L. Deng, D. Yu et al., "Deep learning: methods and applications," Foundations and TrendsR in Signal Processing, vol. 7, no. 3–4, pp. 197–387, 2014.
19. D. Shankar, G. Victo Sudha George, Janardhana Naidu J N S S, P Shyamala Madhuri, "Deep Analysis of Risks and Recent Trends Towards Network Intrusion Detection System" (IJACSA) International Journal of Advanced Computer Science and Applications, Vol. 14, No. 1, 2023, pp. 262–276.

Automatic Identification of Medical Plant Species Using VGG-19 Model

Kompella Bhargava Kiran, Sivala Srinu, B. C. H. S. N. L. S. Sai Baba, Venkata Durgarao Matta, and Immidi Kali Pradeep

Abstract Nowadays it is vital to automatically identify and recognise medicinal plant species in settings like forests, mountains, and dense areas in order to be aware of their existence. The shape, geometry, and texture of various plant parts, such as leaves, stems, flowers, etc., are now used to identify plant species. Systems for identifying plant species based on their flowers are frequently employed. Even though contemporary search engines offer tools for visually searching for a query image that includes flowers, these methods lack robustness due to the intra-class variance among the millions of flower species found worldwide. Therefore, a Deep Learning technique using Convolutional Neural Networks (CNN) is applied in this proposed study work to accurately identify flower species. The cellular phone's built-in camera module is used to capture images of the different plant species. A Transfer Learning technique is used to extract complicated characteristics from pre-trained networks for floral image feature extraction. Generally, to increase accuracy, a machine learning classifier like Logistic Regression or Random Forest is added on top of it. This strategy aids in reducing the amount of hardware required to complete the computationally demanding task of training a CNN. It has been found that employing the VGG-19 pre-trained model architecture and CNN paired with Transfer Learning methodology as a feature extractor outperforms all manually created feature extraction techniques like Local Binary Pattern (LBP).

Keywords Local binary pattern · Transfer learning · Logistic regression · Convolutional Neural Network · Deep learning · Machine learning · VGG-19

K. B. Kiran (✉) · S. Srinu · B. C. H. S. N. L. S. Sai Baba · V. D. Matta · I. K. Pradeep
Department of CSE, Vishnu Institute of Technology, Bhimavaram, India
e-mail: durgarao.m@vishnu.edu.in

A. Patel et al. (eds.), *Advances in Machine Learning and Big Data Analytics II*,
Springer Proceedings in Mathematics & Statistics 442,
https://doi.org/10.1007/978-3-031-51342-8_18

1 Introduction

A systematic division into groups and categories based on an object's characteristics is called classification. By using data to teach the computer, image classification was created to close the gap between computer vision and human vision. The classification of an image is accomplished by classifying the image according to the vision's content into the appropriate category. In this research, we investigate the topic of picture categorisation using deep learning, which is motivated by Saitoh and Kaneko [1]. The traditional techniques for classifying images are a component of the branch of Artificial Intelligence (AI) known as machine learning. Machine learning is composed of a feature extraction module that collects the key features, such as edges, textures, etc., and a classification module that assigns classes based on the extracted features. The fundamental drawback of machine learning is that it is only able to extract a limited number of features from images during separation and is unable to do so from the training set of data. Deep learning is used to address this drawback [2]. Deep learning (DL), a branch of machine learning, is able to learn using its own computational technique.

The use of a deep learning model allows for the persistent homogenous structure breakdown of data, just as a person would do while making decisions. Deep learning uses a layered structure of many methods described as an artificial neural system (ANN) to do this. The biological neural network of the human brain is used to simulate the architecture of an ANN. Due to this, deep-learning models are more effective than traditional machine learning models [3, 4]. When discussing deep learning, neural networks that recognise an image based on its attributes are taken into account. This is completed to create a comprehensive feature extraction model that can address the challenges posed by traditional approaches. The integrated model's extractor should be able to master accurately extracting the distinctive features from the training batch of photos. The feature descriptors from the image are categorised using a variety of techniques, such as GIST, histograms of gradient-oriented and local binary patterns, and SIFT. All around us, there are flowers. They can be used to feed people, animals, birds, and insects. They are additionally employed as human and animal medications. A thorough knowledge of flowers is necessary to recognise new or endangered species when they are discovered. This will aid in the growth of the pharmaceutical sector. Botanists, campers, and medics can all use the system suggested in the paper. To obtain more information about the topic and tailor your search for the most relevant results, this may be expanded as an image search solution where photos can be taken as input instead of words [5].

Processing the enormous amount of data that is produced for the classification of images is generally exceedingly difficult and time-consuming. This inspired me to create the current application, which classifies various flowers using a deep-learning network. This makes it simple for us to categorise the many kinds of flowers. In this system, we attempt to classify various image types and forecast the precise photos from that group of images using a deep learning model. The report is generated for the end customers once the forecast has been made.

2 Literature Survey

The most important step in the software development process is the literature review. This will describe some preliminary research that was carried out by several authors on this appropriate work and we are going to take some important articles into consideration and further extend our work.

Kayiram Kavitha et al. [6] proposed a research article "Medicinal Plant Species Detection Using Deep Learning." In this article, the authors try to analyse MobileNet, ResNet50, Inception v3, Xception, and DenseNet121 Convolutional Neural Networks (CNN) versions for Indian-origin medicinal plant species detection. We tested many CNN variations to categorise photos of medicinal leaves and found that the Inception v3 model performs better than all other traditional approaches. To optimise and produce superior outcomes, our suggested design uses the Inception v3 model and the stochastic gradient descent technique.

Jafar Abdollahi [7] proposed a research article "Identification of Medicinal Plants in Ardabil Using Deep Learning: Identification of Medicinal Plants Using Deep Learning." In this article, the author tries to discuss about Convolutional Neural Network (CNN)-based methods that can distinguish between Indian leaf species. Recently, a number of Deep Learning frameworks have been applied to distinguish, recognise, and characterise different plants. The identification of therapeutic plants that can be found in rural areas is the main goal of this study. The Transfer Learning method chose the mobile net v2 architecture, a well-known pre-trained CNN. The 3000 pictures of medicinal plants from 30 distinct classes were used to create the medical plant dataset, and models were evaluated using pre-trained weights. The trained model's accuracy on a held-out test set was 98.05%, proving the usefulness of this strategy.

C. Sivaranjani et al. [8] proposed a research article "Real-Time Identification of Medicinal Plants Using Machine Learning Techniques." In this article, the authors provide a method for determining the species of a plant from a sample of its leaves. ExG-ExR, a more advanced vegetation index, is utilised to extract more vegetative data from the photos. There is no need to use Otsu or any other threshold value chosen by the user because it fixes built-in zero thresholds. Although the Otsu approach produces more vegetative information in ExG, our ExG-ExR index functions effectively regardless of the lighting context. Therefore, a binary plant zone of interest is identified by the ExG-ExR index. The binary image's original colour pixel is used as the mask to separate the leaves into separate images. The colour and texture characteristics of each extracted leaf are used to identify the plant species.

Kayiram Kavitha et al. [9] proposed a research article "Medicinal Plant Species Detection Using Deep Learning." In this article, based on a leaf image and cutting-edge computer vision, we suggest a deep learning model to identify the species of medicinal plants. To identify medicinal plant species of Indian origin, this study analyses the Convolutional Neural Network (CNN) versions MobileNet, ResNet50, Inception v3, Xception, and DenseNet121. We tested many CNN variations to categorise photos of medicinal leaves and found that the Inception v3 model performs better than all other traditional approaches. To optimise and produce superior outcomes, our suggested design uses the Inception v3 model and the stochastic gradient descent technique. Our test findings reveal that the Inception v3 model classified medicinal plants of Indian provenance with 95% accuracy.

Haryono et al. [10] proposed a research article "A Novel Herbal Leaf Identification and Authentication Using Deep Learning Neural Network." In this article, the authors try to introduce plants known as "herbals" which can be used as an option to treat illnesses naturally. The general public is still unaware of the presence of herbal plants. Due to the wide variety of medicinal plants, it takes specialised knowledge to identify them. To get around this, a clever and precise herbal leaf recognition system is required. This study uses convolutional neural networks and long short-term memory (CNN-LSTM) techniques to recognise and verify herbal plants. Nine different kinds of herbal leaves were separated into two-thirds of the training data and one-third of the testing data for identification. Other data not present in the training data, testing data, or leaf data were used to validate the outcomes of the identification process.

Saurabh Kumar Sinha [11] proposed a research article "Plant Identification Using Machine Learning." In this article, the author tries to discuss about Ayurveda and other plant-based medical systems employ the identification of plants by their leaves for a variety of purposes, including ecology, horticulture, disease detection, rare plant preservation, and Ayurvedic medicine. Our goal in this project is to use neural networks to digitally identify different plant species from an image of a single leaf. Convolutional neural networks, Tensorflow, and Keras will be used to approach our project. The aforementioned method produces outcomes that are satisfactory and highly accurate.

3 Existing System

The existing system lacked a proper way to classify medicinal plant species into several categories based on the picture quality of the plants, such as size, shape, species, colour, and more. The main drawbacks of the current system are listed below.

1. Finding the species type from a large number of photos takes more time.
2. Each matched set of flowers lacks a description.
3. Once a flower has been identified, there is no technique that can forecast the accuracy of the identification.
4. Using any technique, it is not possible to categorise plants or flowers and determine their names automatically.

From Fig. 1, we can clearly identify the proposed model using the raw data as a sampledataset. Here, the raw data is taken from the KAGGLE website https://www.kaggle.com/alxmamaev/flowers-recognition dataset, and then data pre-processing

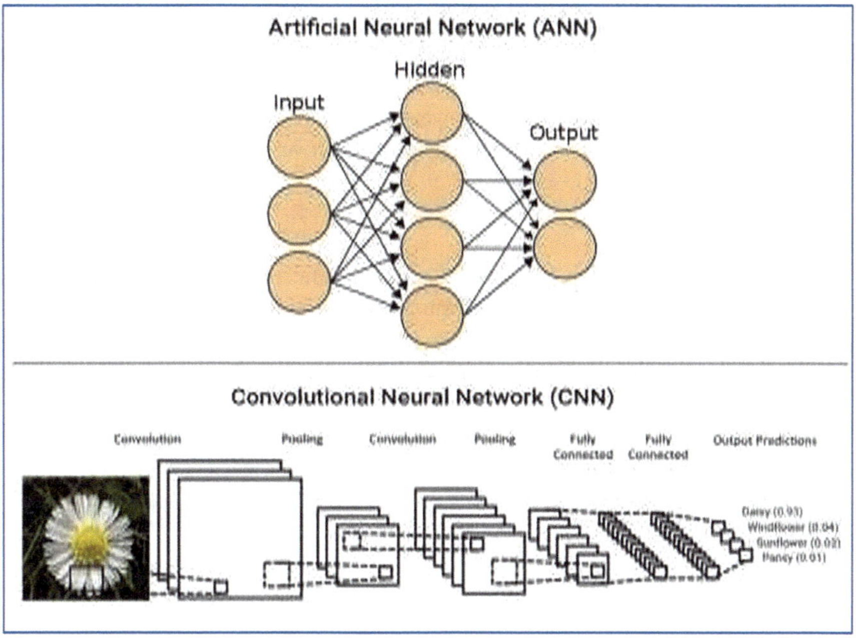

Fig. 1 Representation of the proposed model for flower species identification

is applied to that raw dataset. Once data pre-processing is completed, we now apply data analysis and then tokenisation is applied to that data. Here, we apply the classification algorithm and finally evaluate the performance of the CNN model.

4 Proposed System

To categorise various flowers, we have created a deep-learning network. For this, we used the 4242 floral image dataset from the Kaggle category. The data was gathered using information from Flickr, Google Images, and Yandex Images. This information can be used to identify the plants in the image. Five categories—chamomile, tulip, rose, sunflower, and dandelion—are used to categorise the images. There are roughly 800 images for each class. Photos have a resolution of only 320 × 240 pixels. Photos have varying proportions and are not scaled down to one size. This approach of classifying flowers can be utilised in real-time applications and will be useful to both campers and botanists conducting research [12].

The benefits of the proposed system are:

1. It takes less time and is more accurate to classify flowers or plants using data mining techniques.
2. The ability to predict the type of flower is highly accurate and effective.
3. This will assist botanists in their quest to discover new species from various flowers.

Proposed Dataset
In this module, we try to load the dataset which is collected from the Kaggle website, and then try to give that Excel file information as input to the next module.

Dataset URL: https://www.kaggle.com/alxmamaev/flowers-recognition

From the above description, we can see 4242 different flowers of five categories such as:

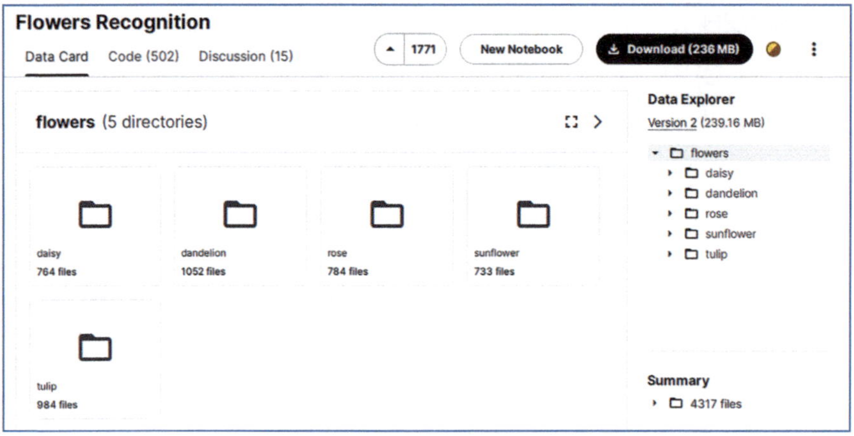

5 Proposed Algorithms

In this application, we attempt to employ a pre-trained CNN model such as VGG-19 to classify the flower species and its importance based on sample input. We obtained the dataset from the KAGGLE website for this application. There are about 4242 samples of nearly five different flower species which are used for medicinal purposes [13].

VGG-19 Model
The VGG-19 is a member of the visual geometry group (VGG) model, which has a total of 19 layers (SoftMax is 1 layer, MaxPool is 5, completely linked layers are 3, and convolution layers are 16). VGG-11, VGG-16, and more versions of VGG

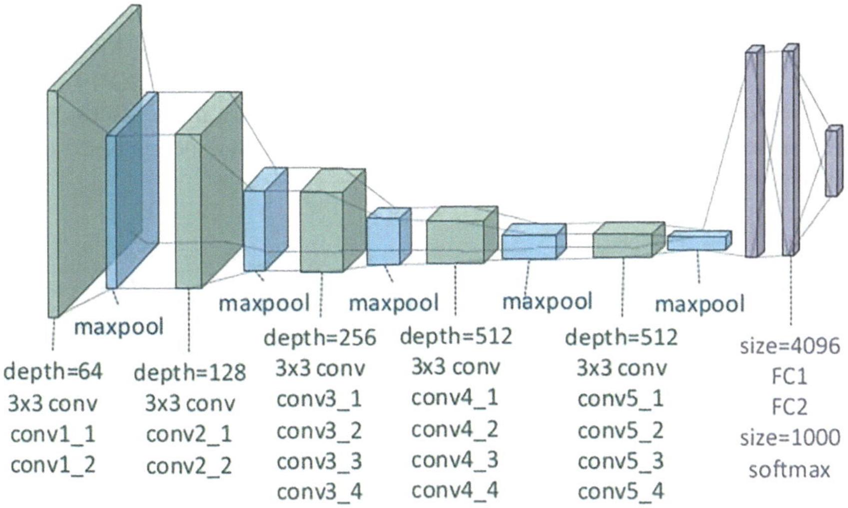

Fig. 2 Representation of the proposed VGG-19 model for flower species identification

are available. In VGG-19, there are 19.6 billion FLOPs. As 3 × 3 convolutional layers are placed on top of one another to enhance the depth level, the VGG-19 model becomes more straightforward and practical [14]. Max pooling layers were used in VGG-19 as a handler to reduce volume size. There was only one FC layer used. As seen in Fig. 2, the resized photos served as the VGGNet DNN's input data. The suggested model takes as input a 256 × 256 × 3 RGB image. In CNN, information is extracted from an image using four layers: the convolution layer, the reLU layer, the pooling layer, and the fully connected layer. Many feature filters in the convolution layer conduct a convolution operation. When two small sections of larger photographs are compared, these traits will identify the image accurately if they match. Four steps are necessary for this layer: It must first align the feature filter in the image before multiplying each image's pixel by the corresponding feature pixel. Next, add the values and divide them by the feature's total number of pixels to obtain the sum.

The final observed value of the filtered image is shown in the centre. Similar to the previous phase, the feature filter also sweeps all over the image. For each feature filter, this method is used to produce the convolution output. ReLU is a different kind of rectified linear activation function that, as illustrated in Fig. 5(c), outputs the input directly if it is positive and 0 otherwise. Due to its quicker and better performance, it is the default activation function for many different types of neural networks. It is used to produce the output by applying it to all the feature maps (feature pictures) that have been corrected.

6 Experimental Results

From the below two figures, it can be seen that the proposed VGG-19 models performed well on the dataset and the data is trained on all the VGG-19; finally, we can see the accuracy of our model and check the flower species based on the sample input image.

Load Dataset

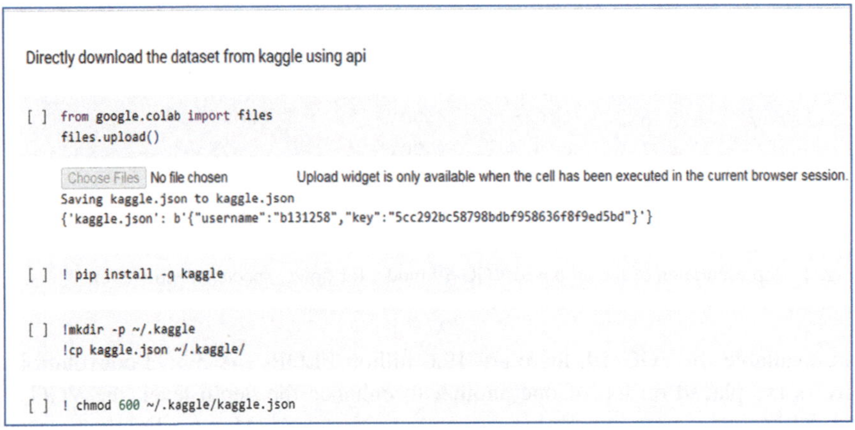

Directly download the dataset from kaggle using api

```
[ ] from google.colab import files
    files.upload()

    Choose Files  No file chosen        Upload widget is only available when the cell has been executed in the current browser session.
    Saving kaggle.json to kaggle.json
    {'kaggle.json': b'{"username":"b131258","key":"5cc292bc58798bdbf958636f8f9ed5bd"}'}

[ ] ! pip install -q kaggle

[ ] !mkdir -p ~/.kaggle
    !cp kaggle.json ~/.kaggle/

[ ] ! chmod 600 ~/.kaggle/kaggle.json
```

Explanation: From the above window we can see the dataset is loaded and preprocessed so that we can apply models for training the model.

Matplot Window

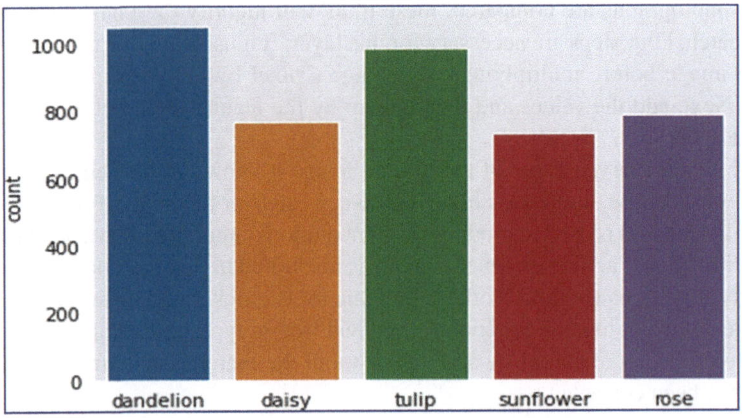

Explanation: From the above window, we can see that five different types of flower species are present.

Apply VGG-19 Model

```
from tensorflow.keras.applications.vgg19 import VGG19
from tensorflow.keras import Model
from tensorflow.keras.layers import MaxPool2D,Dense,Flatten,Dropout
from tensorflow import keras
from tensorflow.keras.models import Sequential


pre_trained_model = VGG19(input_shape=(224,224,3), include_top=False, weights="imagenet")

for layer in pre_trained_model.layers[:19]:
        layer.trainable = False

model=Sequential()
model.add(pre_trained_model)
model.add(MaxPool2D((2,2),strides=(2,2)))
model.add(Flatten())
model.add(Dense(5,activation='softmax'))
```

Explanation: From the above window, we can see VGG-19 model is applied to our input dataset and now we train our model using VGG-19.

Test and Training Validation

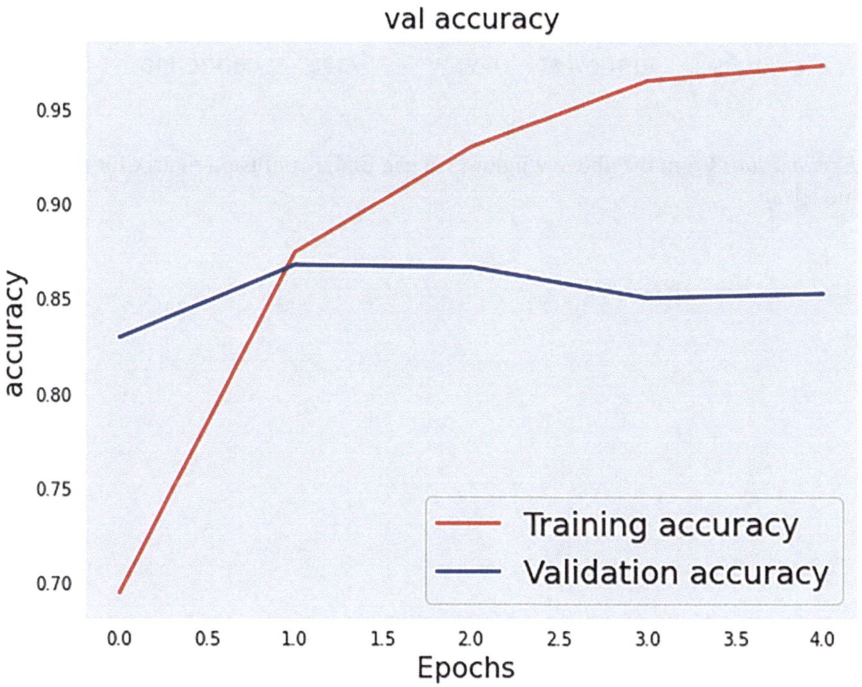

Explanation: From the above window, we can see that the VGG-19 model is applied to our input dataset and now we can see training and validation accuracy.

Confusion Matrix

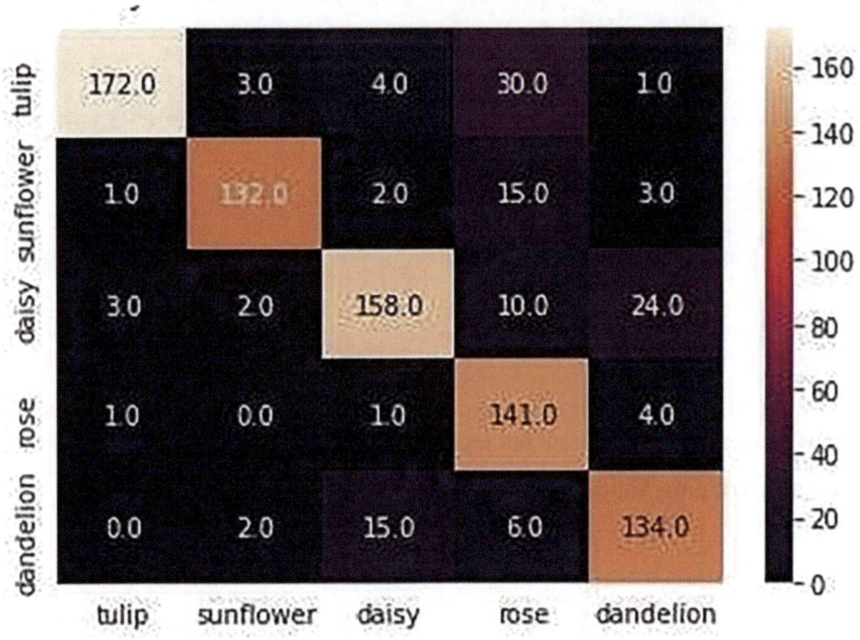

Explanation: From the above window, we can find a confusion matrix for the given model.

Prediction of Sample Input

```
pred=model.predict(x1) # for predicting class
# print(pred)
# prob=model.predict_proba(x1)  # predicting probability
labels_pred=np.argmax(pred,axis=0)
# labels=get_labels(labels_pred)
flowers = ["dandelion","daisy","tulip","sunflower","rose"]
pred_results=pd.DataFrame(data=pred,columns=flowers)
# print(pred_results.head())
import seaborn as sns
# sns.set_theme(style="darkgrid")
ax=sns.barplot(data=pred_results)
plt.show()
# pred_results.head()
```

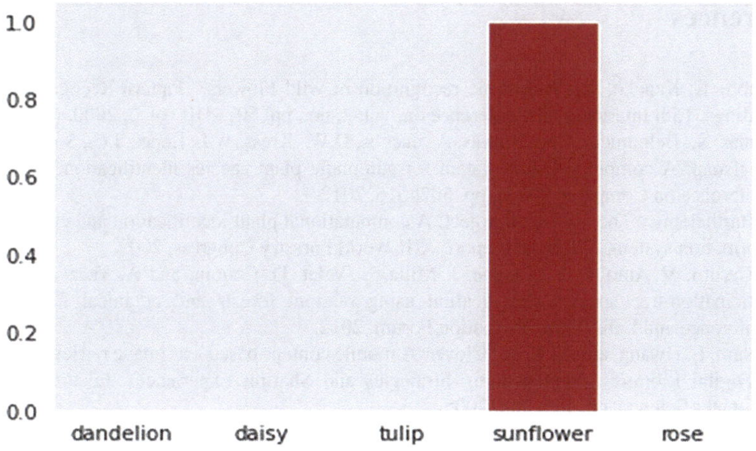

Explanation: From the above window, we can get a prediction of sample input as sunflower based on the test image.

7 Conclusion

With a smaller dataset and constrained computational resources, like a CPU, the suggested method is a quicker way to train a convolutional neural network (CNN). Since there are millions of distinct flower species in the world, this system might be readily modified by training more flower species photos to identify various species anywhere. Therefore, future work would involve creating a larger database that included photographs of not only flowers but also of leaves, fruits, bark, etc., that were gathered from various sources all over the world. This technique would be helpful for first-aid situations as well as for identifying plants used for medicinal purposes. To determine whether a plant species may be used for first aid, the user can simply take an image of it and obtain information about it. The training dataset, which must be produced either by personally collecting images of the plants throughout the city or by using public datasets, is a critical component of developing such a system.

Declaration

1. All authors do not have any conflict of interest.
2. This article does not contain any studies with human participants or animals performed by any of the authors.

References

1. Saitoh, T.; Kaneko, T., "Automatic recognition of wild Flowers," Pattern Recognition, Proceedings. 15th International Conference on, vol. 2, no., pp. 507–510 vol. 2, 2000.
2. Kumar, N., Belhumeur, P.N., Biswas, A., Jacobs, D.W., Kress, W.J., Lopez, I.C., Soares, J.V.B. "Leafsnap: A computer vision system for automatic plant species identification," European Conference on Computer Vision, pp. 502–516, 2012.
3. D. Barthelemy. "The pl@ntnet project: A computational plant identification and collaborative information system," Technical report, XIII World Forestry Congress, 2009.
4. G. Cerutti, V. Antoine, L. Tougne, J. Mille, L. Valet, D. Coquin, and A. Vacavant, "Reves participation-tree species classification using random forests and botanical features," in Conference and Labs of the Evaluation Forum, 2012.
5. Y. Nam, E. Hwang, and D. Kim, "Clover: A mobile content-based leaf image retrieval system," In Digital Libraries: Implementing Strategies and Sharing Experiences, Lecture Notes in Computer Science, pp. 139–148, 2005.
6. K. Kavitha, P. Sharma, S. Gupta and R. V. S. Lalitha, "Medicinal Plant Species Detection using Deep Learning," 2022 First International Conference on Electrical, Electronics, Information and Communication Technologies (ICEEICT), Trichy, India, 2022, pp. 01–06, doi: https://doi.org/10.1109/ICEEICT53079.2022.9768649.
7. J. Abdollahi, "Identification of Medicinal Plants in Ardabil Using Deep learning: Identification of Medicinal Plants using Deep learning," 2022 27th International Computer Conference, Computer Society of Iran (CSICC), Tehran, Iran, Islamic Republic of, 2022, pp. 1–6, doi: https://doi.org/10.1109/CSICC55295.2022.9780493.
8. C. Sivaranjani, L. Kalinathan, R. Amutha, R. S. Kathavarayan and K. J. Jegadish Kumar, "Real-Time Identification of Medicinal Plants using Machine Learning Techniques," 2019 International Conference on Computational Intelligence in Data Science (ICCIDS), Chennai, India, 2019, pp. 1–4, doi: https://doi.org/10.1109/ICCIDS.2019.8862126.
9. K. Kavitha, P. Sharma, S. Gupta and R. V. S. Lalitha, "Medicinal Plant Species Detection using Deep Learning," 2022 First International Conference on Electrical, Electronics, Information and Communication Technologies (ICEEICT), Trichy, India, 2022, pp. 01–06, doi: https://doi.org/10.1109/ICEEICT53079.2022.9768649.

10. Haryono, K. Anam and A. Saleh, "A Novel Herbal Leaf Identification and Authentication Using Deep Learning Neural Network," 2020 International Conference on Computer Engineering, Network, and Intelligent Multimedia (CENIM), Surabaya, Indonesia, 2020, pp. 338–342, doi: https://doi.org/10.1109/CENIM51130.2020.9297952.
11. Robert M. Haralick, K. Shanmugam, Its'hak Dinstein, "Textural Features for Image Classification," IEEE Transactions on Systems, Man and Cybernetics, Vol. SMC-3, No. 6, November 1973, pp. 610–621, 1973.
12. P. Sermanet, D. Eigen, X. Zhang, M. Mathieu, R. Fergus, and Y. LeCun, "Overfeat: Integrated recognition, localization and detection using convolutional networks," ICLR, 2014.
13. C. Szegedy, W. Liu, Y. Jia, P. Sermanet, S. Reed, D. Anguelov, D. Erhan, V. Vanhoucke, and A. Rabinovich, "Going Deeper with Convolutions," IEEE Conference on Computer Vision and Pattern Recognition (CVPR), 2015.
14. Paulchamy, B., K. Maharajan, and Ramya Srikanteswara. "Deep Learning Based Image Classification Using Small VGG Net Architecture." 2022 IEEE 2nd Mysore Sub Section International Conference (MysuruCon). IEEE, 2022.

Identification of Fake Job Recruitment Using Several Machine Learning (ML) Models

Immidi Kali Pradeep, Seerla Mohan Krishna, Kompella Bhargava Kiran, B. C. H. S. N. L. S. Sai Baba, and Venkata Durgarao Matta

Abstract In modern society, especially among beginners, education levels are rising in relation to employment experience levels. In the process of finding suitable jobs. They give precedence to some fake jobs and invest time in recruitment processes. So, to discover the fake recruitment, we created the project we were working on. We employ machine learning methodologies that employ classification techniques to detect such bogus recruiting detection processes. The research presents an application that uses machine learning-based categorization algorithms to prevent fraudulent job posts online. To establish the best job fraud detection model, the outputs of various classifiers are compared. These classifiers are used to detect fake web posts. It aids in the detection of bogus job ads among a huge number of postings. Two types of classifiers are used to detect bogus job ads: single classifiers and ensemble classifiers. However, experimental results demonstrate that ensemble classifiers outperform single classifiers in terms of detecting fraud.

Keywords Machine learning · Bogus recruitment detection · Fraudulent jobs · Ensemble classifiers · Fraud detection

1 Introduction

Internet job advertisements have raised concerns about employment fraud. Businesses often post job openings online for applicants to find, but scammers exploit this convenience to create fake job ads and defraud people [1]. Automated systems are necessary to detect and inform individuals about fraudulent job ads, as they can damage the reputation of legitimate companies [2]. Job seekers might also look for openings that match their interests [3]. This online recruitment method streamlines the hiring process for both employers and job seekers. Naukri.com,

I. K. Pradeep (✉) · S. M. Krishna · K. B. Kiran · B. C. H. S. N. L. S. Sai Baba · V. D. Matta
Department of CSE, Vishnu Institute of Technology, Bhimavaram, India
e-mail: durgarao.m@vishnu.edu.in

A. Patel et al. (eds.), *Advances in Machine Learning and Big Data Analytics II*,
Springer Proceedings in Mathematics & Statistics 442,
https://doi.org/10.1007/978-3-031-51342-8_19

with 49.5 million registered users and 11,000 resumes uploaded daily in December 2016, highlights their importance [4]. Online recruitment fraud (ORF) is a new scam that has emerged alongside the growth of online recruiting. Criminals use online recruitment platforms to lure victims with fake job offers, stealing their personal information and money, and damaging their reputations as well as those of the targeted businesses. According to news sources, ORF causes significant financial losses for job seekers, and major companies like ABB have been warned about these scams [5]. Identifying ORF is important but remains understudied by scholars. The class disparity and high fraud rate make detecting fake job offers challenging.

2 Literature Survey

The initial phase of the software development process involves conducting a literature review. This involves reviewing preliminary research conducted by various authors relevant to our work and identifying key articles that will inform and expand upon our project.

Devsmit Ranparia [6] proposed a research article titled "Fake Job Prediction using Sequential Networks." The authors discuss how companies and fraudsters lure job seekers through various means, particularly on digital job search platforms. Their goal is to reduce such fraudulent activities by employing Machine Learning to predict the likelihood of job postings being fake, enabling candidates to make informed decisions. The model analyzes job posting sentiments and patterns using Natural Language Processing (NLP) techniques, trained as a Sequential Neural Network with GloVe embedding. The effectiveness of the model was evaluated using LinkedIn job postings to assess real-world accuracy, and it was refined through multiple enhancements to improve robustness and realism.

Hridita Tabassum [7] proposed a research article titled "Detecting Online Recruitment Fraud Using Machine Learning." This study builds upon previous work, detailing the methodologies and results in developing a model for detecting Online Recruitment Fraud (ORF) using a proprietary dataset focused on the Bangladesh job market, supplemented by a publicly available dataset. Various machine learning techniques including Logistic Regression, AdaBoost, Decision Tree Classifier, Random Forest Classifier, Voting Classifier, LightGBM, and Gradient Boosting were applied. The study found that LightGBM (95.17%) and Gradient Boosting (95.17%) achieved the highest accuracy in detecting fraudulent job postings, aiming to establish a reliable method for combatting this issue.

Sultana Umme Habiba [8] proposed a research article titled "A Comparative Study on Fake Job Post Prediction Using Different Data Mining Techniques." The author explores the increasing prevalence of job advertisements amid advancements in technology and social communication. Predicting fake job postings has become crucial due to its societal impact. The research addresses this challenge using various data mining techniques and classification algorithms such as KNN, Decision Tree, Support Vector Machine, Naive Bayes Classifier, Random Forest Classifier, Multilayer Perceptron, and Deep Neural Network. The Employment Scam Aegean Dataset (EMSCAD) with 18,000 samples was utilized for testing, demonstrating the effective performance of deep neural networks in this classification task.

3 Existing System

Recently, a significant challenge in the realm of Online Recruitment Frauds (ORF) has emerged in the form of employment scams. Many organizations now opt to advertise their job vacancies online for easy and prompt access by job seekers. However, this convenience has been exploited by fraudsters who deceive job seekers by offering employment in exchange for financial gain. These fraudulent job postings can tarnish the reputation of legitimate companies. Detecting such fraudulent job ads has sparked interest in developing automated solutions to identify them and alert potential applicants, preventing them from falling victim to these scams. The current system faces several limitations:

1. In recent years, organizations have increasingly chosen to post job openings online for rapid accessibility by job seekers.
2. However, this practice can be exploited by fraudsters who solicit money from job seekers in exchange for promised employment.
3. Addressing the challenge of identifying fraudulent job advertisements involves exploring supervised learning algorithms, specifically classification techniques.
4. These classifiers map input variables to target classes using training data to distinguish between legitimate and fraudulent job postings.

4 Proposed System

To identify fraudulent job postings, a machine learning method employs various classification algorithms. This approach involves a classification tool that alerts users upon detecting fake job advertisements within a larger pool of job postings. The focus is on exploring supervised learning algorithms as classification techniques to address the challenge of identifying job posting scams [9]. These classifiers utilize training data to map input variables to specific target classes. The study outlines several classifiers used to distinguish fake job postings, categorized into single classifier-based predictions and ensemble classifier-based predictions. The Naive Bayes Algorithm demonstrated the most effective performance in this research.

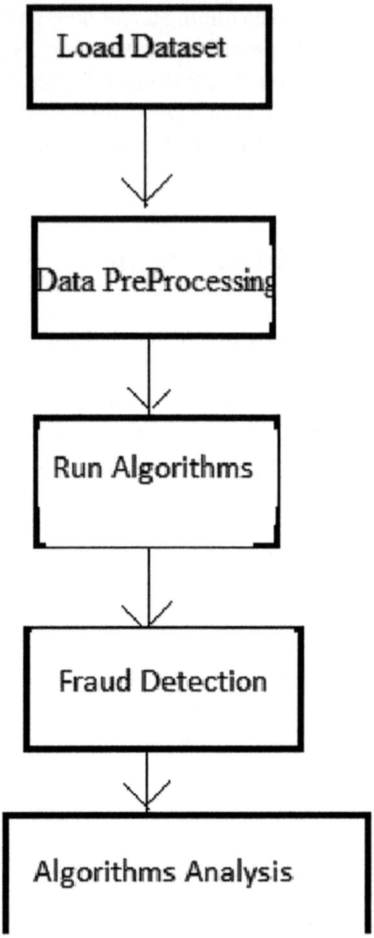

Fig. 1 Proposed model for fake job prediction

Benefits of the proposed system include:

1. Utilization of a machine learning approach with multiple classification algorithms to detect fraudulent job postings.
2. A classification tool that alerts users upon detecting fake job postings within a larger dataset of job advertisements [10].
3. Exploration of supervised learning algorithms to develop classifiers that link input variables to target classes effectively.

Based on Fig. 1 above, it is evident that the proposed model utilizes raw data sourced from the KAGGLE website as its sample dataset. Following acquisition, the raw data undergoes comprehensive data pre-processing procedures. Once the data pre-processing phase is finalized, machine learning algorithms are then applied to discern between fake and legitimate job postings based on specific keywords.

4.1 Proposed Dataset

In this module, we try to load the dataset which is collected from the Kaggle website, and then try to give that Excel file information as input to the next module.

Dataset URL: https://www.kaggle.com/datasets/shivamb/real-or-fake-fake-jobposting-prediction

From the above description, we can see 18K job descriptions are collected as samples from that dataset which contains both genuine as well as fake job profiles.

4.2 Proposed Algorithms

In this application, we attempt to employ multiple classification models before determining which model can reliably identify fake job profiles or normal profiles. We obtained the dataset from the KAGGLE website for this application. There are about 18 K samples in this collection that are a mix of good and fake job profiles.

1. Logistic Regression.
2. SVM.
3. Random Forest.
4. Naïve Bayes.

Logistic Regression
Logistic regression is a widely used machine learning algorithm categorized under supervised learning. Its primary function is to predict the categorical dependent variable based on a set of independent variables. Unlike linear regression, which predicts continuous values, logistic regression predicts the probability of an outcome falling into a specific category. This outcome is typically binary, such as Yes or No, 0 or 1, or True or False. In our current application, logistic regression is employed to accurately classify job descriptions as either fake or normal based on specific criteria.

SVM Classification Model
Support Vector Machine (SVM) is a supervised machine learning technique capable of performing classification, regression, and outlier detection tasks. In classification, the linear SVM classifier draws a straight line between two classes, where data points on one side of the line are assigned to one class and those on the other side to the second class [11]. While conceptually straightforward, the challenge lies in selecting the optimal line for classifying the data effectively. The Linear SVM (LSVM) algorithm addresses this by choosing a line that not only separates

the classes but also maximizes the margin between the classes, thus enhancing robustness against potential errors. In SVM terminology, the "support vectors" are key elements—these are the position vectors drawn from the origin to the decision boundary points, crucial for defining the separation between classes. In our current application, SVM is utilized to achieve highly accurate classification of job descriptions as either fake or normal based on specific features and characteristics.

Naïve Bayes Classification Model
The Naive Bayes classifier is a supervised classification method that applies the concept of Conditional Probability from the Bayes Theorem. Despite potentially inaccurate probability estimates, this classifier's decisions are highly effective in practical applications. It performs optimally when features are either independent or completely functionally dependent. The accuracy of the Naive Bayes classifier is influenced not by the dependencies among features but rather by how much information about the class is lost due to the assumption of independence [12].

Random Forest Classification Model
Random forest is a supervised machine learning technique based on ensemble learning principles. Ensemble learning combines multiple instances of the same or different algorithms to create a more robust prediction model. Specifically, the random forest algorithm integrates numerous decision trees, collectively forming a "forest" of trees, hence the name "Random Forest" [13, 14]. This approach is versatile and suitable for both regression and classification tasks. In our current application, we employ the random forest algorithm to accurately classify job descriptions as fake or normal based on specific criteria.

5 Experimental Results

The two figures below demonstrate that the proposed ML models [15] performed well on the dataset. The data was trained using all three models, and the results show which model achieved the highest accuracy in accurately identifying fake job descriptions from the job profiles.

Word Cloud for Extracting Keywords

```
plt.figure(figsize=(15,15))
plt.imshow(wc, interpolation='bilinear')
plt.axis("off")
plt.show()
```

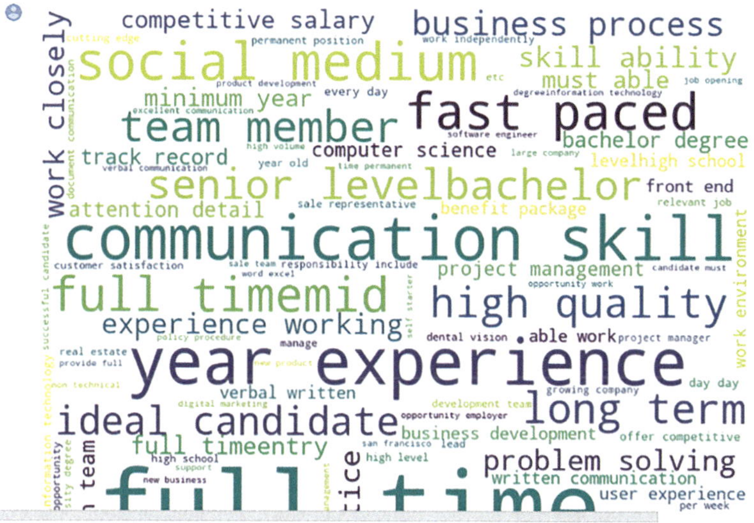

From the window above, we can observe that the dataset has been loaded and pre-processed, allowing us to identify how many relevant keywords match the job descriptions.

Performance Analysis

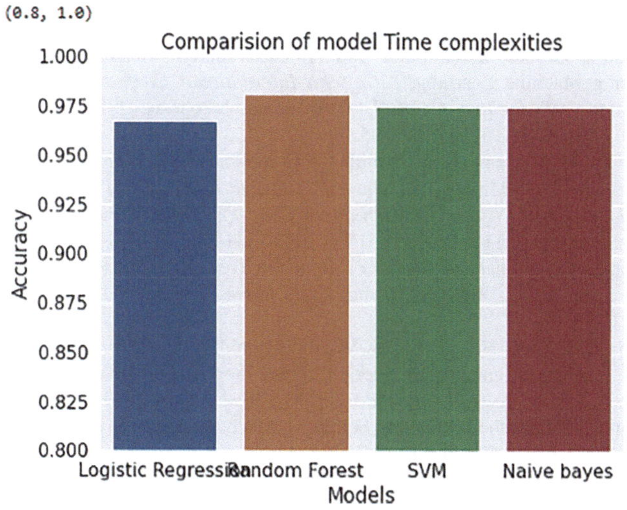

From the information displayed above, we observe that the Random Forest model exhibits higher accuracy compared to the other models in predicting fake job descriptions.

6 Conclusion

Identifying employment scams aids job seekers in securing genuine offers from employers. This research explores various machine learning algorithms as defenses against detecting such scams. A supervised approach employs multiple classifiers to effectively identify fraudulent job postings. According to experimental results, the Random Forest classifier outperformed other classification tools, achieving an accuracy of 98.27%, notably surpassing existing methods.

References

1. B. Alghamdi and F. Alharby, "An Intelligent Model for Online Recruitment Fraud Detection," J. Inf. Secur., vol. 10, no. 03, pp. 155–176, 2019. https://doi.org/10.4236/jis.2019.103009.
2. I. Rish, "An Empirical Study of the Naïve Bayes Classifier An empirical study of the naive Bayes classifier," || no. January 2001, pp. 41–46, 2014.
3. D. E. Walters, "Bayes's Theorem and the Analysis of Binomial Random Variables," || Biometrical J., vol. 30, no. 7, pp. 817–825, 1988. https://doi.org/10.1002/bimj.4710300710.
4. F. Murtagh, "Multilayer perceptrons for classification and regression," || Neurocomputing, vol. 2, no. 5-6, pp. 183-197,1991. https://doi.org/10.1016/0925-2312(91)90023-5.
5. P. Cunningham and S. J. Delany, *K.*, "Nearest Neighbour Classifiers," || Mult. Classif. Syst., no. May, pp. 1-17, 2007. https://doi.org/10.1016/S0031-3203(00)00099-6.
6. D. Ranparia, S. Kumari and A. Sahani, "Fake Job Prediction using Sequential Network," *2020 IEEE 15th International Conference on Industrial and Information Systems (ICIIS)*, RUPNAGAR, India, 2020, pp. 339–343. https://doi.org/10.1109/ICIIS51140.2020.9342738.
7. H. Tabassum, G. Ghosh, A. Atika and A. Chakrabarty, "Detecting Online Recruitment Fraud Using Machine Learning," *2021 9th International Conference on Information and Communication Technology (ICoICT)*, Yogyakarta, Indonesia, 2021, pp. 472–477. https://doi.org/10.1109/ICoICT52021.2021.9527477.
8. S. U. Habiba, M. K. Islam and F. Tasnim, "A Comparative Study on Fake Job Post Prediction Using Different Data Mining Techniques," *2021 2nd International Conference on Robotics, Electrical and Signal Processing Techniques (ICREST)*, DHAKA, Bangladesh, 2021, pp. 543–546. https://doi.org/10.1109/ICREST51555.2021.9331230.
9. H. Sharma and S. Kumar, "A Survey on Decision Tree Algorithms of Classification in Data Mining," || Int. J. Sci. Res., vol. 5, no. 4, pp. 2094–2097, 2016. https://doi.org/10.21275/v5i4.nov162954.
10. E. G. Dada, J. S. Bassi, H. Chiroma, S. M. Abdulhamid, A. O. Adetunmbi, and O. E. Ajibuwa, "Machine learning for email spam filtering: review, approaches and open research problems," || Heliyon, vol. 5, no. 6, 2019. https://doi.org/10.1016/j.heliyon.2019.e01802.
11. L. Breiman, "ST4_Method_Random_Forest," || Mach. Learn., vol. 45, no. 1, pp. 5–32, 2001. https://doi.org/10.1017/CBO9781107415324.004.
12. B. Biggio, I. Corona, G. Fumera, G. Giacinto, and F. Roli, "Bagging classifiers for fighting poisoning attacks in adversarial classification tasks."

13. Abhinav Dayal, Sumit Gupta, Sreenu Ponnada, D Jude Hemanth "Tornado Forecast Visualization for Effective Rescue Planning," 2022 EAI/Springer Innovations in Communication and Computing book series (EAISICC), Vishnu Institute of Technology, Bhimavaram, India. https://doi.org/10.1007/978-3-030-84182-9_7
14. Lakshmanaprabu, S.K., Shankar, K., Ilayaraja, M., Nasir, A.W., Vijayakumar, V., Chilamkurti, N. "Random forest for big data classification in the internet of things using optimal features," vol. 10, 2019, pp. 2609–2618. https://doi.org/10.1007/s13042-018-00916-z.
15. T. Gayathri; T. Madhavi; K. Ratna Kumari "A Prediction of breast cancer based on Mayfly Optimized CNN" 2022, International Conference on Computing, Communication and Power Technology, Shri Vishnu Engineering College for Women, Bhimavaram, India. https://doi.org/10.1109/IC3P52835.2022.00044.

Secure Image Transmission Using Advanced Encryption Standards with Salting, Steganography, and Data Shredding Techniques

Keesara Sravanthi, P. Chandra Sekhar, E. Rishyendra, N. Shiva, P. Aravinda, and V. Sai Varun

Abstract The project suggests using the AES (Advance Encryption Standard) algorithm for image encryption and decryption. As the usage of images is increasing, there is a need to provide security for images. The architecture employs an iterative process with keys that are 128, 192, or 256 bits long and blocks that are 128 bits wide. The round integers for keys with 256 bits are 14, 128 bits are 10, and 192 bits are 12. The complexity and security of the cryptography algorithms both grow by growing secret keys. AES encryption in this paper's technique is to obtain the encrypted image, then AES decryption decrypts it. Various processes secure the facts. Through cryptography, we can prevent unauthorized access to data. In recent years many cryptography methods have been proposed and used to protect confidential data. The popular cryptography methods are AES and DES. Various aspects of various cryptography methods usage on images are presented in this paper. A specification for the encryption of electronic data is called the Advanced Encryption Standard (AES). Compared to well-known cryptographic algorithms and the current data encryption standard, AES is significantly more efficient and secure. A new version of AES with enhanced security is proposed.

Keywords Cryptography · Encryption · Decryption · Cipher text · Plain text · Key

K. Sravanthi (✉)
Department of Computer Science and Engineering, GITAM, Visakhapatnam, AP, India

Department of Information Technology VNR VJIET, Hyderabad, Telangana, India

P. C. Sekhar
Department of Computer Science and Engineering, GITAM, Visakhapatnam, AP, India

E. Rishyendra · N. Shiva · P. Aravinda · V. S. Varun
Department of Information Technology VNR VJIET, Hyderabad, Telangana, India

A. Patel et al. (eds.), *Advances in Machine Learning and Big Data Analytics II*,
Springer Proceedings in Mathematics & Statistics 442,
https://doi.org/10.1007/978-3-031-51342-8_20

1 Introduction

Internet is used widely over the world involving a lot of users. Lots of interactions and communications happen over it. Data can be sent and received through the Internet. This process can be affected by other third sources. It can be dangerous to perform such processes. Better security is needed for the safe transfer of data over the Internet or through other means. If the security is not good enough to protect the data, the data can be misused and the theft of data takes place. Therefore, fear has risen among users regarding their data. Images are also considered as part of the data. So, there is a need to have a secure way to transfer images by using suitable methods.

Basically, good practices should be deployed that offer better security. The security practices protect data from unauthorized access and malicious users. There are many methods to provide security. Cryptography is one of the methods that provide security to data. It is done by performing various operations on plain data [1].

These operations include bit transformations, substitutions, etc. There are different types of cryptography techniques such as symmetric, asymmetric cryptography, and hashing. Symmetric uses a single key, and asymmetric uses different keys [2].

Key is not required for hashing. Hashing involves adding data to the original data or converting original data into other forms of data. The above-mentioned cryptography methods are performed on the text format of data. We cannot implement cryptography directly on images. So, to perform cryptography on images, images are converted to text format. Images are stored in the format of bytes on a computer. By converting the image to byte object format, we can get the text format of the image and we can apply cryptography to the available text format. Images are made of pixels. Another approach to converting images to text is by using pixels. Pixels are represented in RGB values. RGB values of pixels can be represented as text.

Another approach to hiding data is steganography. Steganography is the process of hiding data in another form of data. "stegos" means "to cover" and "grayfia" means "writing" in Greek language. The image steganography is performed to hide the image data in another image file. It is done by altering the values of pixels in the image. Randomly, the encryption algorithm chooses the pixels [3].

Cryptography and steganography can protect secret data. Cryptography converts data to an unreadable format. Steganography protects data by hiding it in other data. However, cryptography is more popular and more used than steganography.

As mentioned, various cryptography methods for images are going to be discussed. DES and AES algorithms are mainly used for cryptography purposes. There are various variations of these algorithms [4]. In today's digital age, the security of data and images has become more critical than ever before. With the vast amount of data being shared and stored on various platforms, the risk of data breaches and theft has increased significantly. Images, in particular, are vulnerable to attacks, as they contain sensitive information that can be misused if accessed by unauthorized parties. To address this concern, various encryption techniques have been developed to secure images from being accessed by unauthorized parties.

The proposed system for AES image encryption and decryption with salting, steganography, and data shredding is designed to provide enhanced security to images. This system incorporates three new features, namely, salting, steganography, and data shredding, to make the encryption process more secure and robust. In this project, we will discuss the architecture of the proposed system and the implementation of these new features to provide better security to images. This system can be used in various applications where image security is of utmost importance, such as medical imaging, military imaging, and surveillance imaging. The use of this system will ensure that images remain secure and can only be accessed by authorized parties, thereby protecting sensitive information from being misused.

2 DES

As a symmetric key block cypher algorithm, DES is used. The Feistel cypher is used as the DES implementation framework. There are 16 cycles of steps in the Feistel structure. DES structure makes use of 64-bit blocks. Despite having a 64-bit key, DES only uses a 56-bit key. The final 8 bits are used afterward but are not employed in the encryption process [5] (Fig. 1).

The biggest DES flaw is the 56-bit key size, and the processors can only implement one million DES encrypt or decrypt operations at a time. Because DES was not intended for application, it operates slowly. AES is recommended over DES.

DDES is a method that applies two separate implementations of DES to the same plain text. By using different keys plain text is encrypted. When decrypting, there is a need for different keys. First DES instance, uses the first key to transform 64-bit plain- text to 64-bit middle text. 64-bit plain text is transformed into 64-bit cypher text by a second DES instance using a second key (Figs. 2 and 3).

Another implementation of DES is called triple DES. Three different DES approaches are applied to simple plain text. In the first approach, the three different keys are used. In the second approach, the two same and one different keys are used. In the third approach, the same type of keys is used (Figs. 4 and 5).

3 RSA

Ron Rivest, Adi Shamir, and Leonard Adleman developed this asymmetric key cryptography algorithm. The developers' surnames served as the basis for the naming. The algorithm can be used to authenticate the information and achieve information security.

Fig. 1 Working of DES

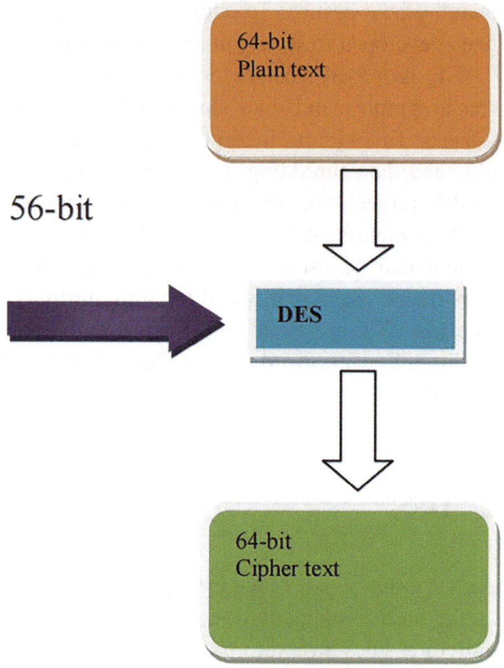

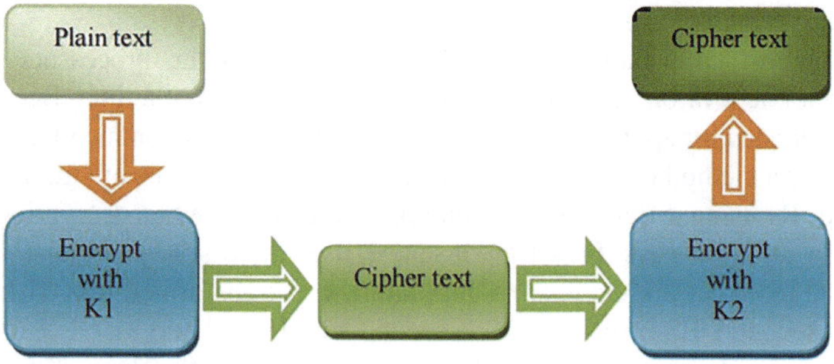

Fig. 2 Encryption in DDES

The majority of public-key cryptosystems are employed to protect data during transmission. An asymmetric key cryptography algorithm called RSA employs a pair of keys, a public, and a private key. As a result, message encryption and decryption take place at the sender and receiver ends, respectively, using the public key of the sender and the private key of the receiver [6].

The creation of keys in RSA is complicated as it uses two prime numbers p and q. But it provides good security as it is not easy to guess these keys as additional concepts like product of prime numbers cannot be cracked easily.

In the RSA algorithm's steps, three main phases are involved in their implementation. They are as follows:

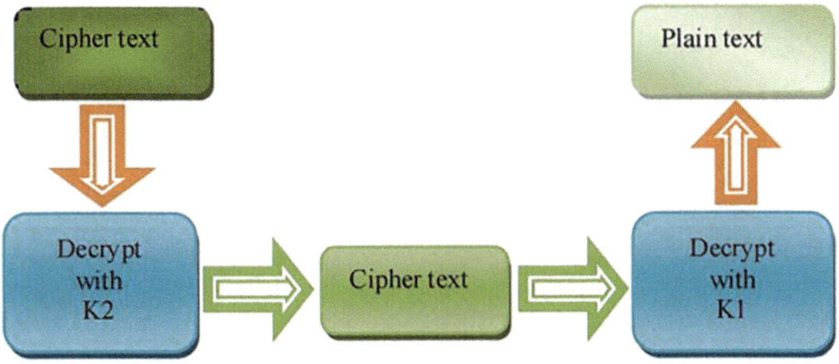

Fig. 3 Decryption in DDES

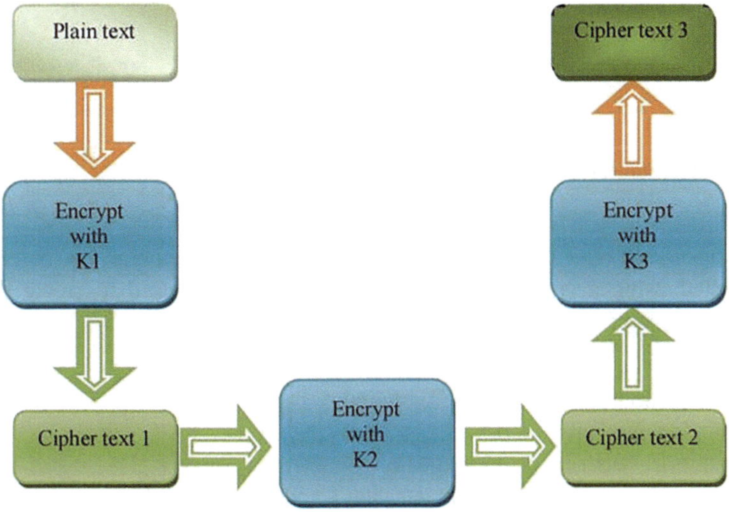

Fig. 4 Encryption in Triple DES

(a) Generation of key.
(b) Encryption.
(c) Decryption.

3.1 Key Generation

The creation of keys, which includes both public and private generation, is the first stage in the RSA algorithm. A public key, as its name suggests, is an open key that is visible to everyone and is used to encrypt messages.

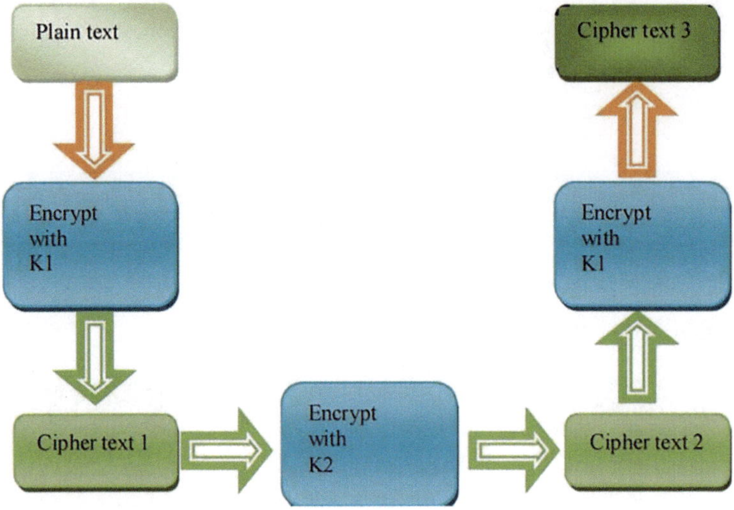

Fig. 5 Decryption in Triple DES

Images sent over the network are encrypted using a public key before being decrypted using a private key. The following steps are used to create the RSA algorithm's keys:

1. Consider two different prime numbers p and q.
2. p and q should be of the same bit length.
3. Calculate n where $n = p * q$.
4. Find $\Phi(n)$, $\Phi(n) = (p-1) * (q-1)$.
5. Choose an integer e that is greater than 1 and coprime to $\Phi(n)$.
6. Public key consists of n and e.
7. A private key is generated by using $\Phi(n)$ and k.
8. K is some integer.
9. Private key $d = (k * \Phi(n) + 1) / e$.

3.2 *Encryption*

$E = T^{1}$
E – Encrypted text
T – Plain text
$1 = e \bmod n$

3.3 Decryption

$^{l}P = E$
P – Decrypted text
E – Encrypted text
$l = d \bmod n$

The biggest flaw in RSA is its suitability for different types of text data. It is best suitable for the numbers format of text. It will be complicated to apply RSA on text that contains characters other than numbers/digits.

4 A Trio Approach

This method is proposed by Sundararaman, Sivaraman, Siva Janakiraman, Har Narayan Upadhyay, and Rengarajan. It uses LFSR and cellular automata. Cellular automata offer both permutation and diffusion. The linear feedback shift register does the job of a pseudo-random generator. The output of a linear feedback shift register depending on its previous state is an essential feature [7]. Because it is so simple to create, cellular automation has many applications in actual systems. The encrypted image is produced by XORing each level of the LFSR with the one before it at the rising clock edge.

5 Using Neural Networks

Nearly every industry today employs neural networks. This is explained by its accuracy and capacity for learning. To encrypt a picture two concepts named chaotic systems, and neural networks are merged, the approach put out by Yousef, Karim, and Nooshin intends to benefit from both of these technologies. To introduce chaos, techniques like Chua and Liu systems are used to take the input. The final encrypted output is then produced by performing a nonlinear mapping on the output using a permutation neural network [8].

Decryption is the opposite of encryption. The result has a high degree of entropy, and the histograms of the unencrypted and original images diverge. Its implementation is difficult.

6 Using Henon Map

The method put forth by Ping, Feng Xu, Yingchi Mao, and Zhijian Wang proposes a way for processing two pixels at once while also changing the pixel value and location. The Henon Map's parameters are chosen in a way that results in chaotic behavior. The map is then discretized. Secret keys are considered from the parameters. Next, discretization of the map is done. Certain parameters are protected by secret keys. Diffusion and permutation are handled by the discrete Henon Map, whereas keystream creation is handled by the classical Henon Map. With this technique, the image must be padded to make it square. Typically, if a secret key is a picture, whenever the encrypted image is generated receiver receives the key. The picture features are, however, introduced into the encryption image using this technique. Therefore, if many photos are encrypted, only one secret key needs to be provided. This technique can turn the provided image into a random cypher image by reducing the correlation between two pixels that are next to one another. This approach is quicker because it tries to alter two pixels at once [9].

7 AES–Proposed System

The U.S. National Institute of Standards and Technology (NIST) established AES in 2001. It is a variant of Rijndael Block Ciphers. AES uses the same key for encryption and decryption. AES involves a number of rounds. Each round includes several processing steps. The number of rounds applied varies as per the AES version [10].

Each round consists of four steps—bytes substitution, shifting of rows, mixed column transformation, and adding round key. The last round does not have a mixed-column transformation. As the name suggests, byte substitution involves the substitution of bytes in input text. Permutation is done by shifting rows and mixed column transformation steps. AES takes input as 128-bit text and returns 128-bit cipher text. AES considers the 128-bit as 4×4 byte matrix and the above-mentioned steps are performed on this 4×4 matrix (Figs. 6, 7 and 8).

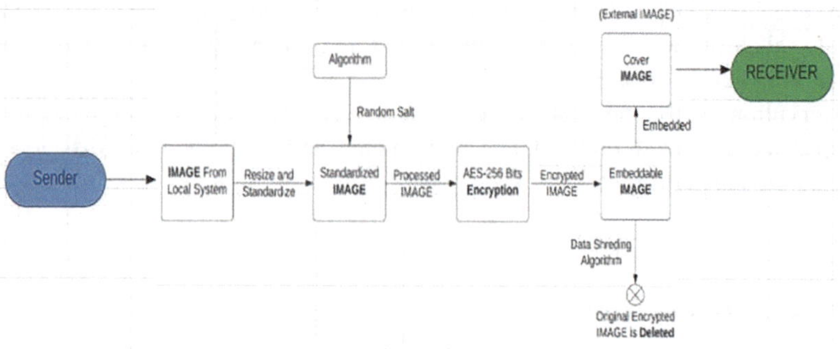

Fig. 6 Encryption process

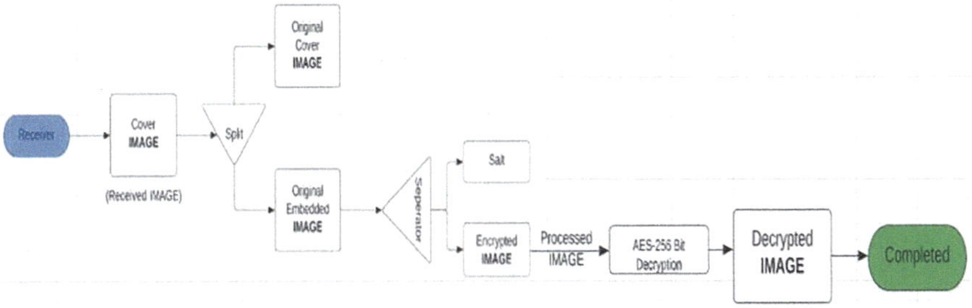

Fig. 7 Decryption process

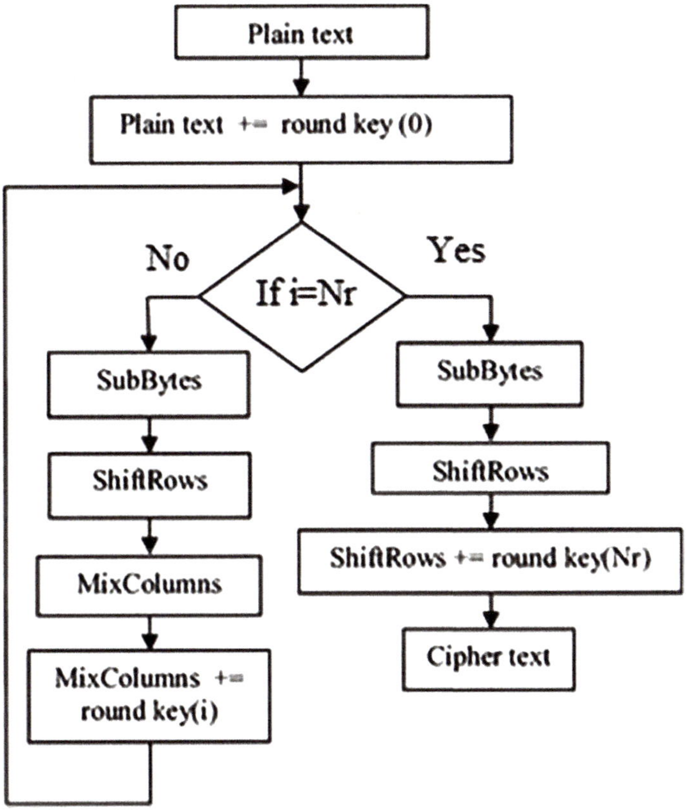

Fig. 8 Overview of AES encryption

Byte substitution involves the substitution of one byte by another byte from a table called S-Box. A byte is not substituted by itself or its complement [11] (Fig. 9).

X \ Y	0	1	2	3	4	5	6	7	8	9	a	b	c	d	e	f
0	52	09	6A	D5	30	36	A5	38	BF	40	A3	9E	81	F3	D7	FB
1	7C	E3	39	82	9B	2F	FF	87	34	83	43	44	C4	DE	E9	CB
2	54	7B	94	32	A6	C2	23	3D	EE	4C	95	0B	42	FA	C3	4E
3	08	2E	A1	66	28	D9	24	B2	76	5B	A2	49	6D	8B	D1	25
4	72	F8	F6	64	86	68	98	16	D4	A4	5C	CC	5D	65	B6	92
5	6C	70	48	50	FD	ED	B9	DA	5E	15	46	57	A7	8D	9D	84
6	90	D8	AB	00	8C	BC	D3	0A	F7	E4	58	05	B8	B3	45	06
7	D0	2C	1E	8F	CA	3F	0F	02	C1	AF	BD	03	01	13	8A	6B
8	3A	91	11	41	4F	67	DC	EA	97	F2	CF	CE	F0	B4	E6	73
9	96	AC	74	22	E7	AD	35	85	E2	F9	37	E8	1C	75	DF	6E
a	47	F1	1A	71	1D	29	C5	89	6F	B7	62	0E	AA	18	BE	1B
b	FC	56	3E	4B	C6	D2	79	20	9A	DB	C0	FE	78	CD	5A	F4
c	1F	DD	A8	33	88	07	C7	31	B1	12	10	59	27	80	EC	5F
d	60	51	7F	A9	19	B5	4A	0D	2D	E5	7A	9F	93	C9	9C	EF
e	A0	E0	3B	4D	AE	2A	F5	B0	C8	EB	BB	3C	83	53	99	61
f	17	2B	04	7E	BA	77	D6	26	E1	69	14	63	55	21	0C	7D

Fig. 9 S-Box

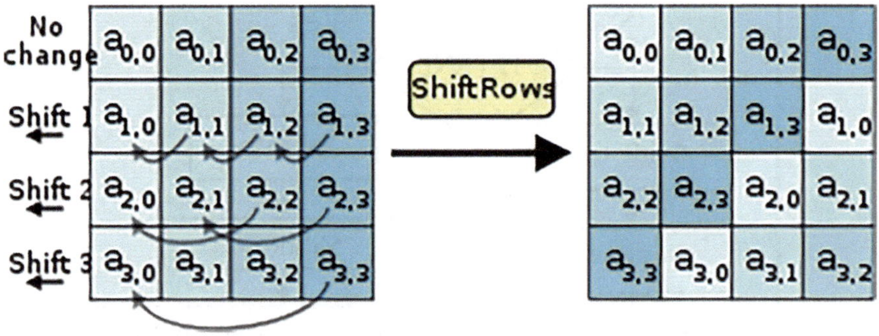

Fig. 10 Shift rows

In the next step, rows in the byte matrix are shifted. The first row is unaltered. Other rows are shifted by some positions to the left (Fig. 10).

The mix columns step includes converting a column of bytes to a transformed column by applying a linear transformation on each column (Fig. 11).

The final step is adding a round key. The round key is generated by expanding the key. In each round, the round key is added to the state or cipher text. The original key is considered as round key for the first round and for other rounds the round key is generated by expanding the round key of the previous round. Adding a round key involves XORing the text byte with the byte of the round key. The resulting text is passed as input to another round. The final text obtained by adding a round key in the final round is the final decrypted text (Fig. 12).

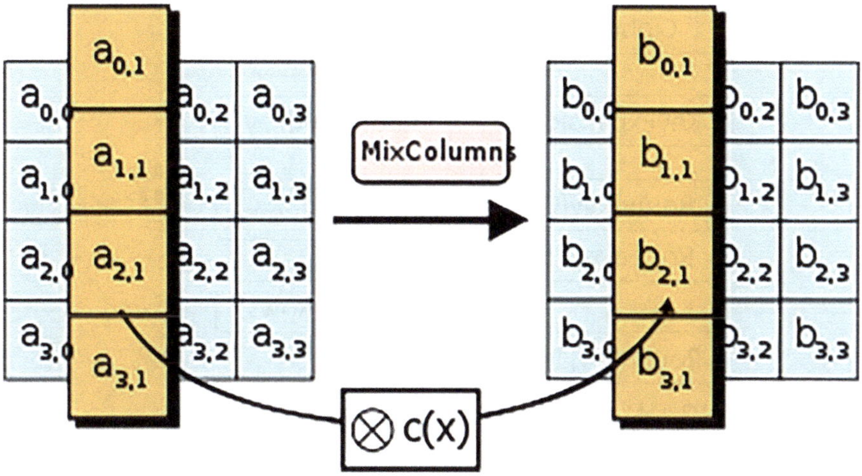

Fig. 11 Mix columns

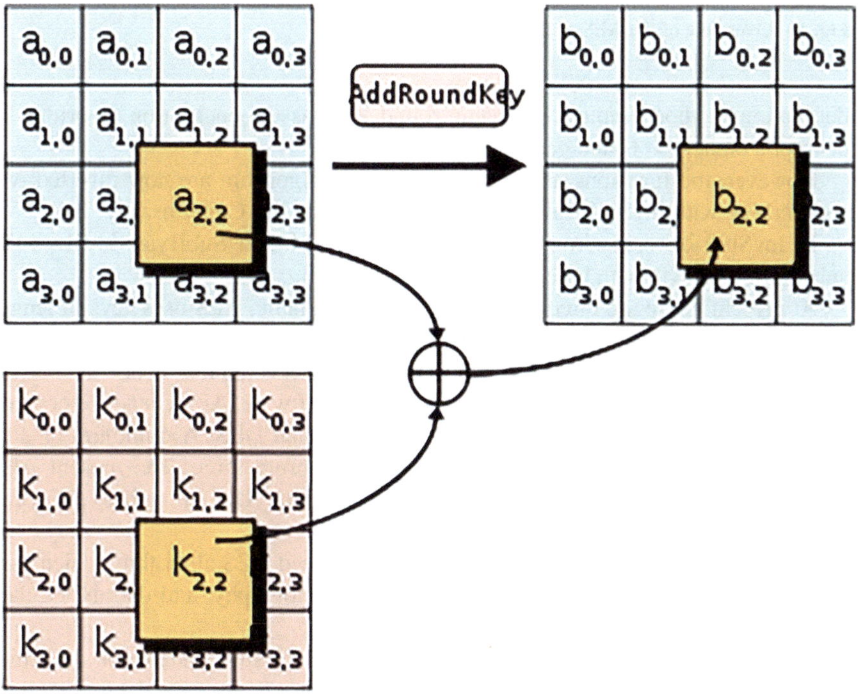

Fig. 12 Round key addition

The above-mentioned process is for the encryption part of AES. To decrypt the cipher text, we need to apply the steps of AES encryption in reverse order. The

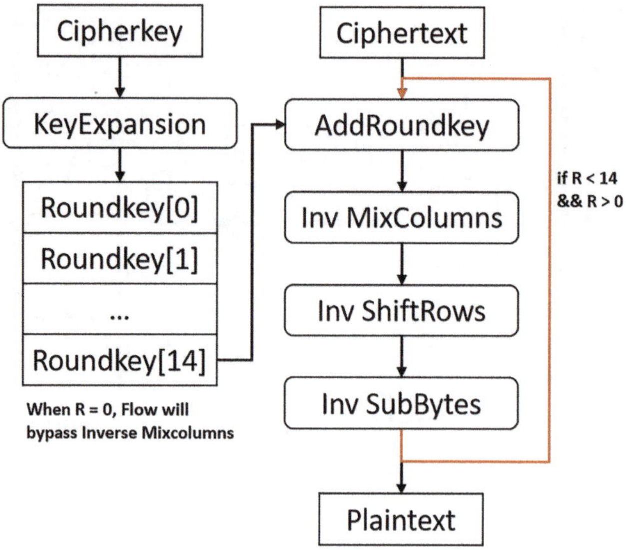

Fig. 13 Overview of the AES decryption

decryption method employs the same round keys as the encryption algorithm to decrypt a ciphertext C and generate a plaintext P.

However, the functions used in the encryption algorithm are now inverted; we are working with InvShiftRows, InvSubBytes, and InvMixColumns (Fig. 13).

In InvShiftRows, the rows are shifted to the right. InvSubBytes involves the substitution of byte with byte from inverse S-Box (Fig. 14).

At present, there are three variations of AES available. AES uses keys of length 128 bits, 192 bits, and 256 bits. The no of rounds applied is dependent on the key size. As key length increases the number of rounds applied is also increased. AES can be easily deployed in both hardware and software. AES cannot be cracked easily [12]. It offers more reliability and security than DES. A minimum of 2^{128} computations are needed to crack AES by using brute force. The present AES methods are sufficient for secure transmission of data. The strength of AES can be further enhanced by increasing key size [13].

A high level of security is provided by the enhanced AES algorithm, which uses a 256-bit key and other features like salting, steganography, and data block data shredding.

Images have increased security thanks to the suggested system for AES picture encryption and decryption with salting, steganography, and data shredding. Steganography conceals the encrypted image in a cover image, then uses salt to provide randomness to the encryption process, and data shredding permanently deletes the original image after encryption. These elements work together to make it difficult for an attacker to access the original image, ensuring that it is preserved.

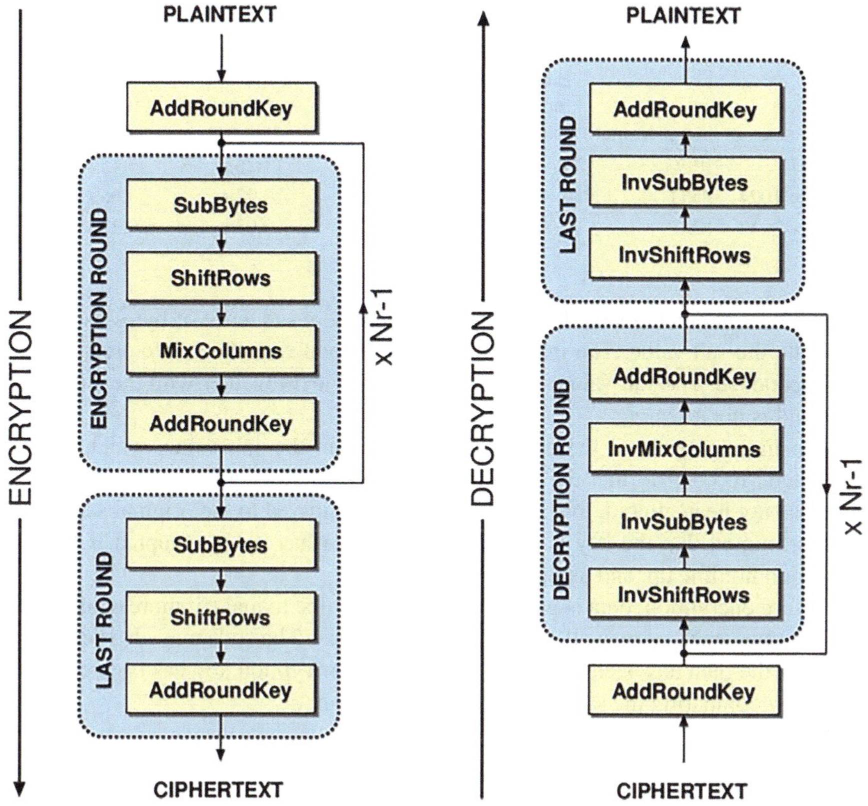

Fig. 14 Overview of the AES process

8 Experimental Results

A new system is proposed that includes AES encryption and decryption along with salting, steganography, and data shredding (Table 1).

8.1 *Salting*

The salting technique involves adding random data to the input data before encryption. This helps to increase the security of the encryption by making it more difficult for attackers to guess the encryption key. Salting is used in the provided code to increase the security of the encryption by making the ciphertext that results more unpredictable. Before encryption, a randomly generated value called the salt is appended to the beginning of the stream.

Table 1 Results for salting

Image size	Image format	Encryption time (without salting)	Encryption time (with salting)	Decryption time (without salting)	Decryption time (with salting)
512*512	JPEG	1.50 s	1.65 s	1.28 s	1.45 s
1024 × 1024	PNG	3.89 s	4.25 s	3.42 s	3.66 s
2048 × 2048	BMP	9.78 s	10.11 s	9.03 s	9.27 s

The os.urandom() method, which creates a random sequence of bytes, is used to generate the salt value. The image data is then joined with the salt to create a new byte sequence. The salt value is often kept in plain text together with the encrypted data and is not a secret.

The first 16 bytes of the decrypted data contain the salt value, which is then extracted. To confirm that the right key was used to decrypt the data and that the data has not been altered, this salt value is then compared to the original salt value. It is presumed that the key is incorrect or the data has been corrupted if the salt values do not line up, and an error is raised.

Before encryption, data is given a random salt value to make it more challenging to predict or crack using well-known plaintext attacks. The code may also determine whether the data has been altered or whether the encryption key has been compromised by comparing the salt value during decryption.

8.2 Steganography

The steganography technique involves hiding the encrypted data inside an image file, making it more difficult for attackers to detect the encrypted data.

The offered code uses the steganography technique to conceal a message inside an image. Utilizing the least significant bits of the RGB color values of each pixel in the image, the message is specifically concealed within the encrypted image data [14, 15].

From the encrypted picture data, a new picture object is first created. The square root of the ciphertext data length determines the size of the image. As a result, a square image made of pixels is produced, with each pixel later storing a single character from the secret message.

The R and G values of the image's initial pixel are used to store the message's length, while the subsequent pixels are used to store the message's characters. The G and B values are set to zero, and the message characters are stored as the pixel's R value. Each row of pixels in the message is used to hold a character from the message in row-major order.

The process is reversed during decryption; the image data is read, and the first pixel is utilized to calculate the message's length. The R value of each remaining

Table 2 Results for steganography

Image size	Image format	Encryption time (without steganography)	Encryption time (with steganography)	Decrypt ion time (without steganography)	Decryption time (with steganography)
512 × 512	JPEG	1.50 s	1.89 s	1.28 s	1.66 s
1024 × 1024	PNG	3.89 s	4.78 s	3.42 s	4.11 s
2048 × 2048	BMP	9.78 s	11.23 s	9.03 s	10.68 s

pixel is then retrieved and used as a character in the message after reading the remaining pixels. After that, the mail can be forwarded to the user.

Although steganography can be a useful way to blend messages with other data, it is not frequently utilized as a major encryption or security approach. In order to add an extra degree of protection or conceal the fact that data is being encrypted at all, it is frequently employed in conjunction with other encryption methods. However, if the encryption is poor or the attacker has access to the original, unencrypted image data, steganography may be subject to attacks that can extract the hidden message [16, 17] (Table 2).

8.3 Data Shredding

The data shredding technique involves securely deleting the original encrypted data to prevent it from being recovered.

The original encrypted data is discarded to prevent its recovery in the aes_decrypt function of the given code after the encrypted image has been successfully decrypted. This is accomplished by overwriting the original encrypted data file with os.urandom() function-generated random data.

Data shredding or file wiping is a typical method for safely deleting sensitive information. It involves overwriting the original data with random data. It guarantees that the information cannot be recovered using specialized tools or procedures that might be able to recover information that has been erased or rewritten [18–20].

Data shredding is used to further secure the file during the decryption process, making it more challenging for an attacker to get the original encrypted data. This helps to protect the data because even if an attacker manages to crack the encryption and read the encrypted file, they will not be able to recover the original content (Tables 3 and 4).

As you can see from the table, each technique adds some processing time to the encryption and decryption process, but overall, the combined use of salting, steganography, and data shredding significantly enhances the security of the encrypted data. The encryption and decryption time increases with larger file sizes and longer messages (for steganography), but the impact of the techniques is

Table 3 Results for data shredding

File size	Encryption time (without data shredding)	Encryption time (with data shredding)	Decryption time (without data shredding)	Decryption time (with data shredding)
10 MB	2.45 s	2.90 s	1.97 s	2.32 s
50 MB	12.05 s	14.23 s	11.34 s	13.20 s
100 MB	25.83 s	30.10 s	23.90 s	28.12 s

Table 4 Results for all techniques

File size	Message length	Encryption time (without techniques)	Encryption time (with techniques)	Decryption time (without techniques)	Decryption time (with techniques)
1 MB	NONE	0.62 s	1.98 s	0.51 s	1.46 s
10 MB	NONE	2.45 s	3.60 s	1.97 s	3.06 s
50 MB	NONE	12.05 s	13.85 s	11.34 s	13.09 s
1 MB	50B	1.06 s	2.52 s	0.93 s	1.88 s

consistent across different file sizes and message lengths. The more time the more difficult to crack the image so it is more secure.

9 Conclusion

Images can be more securely encrypted and decrypted using the proposed AES image encryption and decryption system with salting, steganography, and data shredding, which makes it appropriate for use in a variety of applications like surveillance, military, and medical imaging.

The data shredding feature permanently deletes the original image after encryption, steganography conceals the encrypted image under a cover image, and the salting tool adds a random value to the image. These components increase the proposed system's security and sturdiness, guaranteeing that photos are kept secure and that only authorized users can access them.

References

1. P. Kumar and V. Sharma, "Information Security Based on Steganography & Cryptography Techniques: A Review." International Journal of Advanced Research in Computer Science and Software Engineering, Vol. 4, Issue 10, October (2014):247–250.
2. Stallings W, "Cryptography and network security: principles and practices," Pearson Education India (2006).
3. R. Poornima and Iswarya R. J, "An Overview of Digital Image Steganography," International Journal of Computer Science & Engineering Survey 4.1 (2013): 23–31.

4. Singh G(2013), "A study of encryption algorithms (RSA, DES, 3DES and AES) for information security," International Journal of Computer Applications, 67(19).
5. K. Rabah(2005), "Theory and implementation of data encryption standard: A review," Information Technology Journal, 4: 307–325.
6. N. Priya and M. Kannan, "Comparative Study of RSA and Probabilistic Encryption," International Journal of Engineering and Computer Science, vol. 6, no. 1, pp. 19867–19871, January 2017.
7. Sundararaman Rajagopalan, Sivaraman Rethinam, Siva Janakiraman, "Cellular automata + LFSR + synthetic image: A trio approach to image encryption," 2017 International Conference on Computer Communication and Informatics (ICCCI).
8. Minal Chauhan, Rashmin Prajapati, "Image Encryption Using Chaotic Cryptosystems and Artificial Neural Network Cryptosystems: A Review," International Journal of Scientific & Engineering Research, Vol. 5, Issue 5, May 2014.
9. Ping, P., Mao, Y., Lv, X., Xu, F., Xu, G, "An image scrambling algorithm using discrete Henon map," 2015 IEEE International Conference on Information and Automation, pp. 429–432. IEEE (2015).
10. M. Pitchaiah, Philemon Daniel, Praveen, "Implementation of Advanced Encryption Standard Algorithm," International Journal of Scientific & Engineering Research Volume 3, Issue 3, March – 2012.
11. Rachh, R.R, Anami, B.S, Ananda Mohan, "P.V. —Efficient implementations of S-box and inverse S-box for AES algorithm," TENCON 2009 IEEE Region 10 Conference, Nov. 2009, pp. 1–6.
12. Snehal Kundlik Waybhase & Prashant Adakane, "Data Security using Advanced Encryption Standard," International Journal of Engineering Research & Technology (IJERT). Vol. 11 Issue 06, June-2022.
13. Diaa, S., E, Hatem M. A. K., & Mohiy M. H. "Evaluating the Performance of Symmetric Encryption Algorithms," International Journal of Network Security, Vol. 10, No. 3, (pp. 213–219).
14. Nandhini Subramanian, Omar Elharrouss, "Image Steganography: A Review of the Recent Advances," IEEE Access DOP: January 25, 2021.
15. R. Pakshwar, V.K. Trivedi, V. Richhariya, "A survey on different image encryption and decryption techniques," International Journal of Computer Science, Information Technologies. Volume 4, issue 1, 2013, pp. 113–116, 2013.
16. Mansi S. Subhedar, Vijay H. Mankar, "Current status and key issues in image Steganography: A survey, Computer Science Review," Volumes 13–14, November 2014, pp. 95–113.
17. A. Westfeld, A. Pfitzmann, "Attacks on steganographic systems, International Workshop on Information Hiding," Volume 1768, pp. 61–76, 1999.
18. A. Belazi, M. Khan, A.A.A. El-Latif, S. Belghith, "Efficient cryptosystem approaches: S-boxes and permutation–substitution based encryption," Nonlinear Dynamics. Volume 87, Issue 1, pp. 337–361, 2017.
19. Sikha Bagui et al. (2015), "Database Sharding: To Provide Fault Tolerance and Scalability of Big Data on the Cloud," International Journal of Cloud Applications and Computing, 5(2), 36–52.
20. B. M. Abdelhafiz and M. Elhadef, "Sharding Database for Fault Tolerance and Scalability of Data," 2021 2nd International Conference on Computation, Automation and Knowledge Management (ICCAKM), Dubai, United Arab Emirates, 2021, pp. 17–24. https://doi.org/10.1109/ICCAKM50778.2021.9357711.

Multi Crop—Multi Disease Detection

Srinu Banothu, M. Bhavani, G. Bhadri, and M. UdayKiran

Abstract The financial system in India is heavily dependent on agriculture, making it the foundation of the country's economy. In India, an agricultural nation, more than 55% of people depend on agriculture for their livelihood. The attack of various diseases brought on by bacteria and pests that cannot be seen with the naked eye is causing a lot of problems for the agriculture industry today. In the existing system, the CNN model is built to classify the diseases of paddy crops with 94% accuracy. In this study, an app was created utilizing CNN, which detects the diseases and provides appropriate pesticides. A model that makes use of convolution neural network architecture has been created to make it easier to identify crop diseases from images of leaves. Paddy crop, which is mostly produced in India; tomato crop; and cotton crop, which holds a unique position among all crops and is also referred to as "white gold." These three crops have been taken into consideration for this study. The dataset included six types of diseases. Each crop has two types of diseases. The proposed model achieves 96% accuracy.

Keywords Deep learning (DL) · Multiple crops · Convolution neural network (CNN) · Remedies

1 Introduction

Knowledge of agriculture sectors is critical for contributing to the country's development. Agriculture is a one-of-a-kind source of riches that develops farmers. Agriculture growth is both a need and a requirement in the global market for a strong country. Population in the world is increasing at an exponential rate, demanding tremendous food production during the next 50 years [1]. Farmers are facing many losses. These losses are the result of many issues. The issues are things such as

---

S. Banothu (✉) · M. Bhavani · G. Bhadri · M. UdayKiran
Computer Science and Engineering, Vignan Institute of Technology and Sciences, Hyderabad, Telangana, India

A. Patel et al. (eds.), *Advances in Machine Learning and Big Data Analytics II*, Springer Proceedings in Mathematics & Statistics 442,
https://doi.org/10.1007/978-3-031-51342-8_21

Fig. 1 Brown spot

climate change, decreasing groundwater levels, various diseases, etc. The biggest issue for farmers is the incidence of numerous diseases on their crops. Diseases, once born, slowly spread throughout the crop and damage the entire crop, causing farmers to lose money and decreasing the amount of food available. A crop disease causes a significant loss in the quantity and quality of agricultural produce. Disease categorization and analysis is a critical problem for agriculture's maximum food output. Information regarding different types of crops and illnesses occurring at each level, as well as its analysis at an early stage, plays an important and dynamic role in the agriculture industry. Frequent monitoring may aid in early disease discovery and, as a result, crop loss reduction. An autonomous plant disease diagnosis system based on visual signs can provide a smart agriculture solution to such difficulties. With help of the image processing, deep learning, and machine learning approaches researchers have developed a variety of solutions for plant disease diagnosis.

Rice contains carbohydrates and it is the source of energy. As the population of the world increases, the consumption of rice also increases. Thus, rice production needs to be increased. In paddy crops, two major diseases are the reason for less production. The two diseases are bacterial blight and brown spot.

Bacterial blight is a disease of the paddy leaf caused by the bacterium Xanthomonas oryzae pv. oryzae. The signs of this disease include leaf drying, yellowing, and seedling wilting. Bacterial oozing may be spotted on the leaves in the early morning. This is one of the most devastating diseases, causing a 70% yield reduction. The disease of brown spot is due to the Cochliobolusmiyabeanus fungus, which mostly affects the leaf sheath, spikelets, and leaves.

A disease symptom begins with a spot of brown color and progresses to oval-shaped patches. This type of illness results in paddy loss quantitatively and qualitatively. Figures 1 and 2 show unfolded paddy leaves suffering from two different kinds of diseases [6].

[2] Tomatoes are the most common produce in the world, and they can be found in a variety of ways in any kitchen, regardless of cuisine. India was rated second in tomato output. Unfortunately, the quality and quantity of tomato crops suffer as a

Fig. 2 Bacterial blight

result of different illnesses. The diseases that mostly occurred were early blight and leaf mold shown in Figs. 3 and 4.

The cotton is one of the most predominant fibers. Cotton has a principal influence on the economies of agriculture and industry in the world. Cotton is a principal element in the building of the textile industry. The cotton plant disease often occurs on the leaves of 80-90% of cotton plants. Many cotton leaf diseases, such as bacteria blight, Alternaria, Foliar disease, and others, have been identified, these reduce the volume and standard of cotton produced in substantial quantities. The most occurred diseases are bacterial blight and cotton curl shown in Figs. 5 and 6.

2 Literature Survey

[1] The article describes a solution for illness detection using deep learning. To identify and categorize illnesses, an approach using convolutional neural networks is utilized. In this model, three convolutional layers and three maximum pooling layers come before the following two fully connected layers. The experimental results show that the proposed model outperforms VGG16, InceptionV3, and Mobile Net,

Fig. 3 Tomato early blight

Fig. 4 Leaf mold

three pre-trained models. The average classification accuracy of the proposed model for the nine sickness classes and one healthy class is 91.2%, ranging from 76% to 100% depending on the class.

[2] Computerized tomato leaf illness detection is first accomplished using traditional techniques such as machine learning and image processing, it gives

Fig. 5 Cotton bacterial blight

Fig. 6 Cotton curl

smaller accuracy. Deep learning-based classification is introduced to increase prediction accuracy. This article offers a comprehensive assessment of recent work on the subject of identifying tomato leaf diseases utilizing machine learning, image processing, and deep learning techniques. Also, talk about the techniques used and the adopted deep learning frameworks, as well as the private and public datasets that are accessible to recognize tomato leaf diseases. As a result, advice is given on how to identify high-quality methods to improve forecast accuracy. The struggle found in keeping the machine learning and deep learning approaches in practice is then highlighted.

[3] The system for recognizing rice plant maladies is a computer vision-based strategy that uses deep learning, machine learning, and image processing techniques to lessen the need for conventional approaches in preventing diseases including brown leaf spot, bacterial leaf blight, and fake smut in rice crops. To establish the sick portion of the paddy plant, picture pre-processing, and image segmentation are used, with the illnesses described above being diagnosed only based on their visual contents. To detect and categorize various types of paddy plant diseases, a combination of a support vector machine classifier with convolutional neural networks is utilized.

[4] In this study, a disease diagnosis app was created utilizing CNN and a dataset of 2000 leaf photos that takes the parameters into account. The suggested system will act as a visual interface for keeping track of cotton leaf disease. The plant illnesses were correctly recognized by the app with an accuracy rate of 92.5%.

[5] G. Jayanthi suggested a model for classifying rice diseases automatically using image processing methods. This paper discusses in detail different image classification algorithms.

[6] In this work, we suggest a CNN-based classification scheme for periodontal disease. The periodontal and non-periodontal pictures were the data that were to be used. To increase the effectiveness of data learning, data processing methods including zero centralizing, cropping, and resizing are applied. This research proposes a CNN structure with four outputs based on periodontal status and a size picture as input data.

[7] Arumugam, A. suggested a predictive approach-based model. This model uses data mining techniques to increase the productivity of paddy crops. Their research attempts to offer a predictive modeling technique that will assist farmers in achieving high yields from their paddy crops. They have employed decision trees and clustering as machine-learning techniques.

R. Rajmohan et al. [8] have suggested a method that uses CNN and SVM classifiers to identify the disease that the paddy crop is afflicted with. Their model makes use of the SVM classifier for feature extraction and image removal. They have taken datasets consisting of 250 images each for training and testing.

K. Jagan Mohan et al. [9] provide a model for determining the plant disease that it is afflicted with. They have extracted the features with the use of image processing. After that, the features are collected, and the model uses the SVM and K Nearest Neighbors classifiers to use the characteristics to identify the disease.

[10] In this paper, a novel fast and accurate image processing-based method for grading plant disease is developed. They segmented leaf regions using Otsu segmentation. The plant diseases are graded by calculating the quotient of disease spots and leaf area.

3 Proposed Work

An automated system is a web app that CNN created to offer a more effective method of disease identification and to recommend the best pesticide. The graphic depicts the approach as being illustrated by a block diagram shown in Fig. 7. Fundamentally, it starts with the phases of picture capture and image pre-processing. Then the features are extracted from the images and fed to the CNN model. It involves mainly three phases: the training phase, the validation phase, and the testing phase for categorization.

3.1 Dataset

We collect the data from Kaggle. The name of the dataset that we collect is the Plant Disease Classification Merged Dataset. This dataset, which consists large number of images, belongs to various plant diseases. We collect diseased leaf image data from the above dataset which is required for this research. A total of 2974 photos were taken, including the following crops and their respective diseases shown in Table 1.

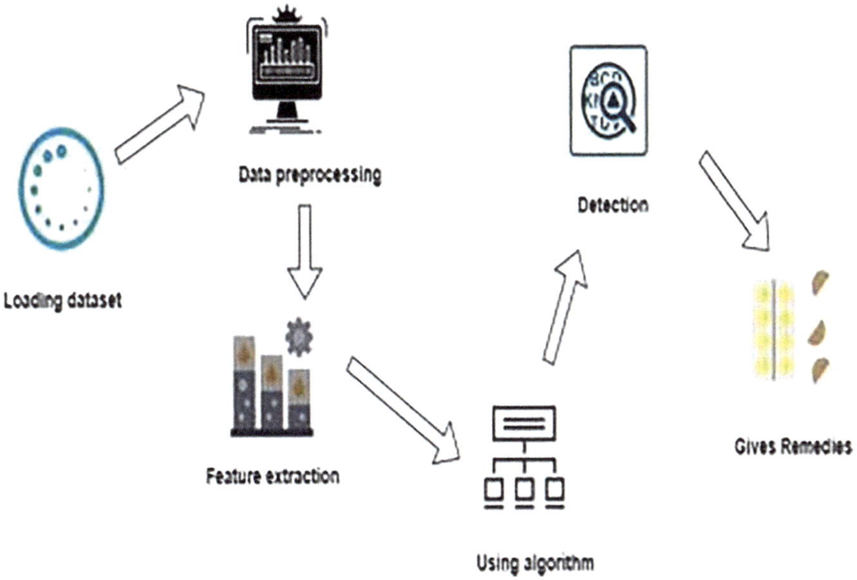

Fig. 7 Block diagram

Table 1 Crops and their respective disease

S. no.	Crop name	Disease name
1	Paddy	Bacterial blight
		Brown spot
2	Cotton	Bacterial blight
		Curl virus
3	Tomato	Early blight
		Leaf mold

3.2 The Architecture of CNN (Convolutional Neural Network)

Deep learning is a part of machine learning, and these two belong to Artificial Intelligence. The convolutional neural network is an algorithm that belongs to deep learning. The deep learning algorithms consist of many layers. Deep learning algorithms are mainly used to work with the input data like images, videos, audio, etc.

The Convolutional Neural Network (CNN) algorithm basically consists of three layers. There are three of them: the input layer, the hidden layer, and the output layer. There can be any number of hidden layers in the CNN algorithm. The hidden layers include a convolution layer, pooling layers, and fully connected layers, as shown in Fig. 10.

3.3 Convolution Layer

To extract features, the convolutional layer makes use of the image's information's local correlation.

The convolution operation process is shown in Fig. 8.

There is a kernel in the upper-left corner of the image. The pixel values covered by the kernel are multiplied by the kernel values, the products are added, and finally, the bias is applied. As seen in Fig. 8, the process is repeated until all potential areas within the image are filtered. One pixel is moved over the kernel.

3.4 Activation Function

An activation function is the final component of the convolutional layer that contributes to the increase in the nonlinearity of output. There are numerous activation functions, some of which are linear and some of which are nonlinear. ReLu (Rectified Linear Unit), Softmax, and tanh functions are frequently employed as activation functions in convolution layers. The Softmax function is described as a combination of multiple sigmoids. In the case of multi-class classification, it is

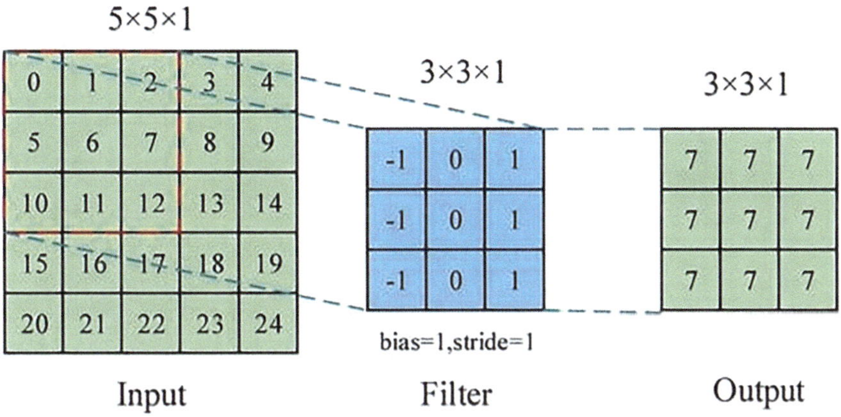

Fig. 8 The convolution operation procedure

typically utilized as an activation function for the final layer of the neural network. The probability for each class is returned by the SoftMax function.

Mathematically, RELU can be represented as:

ReLU

$f(x) = \max(0, x)$

Mathematically, Softmax can be represented as:

Softmax

$$\text{soft}\max(z_i) = \frac{\exp(z_i)}{\sum_j \exp(z_i)}$$

3.5 *Pooling Layer*

The pooling layer simultaneously makes the model invariant to translation, rotation, and scaling by sampling features from the upper layer feature map. Maximum pooling, minimum pooling, and mean pooling are three distinct types of pooling. In the proposed work, we used maximum pooling.

The pooling operation process is shown in Fig. 9.

Maximum pooling is the process of dividing the input image into multiple rectangular regions according to the filter's size and delivering the maximum value for each region. As for average pooling, the output is the average for each region. Convolutional and pooling layers frequently appear alternately in applications. Every neuron in the completely associated layer (fully connected layer)is associated with the upper neuron, and the multi-layered highlights are coordinated and changed

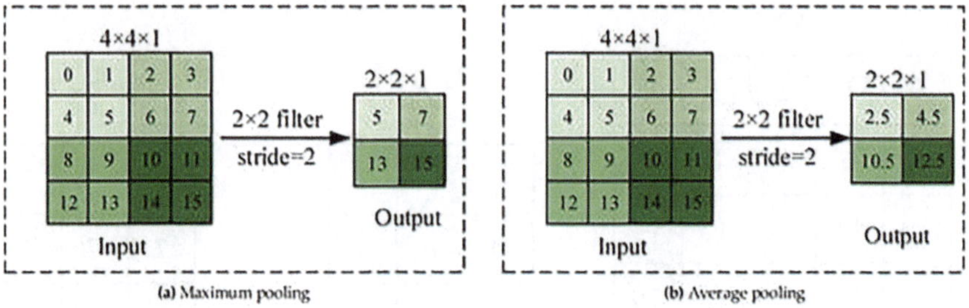

Fig. 9 The pooling operations process

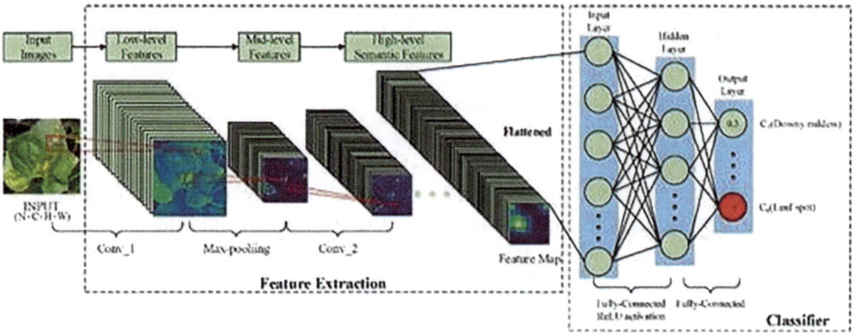

Fig. 10 Convolution Neural Network

over into one-layered highlights in the classifier for characterization or detecting the tasks.

The basic process of Deep Learning is shown in Fig. 10, using the task of disease classification on the surface of cotton leaves as an example. In Fig. 10, to extract features, primarily convolutional, max-pooling, and full connection layers, we employ a CNN-based architecture. The core technique for extracting features from images of cotton plant leaves is the convolutional layer. Some texture and edge information are extracted using the shallow convolutional layer, complicated texture and some semantic information are extracted using the middle layer, and high-level semantic features are extracted using the deep layer. After the convolutional layer, a max-pooling layer is employed to preserve the crucial information in the picture. At the end of the architecture is the classifier. There are complete connection layers in a classifier. The significant level semantic items extracted by the component (feature) extractor are arranged using this classifier.

In Fig. 10, to obtain the features, a set of images is fed into the feature extraction network, and the feature map is then flattened into the classifier for disease classification.

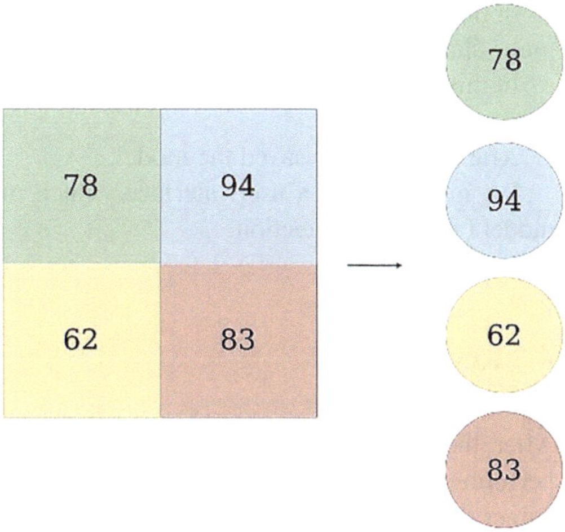

Fig. 11 Flattening

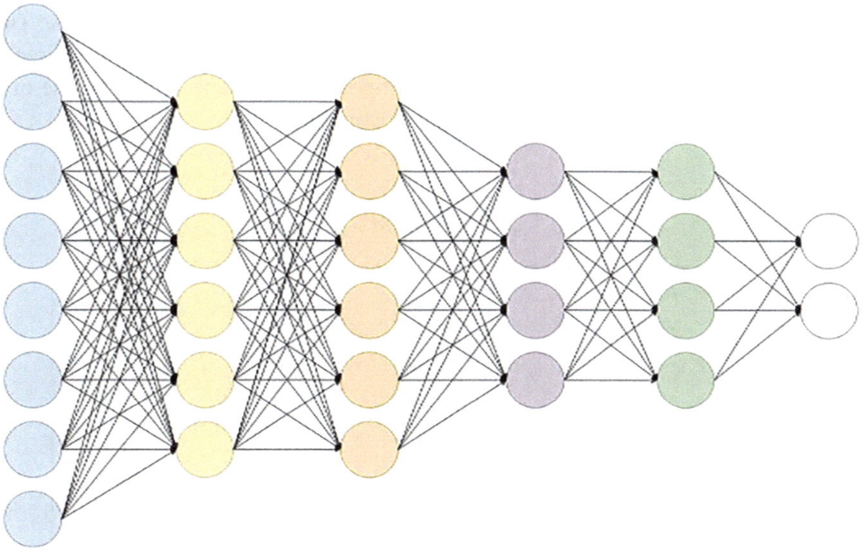

Fig. 12 Fully connected layer

3.6 The Fully Connected Layer

The term "fully connected layers" refers to the network's last few layers.

The earlier layer's feature map is flattened to a vector. Flattening means converting the NXN matrix into one dimension shown in Fig. 11. This flattened vector is then connected to a few fully connected layers.

Complex correlations between high-level features are captured by feeding this vector into fully connected layers shown in Fig. 12. This layer produces a feature vector with only one dimension as its output.

In this proposed method, we use the Softmax function in the last layer as it is a multi-classification model.

In this proposed method, we use 13 layers in CNN and train the model with 30 epochs.

After training, we saved the model.

We implemented a user interface, which makes it easy for users to utilize the model for disease detection.

4 Results

After the building and training of the CNN model, validation and testing are performed. The training and validation accuracy and training and validation loss values are plotted in a graph as shown in Figs. 13 and 14.

The accuracy of the proposed model is 96.875%. The accuracy of the existing CNN model, which classifies diseases of a single crop (paddy crop) is 94.12%. The user can easily upload the leaf image through the user interface to know the disease of a leaf and also provides remedies. The proposed model detects the disease of a leaf with higher accuracy compared to the existing one.

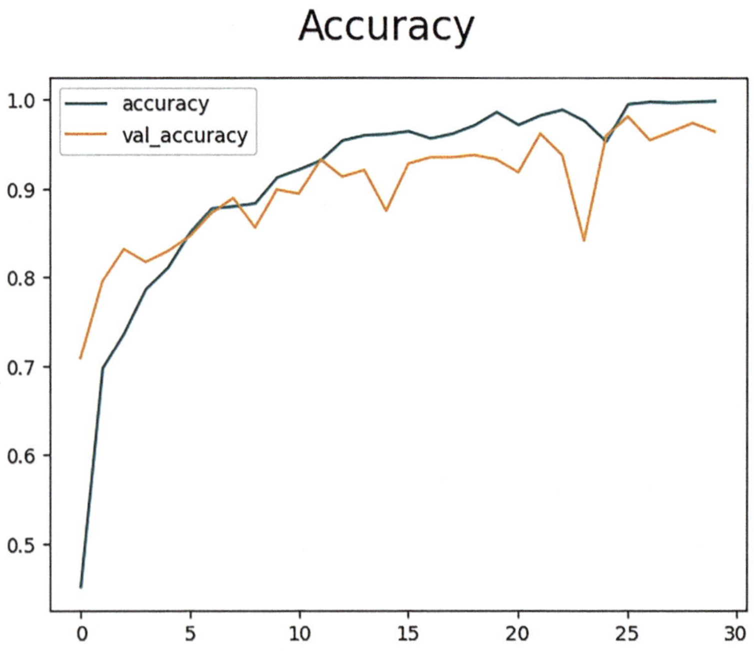

Fig. 13 Accuracy graph

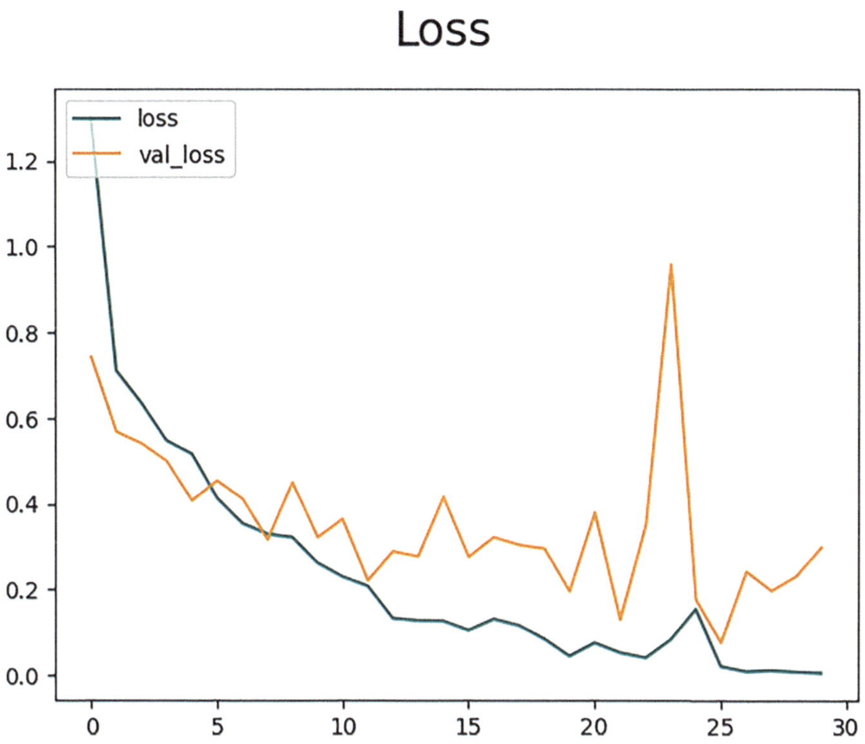

Fig. 14 Loss graph

5 Conclusion

Agriculture is important to the survival of all people and wildlife, so it must be protected and preserved. Lack of technical skills and knowledge in agriculture can cause serious plant health problems. In recent years, the significant involvement of the IT industry in the agricultural sector has transformed the process of detection and diagnosis of several deadly diseases.

This paper followed a CNN-based approach to classifying and detecting plant diseases from leaf images. The experimental results achieved the prediction of disease of the given test dataset. The model successfully detected diseased leaves. The best results were attained with a minimum amount of computing work, demonstrating the effectiveness of the suggested algorithm for identifying and categorizing leaf diseases. The ability to detect plant diseases at an early or starting stage is another benefit of this approach.

References

1. Wan-jie Liang, Hong Zhang, Gu-feng Zhang et al., Rice Blast Disease Recognition Using a Deep Convolutional Neural Network. Sci Rep 9, 2869 (2019). https://doi.org/10.1038/s41598-019-38966-0
2. International Conference on Computational Intelligence and Data Science (ICCIDS 2019)
3. ToLeD: By using a Convolution Neural Network, Tomato leaf disease detection https://www.sciencedirect.com/science/article/pii/S1877050920306906
4. Rajasekaran Thangaraj, S. Anandamurugan, P Pandiyan, Vishnu Kumar Kaliappan, Artificial intelligence in tomato leaf disease detection: a comprehensive review. J Plant Dis Prot 129, 469–488 (2022). https://doi.org/10.1007/s41348-021-00500-8
5. Kabir, Muhammad Mohsin, Abu Quwsar Ohi, and M. F. Mridha "Diagnosis of A Multi- Plant Disease Method Using a Convolutional Neural Network." arXiv preprint 2011.05.151 (2020).
6. Haridasan, A., Thomas, J., and Raj, E.D. Paddy plant disease detection and classification using Deep Learning System . *Environ Monit Assess* **195**, 120 (2023). https://doi.org/10.1007/s10661-022-10656-x
7. 2020 International Conference on Recent Trends in Electronics, Information, Communication, and Technology (RTEICT) | 978-1-7281-9772-2/20/$31.00 2020 IEEE. https://doi.org/10.1109/RTEICT49044.2020.9315634
8. K. Dakshinya, M. Roshitha, P. A. Raj, and C. Anuradha, "Cotton Disease Detection," International Conference on Artificial Intelligence and Smart Communication (AISC)- 2023, Greater Noida, India, 2023, pp. 361-364. https://doi.org/10.1109/AISC56616.2023.10084992.
9. Suraksha, I.S., Sushma, B., Sushma, R.G., Susmitha, K., and Uday Shankar, S.V., "Paddy Crops Disease Prediction Using Data Mining and Image Processing Techniques", International Journal of Advanced Research in Electrical, Electronics, and Instrumentation Engineering, Vol. 5, Issue. 6, May 2016.
10. Rajmohan, R.,Pajany, M., Rajesh, R., Raghuraman, D., and Prabu, U., "Crop Disease Identification Using Deep Convolutional Neural Networks and SVM Classifier", International Journal of Pure and Applied Mathematics, Vol. 118, No. 15 (2018),
11. I Farmer. Available online: https://ifarmer.asia/ (accessed on January 11, 2022).
12. V. Moysiadis, P. Sarigiannidis,V.Vitsas, A. Khelifi, Smart farming in Europe. *Comput. Sci. Rev.* **2021**, *39*, 100345.
13. S. Skawsang, M. Nagai, N. K. Tripathi; Soni, P. Predicting rice pest population occurrence with satellite-derived crop phenology, ground meteorological observation, and machine learning: A case study for the Central Plain of Thailand. *Appl. Sci.* **2019**, *9*, 4846.
14. G.Fenu,Malloci, F.M. LANDS DSS: A Decision Support System for Forecasting Crop Disease in Southern Sardinia *Int. J. Decis. Support Syst. Technol., IJDSST* **2021**, *13*, 21– 33.

Avian Soundscape Analysis Using Machine Learning

V. S. S. P. L. N. Balaji Lanka, A. Phani Krishnaja, A. Vaishnavi, and M. Yoshitha

Abstract Bird species identification is a fundamental task in ornithology research, it sometimes can be a challenging and time-consuming process, particularly with large collections of audio signals. In this paper, we propose a mechanism based on audio signal processing, and machine learning to facilitate the identification of bird species. Our proposed mechanism involves two stages. In the first stage, we constructed an ideal dataset of sound recordings of different bird species which were subjected to various sound preprocessing techniques such as pre-emphasis, framing, silence removal, and reconstruction. In the second stage, the input is sent to our CNN model, and predictions are made.

Keywords Auditory bird analysis · Bird sounds · Avian identification · Convolutional neural network · Spectrograms

1 Introduction

The conservation of various species worldwide is crucial for humanity, yet information on many species remains insufficient, particularly in countries with high biodiversity such as tropical regions. The task of classifying birds within new or existing species is typically performed by many bird researchers, rangers, or ecological consultants, and is a time-consuming and challenging task due to the large volume of birds and species. Visual analytics, which involves using interactive visual interfaces to facilitate analytical reasoning, is a growing field that is being used in many areas to help understand large volumes of data. Ornithologists are potential users of visual analytics systems for ecological surveillance, biodiversity conservation, and species identification. Audio records, rather than bird photos, are commonly used in current practices, as birds can be difficult to photograph due

V. S. S. P. L. N. Balaji Lanka (✉) · A. P. Krishnaja · A. Vaishnavi · M. Yoshitha
Department of CSE, Vignan Institute of Technology and Science, Hyderabad, Telangana, India

A. Patel et al. (eds.), *Advances in Machine Learning and Big Data Analytics II*,
Springer Proceedings in Mathematics & Statistics 442,
https://doi.org/10.1007/978-3-031-51342-8_22

to their elusive nature, quick movements, and tendency to hide in trees or become frightened by human presence. This paper proposes a mechanism for visualizing large collections of bird records to enable ornithologists to analyze the relationships between bird species. The paper proposes a procedure with the help of a number of steps for building visualizations. The paper is organized into sections: the proposed method, flowchart, obtained results, conclusions, and future work.

2 Related Work

Automated bird species identification using sound is an area of key interest in bird watching. Four relevant papers provide a wonderful overview of this domain:

Ferreira, Cunha, and Gomes (2019) [6] review the methods and ideologies employed for bird species identification using sound analysis. They discuss different features used in signal processing and machine learning algorithms for classification, such as neural networks, support vector machines, and random forests.

Stowell and Plumbley (2014) [7] present an approach to bird species identification using unsupervised feature learning, allowing the system to learn discriminative features from the audio signals without manual annotation.

Potamitis and Ganchev (2016) [8] provide details of different techniques used for bird sound classification, viz. feature extraction, feature selection, and classification algorithms. They also cover the different types of datasets used for training and testing, as well as the evaluation metrics used to assess classification system performance.

Fagerlund, Van Dijk, and Waldenström (2020) [5] review recent advances in deep learning techniques for bird sound classification, including the different types of deep neural networks used, challenges, and limitations of using these techniques in real-world applications.

3 Existing System

In the existing system, we use Deep Learning techniques like LSTM. Deep learning techniques on sounds take a lot of time to train the model with unsatisfactory accuracy and high computation requirements.

3.1 Disadvantages

- These systems only provide the name of the birds, they are unable to identify further information about the species.
- The efficiency of LSTM algorithms is not suitable for larger datasets.

4 Proposed System

In this project, we propose a new model using ResNet, a CNN, to train the model, by extracting only the relevant parts of the signal and applying optimizations like Transfer Learning.

4.1 Advantages

- Neural networks can train faster and give more optimal results.
- Use of ResNet and Transfer Learning allows the network to learn fast and makes computation easier, enabling training on weaker hardware.

5 Methodologies

The procedure we followed in this paper is represented as a block diagram in Fig. 1.

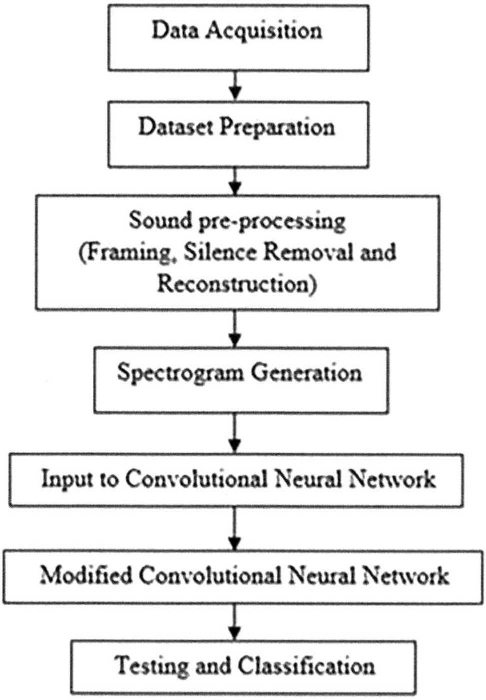

Fig. 1 Flowchart of the methodology. (Source: Chandu B, Munikoti A, Murthy KS, et al. (2020) Automated bird species identification using audio signal processing and Neural Networks. 2020 International Conference on Artificial Intelligence and Signal Processing (AISP). doi: 10.1109/aisp48273.2020.9073584)

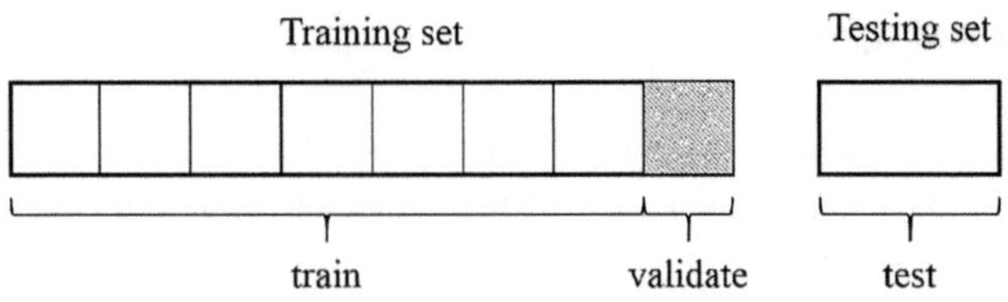

Fig. 2 The dataset partitions. (Source: Example of the division of the dataset for cross-validation and testing https://www.researchgate.net/figure/Example-of-the-division-of-the-dataset-for-cross-validation-and-testing_fig2_312255358)

5.1 Dataset

The quality of a dataset is crucial for machine learning-based approaches to problems. It should be accurate, precise, reliable, consistent, complete, comprehensive, relevant, granular, and unique. The dataset has been segregated into three that is a trained, validated, and finally a tested dataset which is of the ratio 70:10:20, then a cross-validation strategy is employed in order to ensure equal data contribution from all species (Fig. 2).

In addition, the authors collected a subset of the Xeno-canto [11] collaborative database, consisting of data of different bird species from throughout the world, contributed by. The BirdCLEF dataset [4] was also used to obtain audio files for birds missing from the Xeno-canto database.

5.2 Feature Extraction

The audio feature extraction process refers to selecting the relevant and useful information of any given audio signal. When we record sound using a microphone, the resulting audio signal can contain a wide range of frequencies. Depending on the source of the sound, such as bird vocalizations or other types of audio signals, there may be a preponderance of higher frequency energies. To ensure that we can capture and analyze these high-frequency components accurately, we often use a technique called predistortion. This method amplifies or enhances the high-frequency components while reducing the energy of other frequencies.

To get the desired result for us, it is necessary to emphasize the high frequency energy for the interest, while suppressing other frequencies that may interfere with our analysis. This is accomplished through the use of a simple first-order highpass filter, which allows us to remove lower-frequency components from the required signal. This filter is obtained for us with the following equation

$$y\ [n] = x\ [n]\ S^{*}\ x\ [n1]$$

which enables all of the given data points to be processed via the filter [2].

Utilizing this technique, we can isolate the high-frequency components of interest and analyze them more effectively, which can be especially useful in applications such as bird song detection and analysis, speech recognition, and music analysis.

5.3 Audio Signal Framing and Removal of the Silence

The audio signals or sound clips we get are basically dynamic signals that are subjected to change over time and contain various statistical properties. To extract useful information from the signal, we need to partition it into small pieces called frames based on their sizes and remove any unwanted periods of silence. The length or size of these frames can be obtained when you take the whole length or sizes of the frames of the signals and the time interval between two successive sampling points and we assume that each frame would be about 2.5% or 3% of the total amount of time of the sound clips [3].

Once these audio signals are framed, a threshold function is applied to remove silence periods. The threshold function is designed to separate desired audio signals from background noise or periods of silence, and any signal less than the threshold is taken as noise. This is repeated for all frames, and the threshold can dynamically vary to 7% of the peak amplitude of each frame.

By performing this silence removal process, we can effectively extract the desired audio signals from the overall signal and eliminate any periods of silence or noise that may interfere with our analysis. This can be especially useful in applications such as speech recognition, where accurate detection and analysis of spoken words is critical.

5.4 Understanding Transfer Learning

Transfer or Multi-task Learning is one of the techniques that is employed to improve performance by knowledge transfer from multiple related sources. This allows us to refrain from depending on a large group of targets for the data. It is akin to the human ability to learn something and being able to transfer that learned knowledge by experience to another person. It utilizes knowledge from known domains to maximize learning. It is not always useful, as the knowledge of the two domains might not overlap. Additionally, it might not always be constructive, and can also cause slowing down in new learning. It is most effective if used on small datasets with data/knowledge of similar domains [12].

5.5 Introduction to ResNet

ResNet and other trained neural networks and CNN models are used for classifying the images or the audio signals. On a GPU rather than a CPU, a CNN is quickly built. To train this network to categorize various types of photos, more than a million images were employed. From a picture as the input, a category output is generated. Dropout is used before the first and second layers are fully joined. The foundation for classifying an image or audio signal is by extracting its features which is known to be one of the simplest ways to train a classifier that is based on images or audio. The accuracy rate of Resnet is known to be high and may be utilized with Tensorflow, MATLAB, and Python. This application's main goal was to categorize bird species using the graphical representation images of the sounds of numerous types of birds. ResNet was used as a result (Fig. 3).

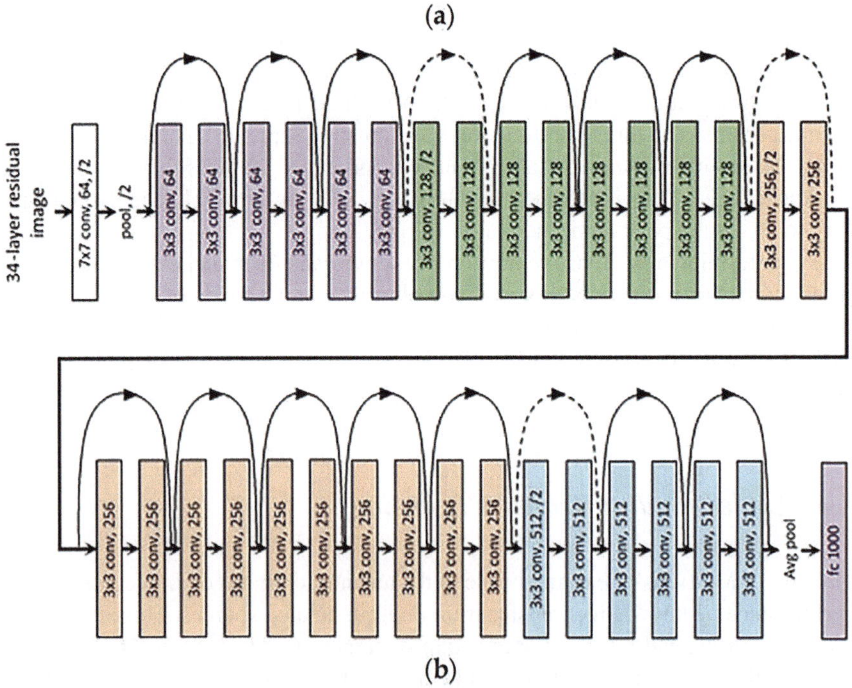

Fig. 3 ResNet architecture. (Source: Resnet architecture explained. In: Medium.https://medium.com/@siddheshb008/resnet-architecture-explained-47309ea9283d)

5.6 *Transfer Learning Using ResNet*

For classification, a trained CNN viz. ResNet was employed. For our application domain (Sounds of several bird species), we needed to re-train ResNet to recognize new types of data objects. Transfer learning is used here along with the ResNet to avoid the pain of building the neural network from the start.

We started by initially recording the audio clips of two different birds viz. American robins and doves. The graphical representation of these sounds was created and saved. The network is trained in the opposite direction because the classification objects differ from those on which ResNet was trained. The number of audio classes in the fully connected layer is from almost a thousand to 2175 and the categorical output we get after classifying is saved in the last or the output layer. The model was validated randomly with a large set of audio recordings after training was completed. Variations were made to parameters such as the rate of learning, how many epochs were required, the size of the batch, and the division of the trained and the tested data to improve accuracy and performance.

6 Practical Implementation

Once the convolutional neural network (ResNet) has been trained, we can finally obtain the bird species from the input of the audio clip of that bird. As a result, the network has been trained on a dataset that could work on a specific set of species when the input data are free of noise or perturbations. However, this is unusable when recording outdoors due to the addition of ambient noise. Numerous factors, including overlapping human voices, vehicular noise, and natural phenomena, can cause the noise. When predicting birds in a clean dataset, it is necessary to make sure that the model performs at the same accuracy as it did previously in a pure environment.

To ensure diversity and avoid the issue of overfitting the data, the dataset's audio clips of various bird species were selected at random and were converted to a fixed sampling rate instead of a variable sample rate. Additionally, the standard bit rates of 128 kbps and 320 kbps are used to ensure that the audio contains a lot of information at a small file size. Spectrograms are produced for each sample following the conversion of all of the clips into the bit streams and sampling rates that are desired [9]. The neural network is retrained with the help of these spectrograms. The model can be reused over iterations to classify the audio by transforming the signal into spectrograms after transfer learning is finished. Real-time testing of the system yielded classification results with an accuracy of up to 95% [10]. To operate the aforementioned system, a Graphical User Interface (GUI) was developed that included all of the functionalities to be carried out easily.

7 Results

In Fig. 4, we display an output screen that asks the user to upload any of the audio files of the birds. The user has to upload mp3 files only and then he needs to click on the submit button.

In Fig. 5, we display the output. The process is done in the backend which cannot be shown to the users as it is not necessary. In the output screen, we display the picture of the bird, its origin as well as its scientific name.

Fig. 4 Application screen

Fig. 5 Output screen

8 Conclusion

This study proposes a methodology for visual analysis of bird species by means of their sounds, which can assist ornithologists in species identification without the use of pictorial representations. The approach was tested on a set of 5 species, demonstrating its potential. It can be extended to a larger number of species with a higher rate of precision with more capable hardware and fine-tuning the parameters.

The study involved identifying five contrasting bird species by analysis of their sounds and creating spectrograms of them, which were further used to create a classification model. It consisted of real bird sounds from their natural habitat, including samples consisting of human and ambient noises. The accuracy of the system for five birds was 80%, which could be improved further by tuning all the parameters and by allowing for more dataset input with more capable hardware and using GPUs instead of CPUs.

9 Future Scope

There are enormous possibilities for future advancements and discoveries in both economic and scientific domains with this project. A mobile application could be developed, allowing users to record bird sounds with their smartphones, which would then be processed and analyzed by the app to provide the user with bird images, descriptions, and population distributions. Alternatively, the recording could be offloaded for processing on an HPC with a convolutional neural network (CNN) for better accuracy, performance, and results. This can alternatively be installed on compact computing devices such as an Arduino Uno or a Raspberry Pi, which could be carried by bird enthusiasts to track every bird they encounter, even the ones they do not spot. This data can also help in predicting bird migration and stay patterns, and it can also be used to identify locations of endangered species and provide a safe habitat for them.

References

1. *Kaggle: Your Machine Learning and Data Science Community*, http://www.kaggle.com/.
2. B, Chandu, et al. "Automated Bird Species Identification using Audio Signal Processing and Neural Networks." *International Conference on Artificial Intelligence and Signal Processing (AISP)*, 2020.
3. Bhor, Samruddhi, et al. "Automated Bird Species Identification using Audio Signal Processing and Neural Network." *International Conference on Sensing, Diagnostics, Prognostics, and Control (SDPC)*, 2019.
4. "BirdCLEF." *IUCN*, http://www.iucn.org/theme/species/our-work/birds.
5. Fagerlund, A., et al. "Automated bird species identification: A review of methods and applications." *Avian Biology Research*, vol. 13, no. 4, 2020, pp. 231–246.

6. Gomes, J., et al. "Bird species identification: A review of the methods and strategies." *Expert Systems with Applications*, 2019, 116, 363–381.
7. Plumbley, M. D., and D. Stowell. "Automatic large-scale classification of bird sounds is strongly improved by unsupervised feature learning." *PeerJ*.
8. Potamitis, I., and T. Ganchev. "A survey on bird sound classification techniques." *Applied Computing and Informatics*.
9. Reyes, Angie K., and Jorge E. Camargo. "Visualization of audio records for automatic bird species identification." *2015 20th Symposium on Signal Processing, Images and Computer Vision (STSIVA)*, 2015.
10. Sun, Rong, et al. "FFT Based Automatic Species Identification Improvement with 4- layer Neural Network." *13th International Symposium on Communications and Information Technologies (ISCIT)*, 2013.
11. "Xeno Canto." www.xeno-canto.org/.
12. Zhuang, Fuzhen, et al. "A Comprehensive Survey on Transfer Learning." 2019.
13. Chandu B, Munikoti A, Murthy KS, et al. (2020) Automated bird species identification using audio signal processing and Neural Networks. 2020 International Conference on Artificial Intelligence and Signal Processing (AISP). https://doi.org/10.1109/aisp48273.2020.9073584.
14. Example of the division of the dataset for cross-validation and testing https://www.researchgate.net/figure/Example-of-the-division-of-the-dataset-for-cross-validation-and-testing_fig2_312255358.
15. Bangar S (2022) Resnet architecture explained. In: Medium. https://medium.com/@siddheshb008/resnet-architecture-explained-47309ea9283d.
16. Debnath S, Roy P, Ali A, Amin M (2016a) Identification of bird species from their singing. 2016 5th International Conference on Informatics, Electronics and Vision (ICIEV). https://doi.org/10.1109/iciev.2016.7759992.

Smart Security and Surveillance System

G. Harish Reddy, M. Chetan Reddy, L. Uday Kumar Reddy, and R. Ramana Reddy

Abstract The surveillance systems in the majority of the areas are conventional, i.e., they are only capable of recording the footage. If any burglary occurs in a mall, theatre, or even in a house, then at the time of robbery there will not be any intimation by the security systems that something is going wrong, but this smart surveillance system intimates the same with a beep sound so that we can take the immediate action during the robbery. We can train the model by providing images of the person involved in the robbery or the person whom we want to identify. For this purpose, a Local Binary Pattern Histogram (LBPH) algorithm is used; with the help of these images, it can identify the same person in the image or footage we provide during the investigation. Also, there are many features such as face detection, motion detection in a particular area in the frame, lost item detection, and simple recording features. We are using the OpenCV module in Python. OpenCV is an open-source computer vision and machine learning software library, it has many algorithms that are useful to detect and recognize faces, identify objects, classify human actions in videos, and track camera movements.

Keywords Face detection · Motion detection · Stolen Detection · Security

1 Introduction

Surveillance refers to the supervision or keeping an eye on specific events or gatherings. So it is believed that the greatest alternative for monitoring and observing is video surveillance. Manual surveillance takes too much time and effort. We can quickly ascertain what is occurring at a certain location by employing this surveillance system, and we can simultaneously monitor numerous locations remotely [1].

G. Harish Reddy (✉) · M. Chetan Reddy · L. Uday Kumar Reddy · R. Ramana Reddy
Vignan Institute of Technology & Science, Hyderabad, Telangana, India

A. Patel et al. (eds.), *Advances in Machine Learning and Big Data Analytics II*,
Springer Proceedings in Mathematics & Statistics 442,
https://doi.org/10.1007/978-3-031-51342-8_23

It is challenging to keep an eye on the security of our homes and workplaces in the fast-paced environment we live in today. Consequently, the need for camera surveillance systems has skyrocketed. It is feasible to continuously monitor homes and workplaces for security reasons and preserve the data for later use by using these systems. However, these systems' primary flaws are manual monitoring and significant storage requirements.

Security and surveillance have a big impact on a lot of different areas of life. Security systems are strongly advised everywhere due to the rise in shady activities and thievery. The current surveillance system in many places is conventional, i.e., they are only capable of recording footage. The commonly used security devices are Retina sensors, Infrared sensors, RFID readers, and cameras. These are all less efficient and sometimes expensive methods for implementation. They do not have any extra security features such as an alarm when an object is lost or when they identify motion in a restricted area or a person or maybe a most wanted criminal or robber. So we developed a security system called the "Smart Surveillance System" which is the solution to the above-mentioned loopholes [2].

The "Smart Security and Surveillance System" is our proposed system. It is the solution to all the problems which were present in the existing surveillance system. This new surveillance system has several security features which are used to enhance the current security systems [3].

The features of this app are Stolen Detection, Motion Detection, Face Detection, Motion Detection in a particular area, and Normal Recording. The Stolen Detection security feature is used to detect which item is stolen from a frame in which the camera is installed. This feature warns immediately with the alarm sound right after the theft of an item. This feature can be used in jewelry shops and museums. Motion Detection feature is used to detect the motion in the entire frame where the camera is being placed. That is, it detects the motion anywhere in the frame. The Face Detection feature is used to detect the person whom we had trained for the application. Suppose we had missed the robber in a theft, then his photo is used to train the application, after that whenever that person is caught in the camera, it identifies and shows his name on the monitor. Motion Detection in a feature area is used to detect the person whom we had trained for the application. Suppose we had missed the robber in a theft, then his photo is used to train the application, after that whenever that person is caught in the camera, it identifies and shows his name on the monitor. Normal Recording is the same as the present surveillance system [5].

2 Methodology

The different features that can be performed using this project are:

1. Stolen Detection
2. Person Detection
3. Motion Detection
4. In and Out Detection

2.1 Stolen Detection

This function is used to identify the object that has been stolen from the webcam-visible frame. Meaning that it continuously examines the frames to see which item or object has been removed by the burglar.

The differences between the two frames are discovered using structural similarity. The first frame was taken before any noise occurred, while the second was taken after the noise stopped occurring in the frame.

The three crucial features are extracted from an image via the Structural Similarity Index Metric (SSIM):

(i) Structure.
(ii) Contrast.
(iii) Luminance.

These features act as the foundation for the comparison between the two photos.

The SSIM, which ranges in value from −1 to +1, is calculated using this system between the two images provided. A number of +1 denotes that the two photographs are identical or extremely similar, whereas a value of −1 denotes that the two images are highly dissimilar (Fig. 1).

2.2 Person Detection

Person Detection is a very useful feature of our project, It is used to find if the person in the frame is known or not. This process takes two steps:

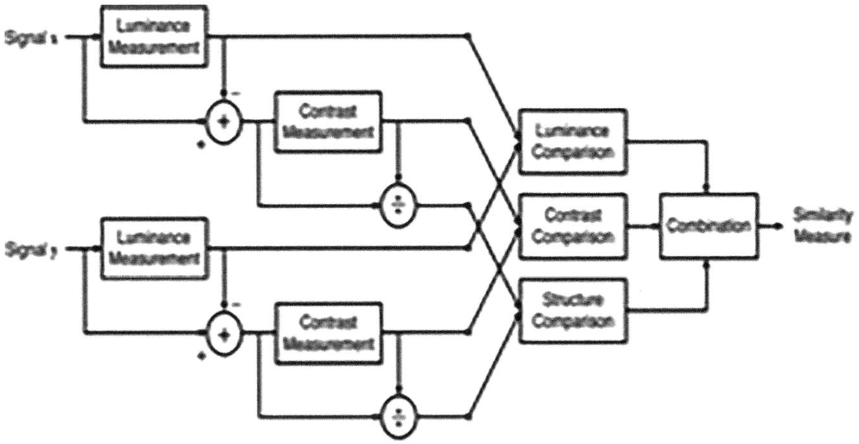

Fig. 1 Similarity measure

frame1

10	90	16	16
0	11	11	11
18	30	33	33
18	18	18	18

frame2

10	90	16	16
0	13	17	11
18	34	31	33
18	17	19	18

frame2 - frame1

0	0	0	0
0	2	6	0
0	4	-2	0
0	-1	1	0

abs (frame2 - frame1)

0	0	0	0
0	2	6	0
0	4	2	0
0	1	1	0

Fig. 2 Frame analysis

First, it recognizes the faces in the images.

Second, by using the LBPH face recognizer technique, use the trained model to predict who the individual is.

2.3 *Motion Detection*

The majority of CCTVs have this feature, which is used to detect noise in the frames; every frame is continuously analyzed and noise-checked. The subsequent frames check for noise. Simply put, we calculate the absolute difference between two frames. In this way, the difference between the two images is analyzed, and the motion's borders are found. If there are no boundaries, there is no motion; if there are, motion is there (Fig. 2).

Since every pixel in a picture has a brightness value, as you are probably aware, all images are simply integer values of pixels.

As a result, we just compute the absolute difference as the negative will be completely meaningless [6].

2.4 *In and Out Detection*

This feature is capable of detecting when someone enters or leaves a room. So, the process is as follows:

1. It starts by looking for noises in the frame.
2. Then, if anything happens, it determines if it comes from the left or the right side.
3. As a last check, if there is a motion from left to right, it will recognize it as entering and take a picture. Or vice-versa.

So, this feature does not involve any intricate mathematics.

Basically, to determine from which side the motion occurred, we first detect motion, then draw a rectangle over the noise, and finally, we check the coordinates to see if those points are on the left side. If they are, then the motion is classified as left motion.

3 Existing System

In the existing system, the CCTVs used in malls, theaters, or any public places are only capable of recording video footage with date and time. Here the CCTVs are connected to the computers with the help of wires which can only be used to watch and record the footage. There is no specialized software or any application with different functionalities or operations. A slightly better version of the wired system is the wireless system. Although the wireless surveillance system also does the same functionality as the wireless does, there is no enhancement in the security.

4 Proposed System

- Stolen Detection
- Motion Detection
- Person Detection
- Motion Detection in a Particular Area
- Normal Recording

Advantages:

- Enhanced security
- Reduced burden for cops
- User-friendly interface
- Easy controls
- Cost-effective

5 Results

5.1 Feature 1—Monitor

Stolen Detection This security feature detects which item is stolen from the particular frame in which the camera is installed. It warns immediately with an alarm sound right after the theft. This feature can be used in jewelry shops and museums.

Figures 3 and 4 show that it detects when a bottle is stolen, which is true.

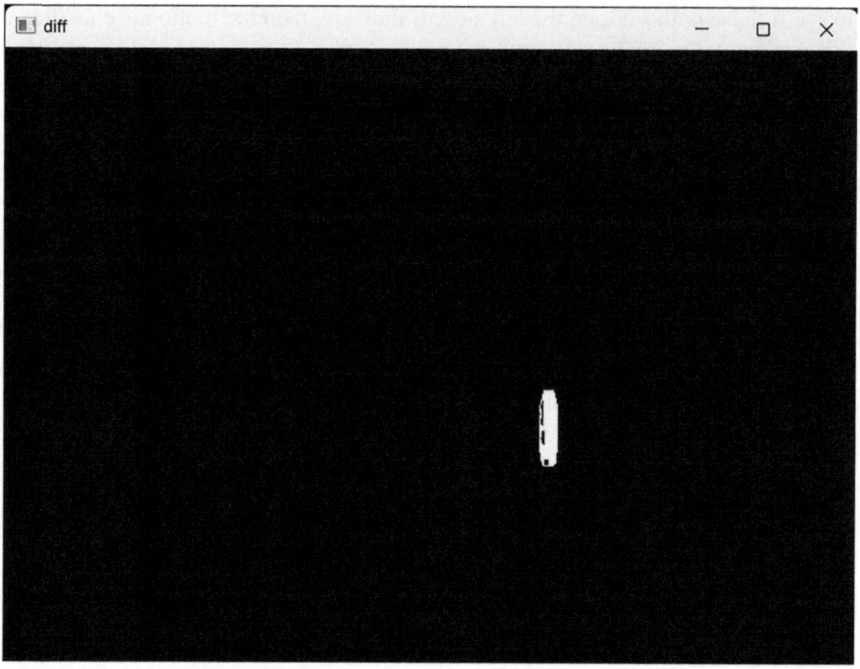

Fig. 3 Monitoring Stolen Detection—bottle stolen

5.2 Feature 2—Noise Detection

Motion Detection This feature is used to detect the motion in the entire frame where the camera is being placed. That is, it detects the motion anywhere in the frame.

This is a working captured output for No-Motion (Fig. 5) and Motion being detected (Fig. 6) by this application.

5.3 Feature 3—In-Out Detection

As shown in Fig. 7, it has detected me entering the room, and being detected as entered, saved the image locally.

Similarly, Fig. 8 shows that it has detected me exiting the room.

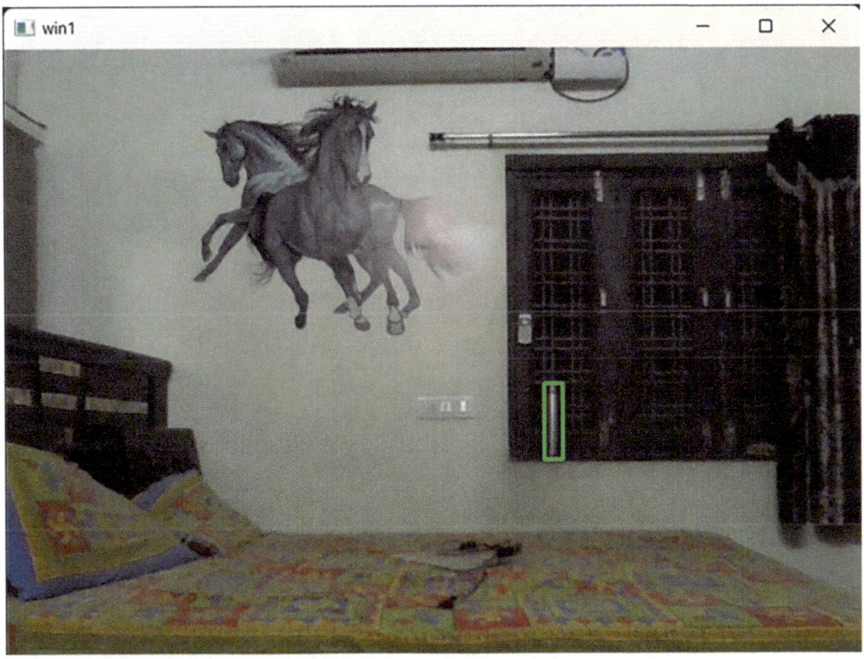

Fig. 4 Marking Stolen Detection—bottle stolen

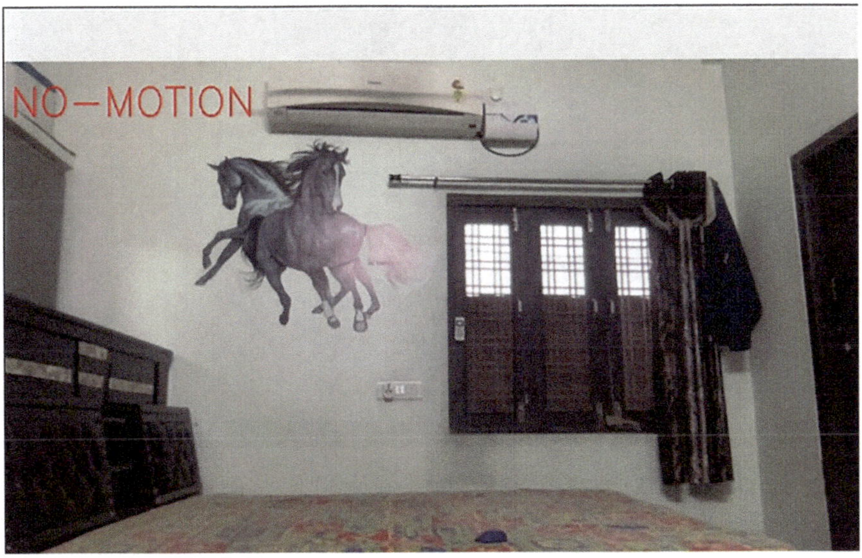

Fig. 5 Detecting the motion

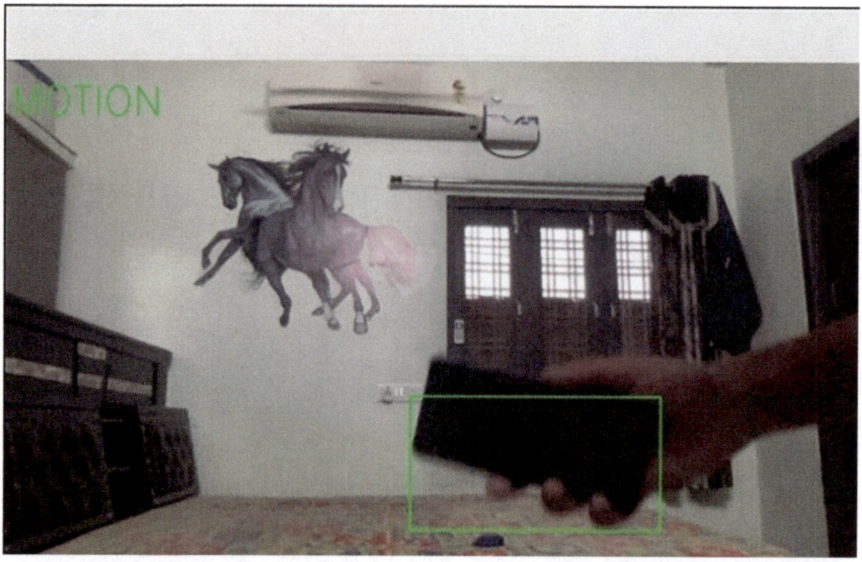

Fig. 6 Capturing No-Motion and Motion

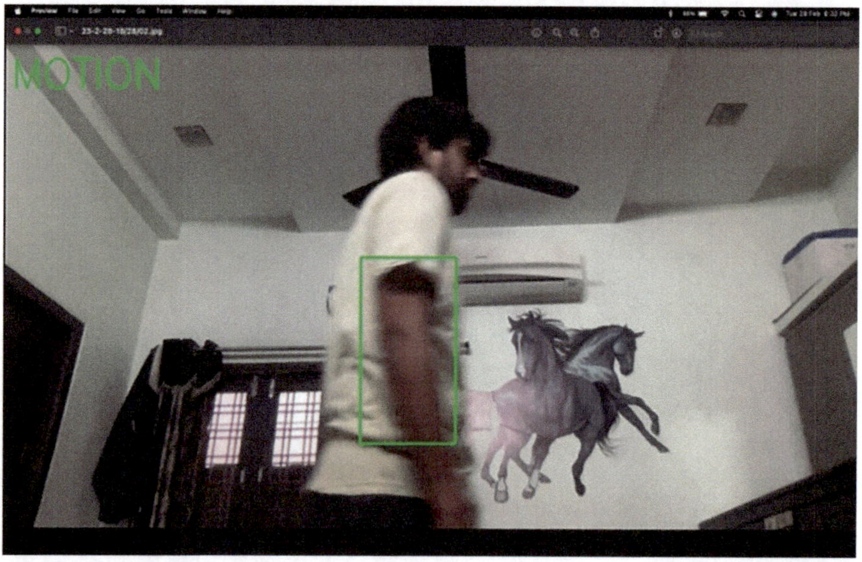

Fig. 7 Capturing In detection

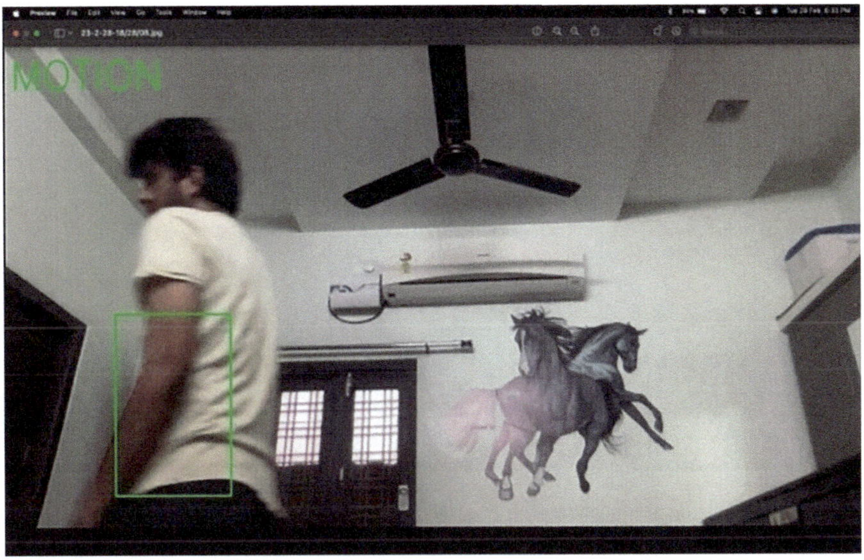

Fig. 8 Capturing Out detection

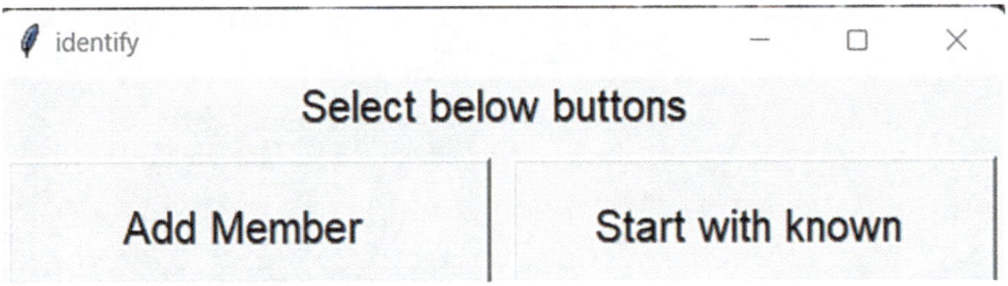

Fig. 9 Buttons for adding new members

5.4 *Feature 4—Face Identification*

Person Detection This feature is used to detect the person whom we had trained for the application. Suppose we had missed the robber in a particular theft, then his photo is used to train the application, after that whenever that person is caught in the camera, it identifies and shows his name on the monitor (Figs. 9 and 10).

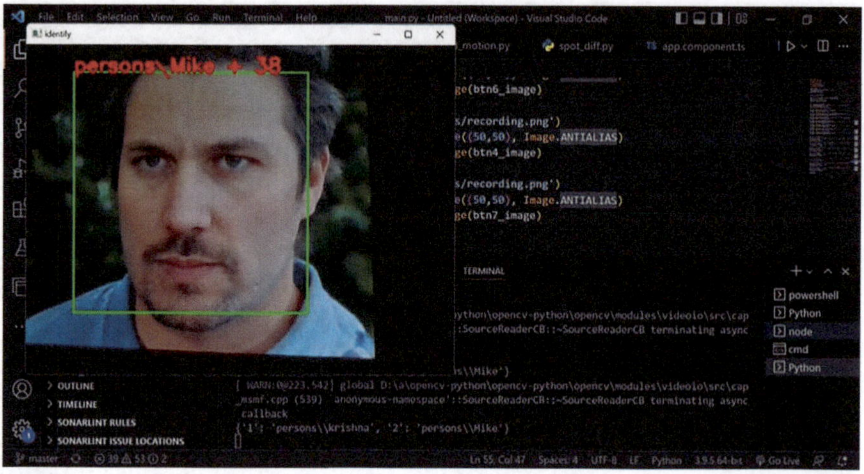

Fig. 10 Recognizing new members

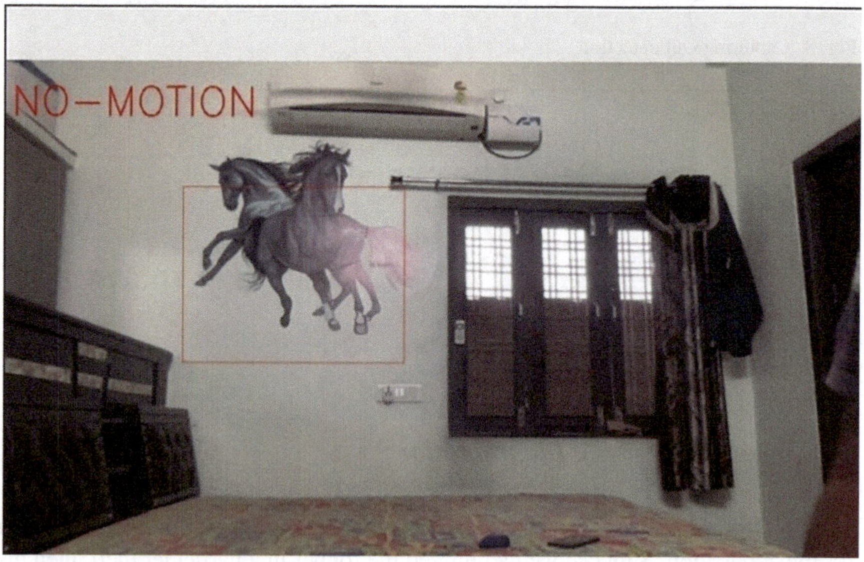

Fig. 11 Detecting No-Motion

5.5 Feature 5—Rectangle

Motion Detection In a particular area: This feature is similar to motion detection, but here we can specify which part of the frame should be motion-restricted or restricted area. Example "power supply room in an exhibition" (Figs. 11 and 12).

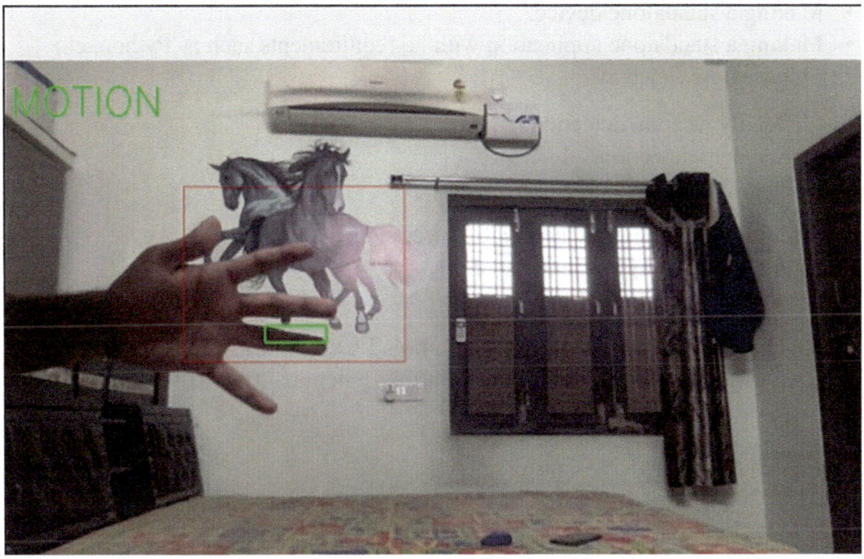

Fig. 12 Detecting Motion

5.6 *Feature 6—Normal Recording*

This is the normal recording like the present surveillance system.

6 Conclusion

The Smart Surveillance System has more security features when compared to the existing surveillance system. With the current surveillance system, so much manpower is needed to identify theft or crime. But with this project, manpower can be reduced to some extent. With the help of Machine Learning, it makes the task of identifying a particular person easier when compared to manually finding them. So, the surveillance systems need to be smart.

6.1 *Future Scope*

Grounded on the technology advancements similar being having the capability of small size but high processing power, this project can be astronomically used. Below are some unborn drills on this project.

- Making a standalone device.
- Making a standalone application with no requirements such as Python, etc.
- More features such as:
  - Deadly weapon detection.
  - Accident detection.
  - Fire detection, etc.
- Adding deep learning with a high-power device.
- Adding in-built night vision capability.
- Creating Portable CCTV.

Adding DL support would produce a broad scope in this project similar to DL, then we would be able to add up much further functionality.

References

1. M. S. D. Gupta, V. Menezes, and V. Patchava, "Surveillance and monitoring system using Raspberry Pi and SimpleCV."
2. B. Pattnaik, M. Sahani, C. Nanda, and A. K. Sahu, "Web-based online embedded door access control and home security system based on face recognition."
3. Muhammad Affan, Syed Umaid Ahmed, and Hamza Khalid "Smart Surveillance and Tracking System."
4. Chang Tsun Li, Leyva, Roberto, and Victor Sanchez. "Fast detection of abnormal events in location features." 2018 IEEE International Conference on Acoustics, Speech and Signal Processing (ICASSP). IEEE.
5. Jung I., Lee M. Smart Video Surveillance System Using Dynamics. In: Lu B.-L., Zhang L., Lee W., Lee G., Ban S.-W., Jung I., Kwok J., editors. Lecture Notes in Computer Science. Springer; Berlin, Germany.
6. Mrunal, Malekar. "Detecting motion of Activities of Surveillance Videos using Deep Learning."

Cyber Money Laundering Detection Using Machine Learning Methods

Deepthi Kamidi, Ravipati Vishnu Priya, Jayanth Alla, and Adivi Srikar

Abstract Today, tax evasion represents a serious danger not exclusively to monetary foundations yet in addition to the country. This crime is turning out to be increasingly refined and appears to have moved from the platitude of medication dealing to funding psychological warfare and most likely not failing to remember individual addition. Most global monetary foundations have been carrying out the enemy of tax evasion answers for battle venture misrepresentation. Be that as it may, customary insightful procedures consume various worker hours. As of late, information mining approaches have been created and are considered also fit procedures for identifying illegal tax avoidance exercises. Inside the extent of a joint effort project to foster another answer for the counter-tax evasion units in a global speculation bank, we proposed a straightforward and proficient information-digging-based answer for hostile to illegal tax avoidance. In this paper, we present this arrangement created as a device and show some fundamental trial results with genuine exchange datasets. We intend to utilize administered AI to characterize the exchanges as fake or not deceitful in light of information such as a change in equilibrium, approaching, and active unfamiliar and homegrown exchanges.

Keywords Money laundering · Criminal activity · Machine learning · Data mining · Tax evasion

1 Introduction

Tax evasion (ML) is a cycle that causes ill-conceived pay to seem genuine; this is likewise the interaction by which hoodlums endeavor to cover the genuine beginning and responsibility for continuing their crime. ML has been described by Genzman as a development that "deliberately partake in a money related trade with the profits

D. Kamidi (✉) · R. Vishnu Priya · J. Alla · A. Srikar
Vignan Institute of Technology and Science, Yadadri Bhuvanagiri Dist, Telangana, India

A. Patel et al. (eds.), *Advances in Machine Learning and Big Data Analytics II*,
Springer Proceedings in Mathematics & Statistics 442,
https://doi.org/10.1007/978-3-031-51342-8_24

of some unlawful activity fully intent on progressing or carrying on that unlawful activity or to stow away or veil the nature region, source, ownership, or control of these profits." Through tax evasion, crooks attempt to change over financial returns got from illegal exercises into "clean" reserves utilizing a lawful medium, for example, huge venture or benefits supports facilitated in retail or speculation banks. Identifying the board extortion is a troublesome errand while utilizing ordinary review methodology. To begin with, there is a lack of information concerning the qualities of the board extortion. Also, given its inconsistency, most evaluators come up short on experience important to recognize it. At last, administrators intentionally attempt to mislead reviewers. For such administrators, who appreciate the impediments of any review, standard evaluating systems might demonstrate deficiencies. These impediments propose that there is a requirement for extra insightful strategies for the compelling discovery of the executive's misrepresentation. It has likewise been noticed that the expanded accentuation on framework evaluation conflicts with the calling's position regarding misrepresentation recognition since most material fakes begin at the high levels of the association. Applying information mining to misrepresentation location as a feature of a routine monetary review can be testing and, as we will make sense of, information mining ought to be utilized when the potential result is high. With regards to extortion location for a given review client, the review group could pursue three significant choices:

- What specific types of misrepresentation (such as income acknowledgment, minimized obligations, etc.) should be kept in mind when planning a review for a certain client?
- What sources of data (such as diary entries, messages, and so on) may be used to demonstrate each type of misrepresentation?
- Which information mining methods (such as coordinated or undirected procedures) could be the most effective for finding anticipated extortion proof in the selected information?

Even if each of these questions has significant consequences on its own, solving them all together is an attempt to answer them.

2 Literature Survey

There are recent papers that dealt with the Hostile to Tax evasion expectation approach such as:

Kumar, S. Das, and V. Tyagi (2020), "Hostile to Tax Evasion Identification Utilizing Innocent Bayes Classifier." The authors used a modified dataset of 10,000 transactions for AML detection, leveraging a Naive Bayes classifier. They focused on data cleaning, statistical analysis, and mining processes to identify money laundering activities. The approach achieved an accuracy of 81.25%.

B. Raeesa and B. Ranjitha (2020), "A Hazard Score Examination Connected with Tax Evasion in Monetary Establishments Across Countries." This paper examines the reduction of financial crimes in the UAE due to anti-money laundering policies that align with the guidelines of the Financial Action Task Force (FATF). It appears to focus more on the regulatory framework and the application of AML in the context of international banking.

C. Arman Z. and James O. (2020), "A Solid Structure for Hostile to Tax Evasion Utilizing AI and Mystery Sharing." The authors proposed a robust framework using machine learning and secure data sharing for AML detection in banks. The system includes three phases: alerting auditors, machine learning-based detection, and generating Suspicious Activity Reports (SARs). This framework emphasizes collaborative data sharing among banks to improve detection.

D. Chih-Hua T. and Tai-Jung K. (2019), "Distinguishing Illegal Tax Avoidance Records." The authors introduced McLay's technique, which is an AI-based method for identifying suspicious money laundering accounts. The technique utilizes a two-stage confirmation process for suspicious accounts, focusing on review and accuracy, with an impressive audit speed of 26.3%.

3 Existing System

Although approaches extraordinarily fluctuate, numerous strategies accept tax evasion cases to be exceptions. Normally, these methodologies utilize unaided oddity discovery strategies to demonstrate licit ways of behaving and track down the occurrences that veer off from it. A few investigations even report that the ML approaches had the option to identify tax evasion designs that were beforehand obscure or not got by rule-based frameworks. Other extortion-related use-cases, for example, Mastercard misrepresentation and organization interruption discovery lead to exploring different avenues regarding AL.

4 Proposed System

We split the information into successive train and test datasets for all investigations. The train set incorporates all marked examples up to the 34th time-step (16,670 exchanges), and the test set incorporates all named tests from the 35th time-step, comprehensive, forward (29,894 exchanges). We train each directed model on the train set utilizing all elements and afterward assess them on the whole test set. To gauge execution over the long run. We utilize the sci-kit-learn execution of Brain Organization Benefits. We have proposed this AML with directed learning and got the most elevated precision for expectation as a condition of the workmanship technique.

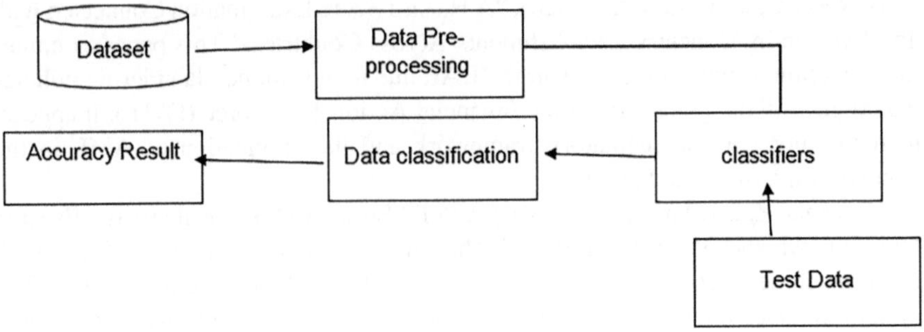

Fig. 1 System architecture

5 System Architecture

Figure 1 helps to understand the process of the architecture of the paper.

6 Methodology

Methodology in machine learning refers to the systematic approach and set of techniques used to design, develop, and evaluate machine learning models. It involves defining the problem, selecting appropriate data, preparing and preprocessing the data, selecting and optimizing the machine learning algorithms, evaluating the model performance, and deploying the models in actual-world applications.

- Collecting the data.
- Pre-processing the data.
- Extracting features.
- Assessment model.
- Flask.

6.1 Collecting the Data

Several pieces of research that were compiled from Visa commerce records provided the data for this essay. We place more emphasis on this stage than on selecting the subset of all pertinent data that you will use. Ideally, stores of data (models or discernments) for which you are absolutely positive that you comprehend the objective response are where ML problems begin. Data that you most definitely comprehend the objective response is referred to as verified data.

Dataset The dataset used by most papers have been old transactions received from financial institutions. That consists of traditional banking transactions. This synthetic dataset will help us test money laundering detection with very sparsely labeled data.

6.2 Pre-processing the Data

Preparing and translating raw data into a format appropriate for analysis and model construction is an important step in machine learning. Typical methods for data preparation include the following: Data pre-processing (Fig. 2) is needed for the entire data.

Data cleaning involves cleaning data. This entails dealing with duplicate values, addressing missing values, and eliminating outliers from the data.

Transforming the data involves scaling, normalization, and encoding the data to prepare it for analysis. For example, feature scaling techniques such as Standardization or Min-Max scaling are used to normalize data so that all features have equal importance during analysis.

By decreasing noise and redundancy, feature selection, which entails choosing a subset of the most important characteristics from the data, can enhance the performance of the model.

6.3 Extracting the Features

Feature engineering is the process of developing new features from already existing ones, which can increase the model's capacity for prediction.

Data integration is the process of merging information from various sources into one dataset.

Data reduction is the process of lowering the dimensions of the data, which can enhance the performance of the model and speed up computation.

The quality of the data preparation stage can significantly affect the performance of the final model, making it a crucial step in machine learning overall.

The process of turning raw data into processable numerical features is known as feature extraction.

6.4 Assessment Model

Model evaluation is a crucial step in the process of improving models. It helps in determining the best model to protect our information and how well the chosen model will work moving forward. It is not advisable to evaluate model performance

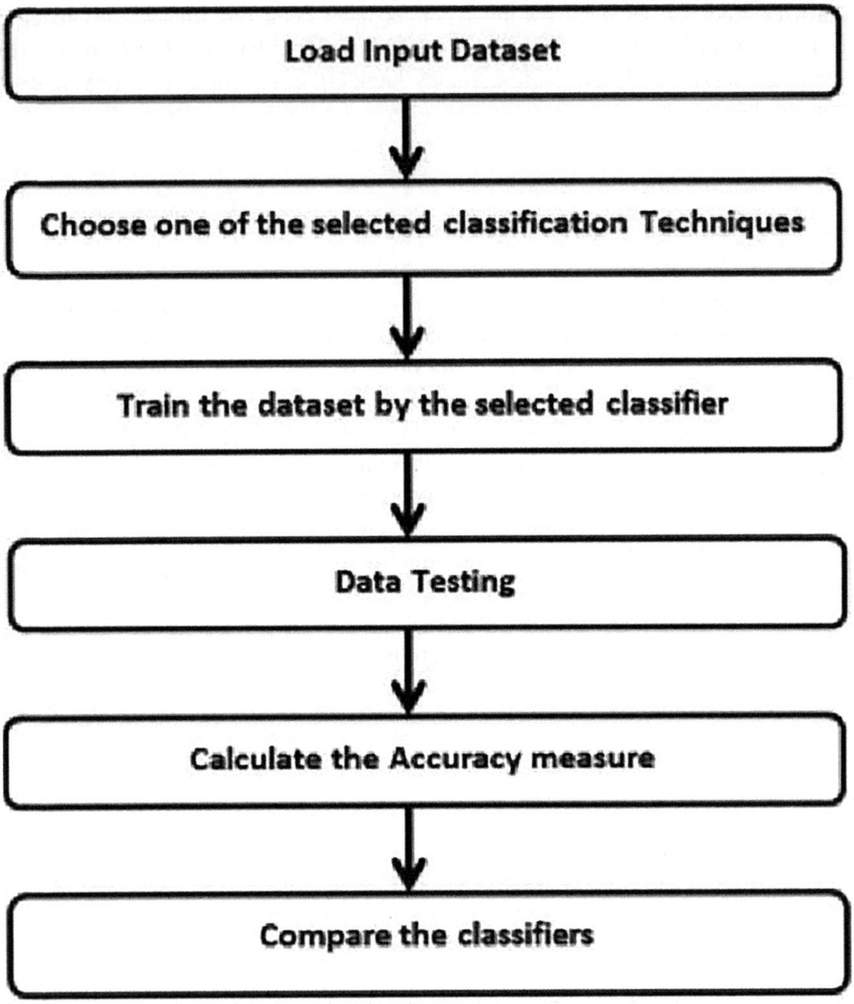

Fig. 2 Data pre-processing

using the data used for preparation because doing so would undoubtedly result in overly optimistic and overfit models. Respite and Cross-Support are two techniques for assessing models in information science. The two approaches compare model execution on a test set that is hidden from the model to prevent overfitting. Each gathering model's effectiveness is evaluated based on how it appeared in the middle. The anticipated outcome will come to pass diagrams that depict synchronized information.

6.5 Flask

Flask is smaller than the conventional Python-written web structure. It is given a microframework because it does not require clear-cut tools or libraries. It lacks the information base layer, structure support, and other elements where conventional limits from previously inaccessible libraries were expected. Nonetheless, Cup continues to release updates that can add application features as if they had already been completed in Container. The object-social mappers, the structure underwriting, the move-making due, the many open insistence motions, and two or three common system-related devices all have improvements.

Layouts are documents that contain static information along with placeholders for dynamic information. A layout is delivered with explicit information to create a final report. Cup utilizes the Jinja format library to deliver layouts.

In your application, you will utilize layouts to deliver HTML which will show in the client's program. In Carafe, Jinja is arranged to auto-escape any information that is delivered in HTML layouts. This implies that it is protected to deliver client input; any characters they have entered that could play with the HTML, for example, < and > will be gotten away with safe qualities that appear to be identical in the program but do not cause undesirable impacts. Jinja looks and acts for the most part like Python. Unique delimiters are utilized to recognize Jinja's sentence structure from the static information in the format. Anything among {{and}} is an articulation that will result in the last archive. {% and %} indicate a control stream explanation like if and for. Not at all like Python, blocks are signified by start and end labels as opposed to space since static text inside a block could change space. Each page in the application will have a similar essential format around an alternate body. Rather than composing the whole HTML structure in every format, every layout will expand a base layout and supersede explicit segments.

6.5.1 Proposed Approach

1. First, we take the transactions dataset.
2. Filter dataset as per necessities which has trait as per investigation to be finished.
3. Split the dataset into preparing and testing.
4. Perform Destroyed to adjust the information on the resultant dataset framed.
5. Analysis should be possible on testing information in the wake of preparing to utilize Calculated Relapse, Arbitrary Woods, and Brain Organization.

At last, you will obtain results as exactness measurements.

6.5.2 Classifiers

1. *Random Forest:*

The random forest algorithm is a learning method that combines many classifiers to improve the performance of a model. A supervised machine-learning system built up of decision trees is called a Random Forest. For both classification and regression issues, Random Forest is used. Coming up next are the essential stages expected to execute the irregular woods calculation.

- Pick N records aimlessly from the datasets.
- Utilize these N records to make a choice tree.
- Select the number of trees you need to remember for your calculation, then, at that point, rehash stages 1 and 2.
- The classification to which the new record belongs in the order issue is predicted by each tree in the timberland.
- The way that there are numerous trees and they are completely prepared utilizing various subsets of information guarantees that the irregular timberland strategy is not one-sided.
- The irregular woods strategy fundamentally relies upon the strength of "the group," which reduces the framework's general predisposition. Since it is extremely challenging for new information to influence every one of the trees, regardless of whether another information point is added to the datasets, the general calculation is not highly different.

2. *Linear Regression:*

One of the easiest and most well-known calculations of human cognition is direct apostatizing. A clear strategy is used for critical assessment. Direct loss of faith raises doubts about unsurprising/real or numerical aspects, such as contracts, pay, age, item costs, and so forth. Straight apostatize, also known as straight lose the faith, is the rapid association between one dependent and one free element. Since direct apostatize indicates a close relationship, it suggests calculating the relationship between the value of the autonomous variable and the value of the dependent variable. The model for prompting people to lose faith provides a crooked straight line that looks for connections between the variables. Consider the image below (Fig. 3). Relation between dependent and independent variables helps to understand the regression mapping of different variables.

3. *Logistic Regression:*

Logistic Lose the Faith is one of the most striking PC-based insight assessments that fall under the category of planned learning. It is used to predict the direct subordinate variable using a predetermined set of free factors. A fixed dependent variable's final result is predicted by a logical fallacy. In this manner, the outcome needs to be a distinct or flat worth. In any instance, rather than providing a particular value like 0 and 1, it provides the probabilistic features that lie a few spots in the

Fig. 3 Relation between dependent and independent variables

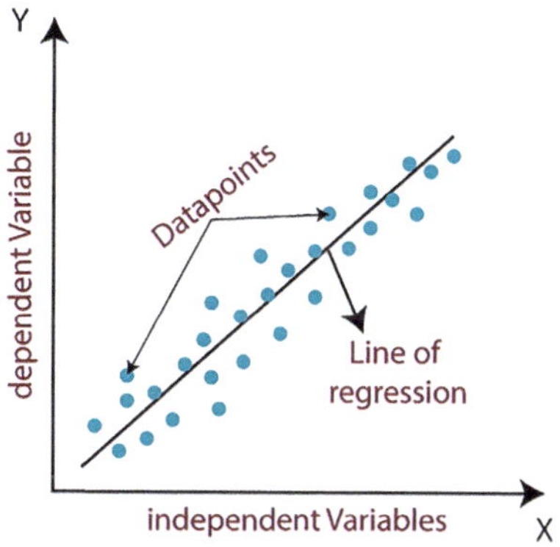

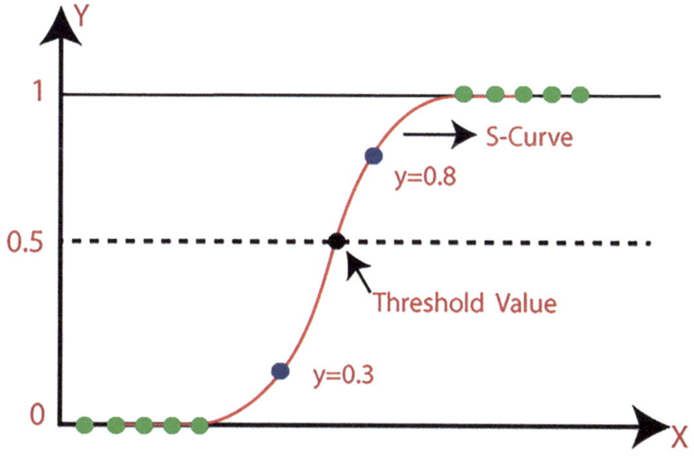

Fig. 4 Representation of threshold value

range of 0 and 1. In the end, it will either be Yes or No, 0 or 1, important or Hoax, and so on. Comparable to the Straight Apostatize in terms of how they are used is the Logistic Lose the Faith. While Fundamental apostatize is used to address portrayal concerns, Direct Fall away from the Faith is used to address apostatize issues. In Decided to Lose the Faith, we fit an "S" framed limit rather than an apostatized line, which predicts the two largest characteristics (0 or 1). The wind from the chosen boundary illustrates the likelihood of something, such as if phones are dangerous, whether a mouse is enormous or does not consider its weight, etc. Given its ability to provide probabilities and request new information using infinite and discrete datasets, the logistic break faith is a significant reenacted knowledge calculation. Calculated Break faith can be used to organize the observations using different types of information and will undoubtedly end up being the best factors used for the presentation.

The crucial ability is displayed in the picture below (Fig. 4). Representation of threshold value using the logistic regression:

4. *SVM*

Support Vector Machines (SVMs) are a sort of regulated learning calculation that can be utilized for grouping or relapse undertakings. The major idea behind SVMs is to find a hyperplane that maximally detaches the different classes in the readiness data. This is finished by finding the hyperplane that has the best edge, which is depicted as the distance between the hyperplane and the nearest bits of information from each class. At the point when they are not absolutely solidly settled, new information can be portrayed by choosing which side of the hyperplane it falls. SVMs are especially significant when the information has many elements, as well as when there is a reasonable edge of the fragment in the information.

Accuracy: The level of accurate presumptions for the test information is construed by accuracy. By confining how many cautious appraisals by the firm number of suppositions, it could not out and out for all time lay outed.

7 Result Analysis

Laundering cash with digital currency imparts a ton of normal qualities to the original tax evasion. The cycle in laundering cash is something very similar as is the outcomes thereof. It is the properties intrinsic to the digital currency—the illegal tax avoidance vehicle that represents the greatest test to controllers. Tax evasion is a harmful wrongdoing that propagates predicate violations where the laundered cash is inferred. It makes an endless loop where criminal elements and hoodlums possess a steady flow of assets. The emergence of digital currency has been a major problem. On one hand, it is simpler to safely execute on the web yet then again it has been taken advantage of to work with a bunch of cybercrimes and help the crooks to wash the returns securely. Because of its decentralization, security, irreversibility, and secrecy, digital currencies like Bitcoin have been exploited. Although Bitcoin is the most unmistakable digital money, advertising volatility is not resistant. Nonetheless, this does not mean that assuming it falls, the utilization of digital currency will fail to be utilized in laundering cash. Bitcoin has quite recently shown the valuable open doors accessible to crooks and as Freedom Save has shown, the potential outcomes are unfathomable.

8 Conclusion

In this paper, a proposed ML identification conspire lays out prerequisites and moves made to have various banks team up in an AML drive. We train each managed model on the train set utilizing a few elements and afterward assess them on the whole test set. To quantify execution over the long run. We utilize the sci-kit-learn execution of Keras-based ML. The civility of the banks is provided cryptographic freedom to conduct continual evaluations handled by a simulated intelligence estimation to ultimately have the outcomes handled in a co-employable way. Further research includes examining how tax evasion recognition responsibilities are shared when banks join a distributed ML without a focal authority. This kind of organization could be researched further with conveyed records where every hub plays out some brain network handling as a commitment.

References

1. Bhattacharyya, Siddhartha, Sanjeev Jha, Kurian Tharakunnel, and J Christopher Westland. 2011. 'Data mining for credit card fraud: A comparative study', Decision Support Systems, 50: 602–13.
2. Bindu, PV, P SanthiThilagam, and Deepesh Ahuja. 2017. 'Discovering suspicious behavior in multilayer social networks', Computers in Human Behavior.
3. Glen L. Gray ., Roger S. Debreceny. (2014). A taxonomy to guide research on the application of data mining to fraud detection in financial statement audits. Journal: International Journal of Accounting Information Systems, Volume 15, Issue 4, pp. 357– 380
4. Han, Jiawei, Micheline Kamber, and Jian Pei. 2011. Data mining: concepts and techniques: concepts and techniques (Elsevier).
5. Keyan, Liu, and Yu Tingting. 2011. "An improved support-vector network model for anti-money laundering." In Management of e-Commerce and eGovernment (ICMeCG), 2011 Fifth International Conference on, 193–96. IEEE.
6. Khan, Nida S, Asma S Larik, Quratulain Rajput, and Sajjad Haider. 2013. 'A Bayesian approach for suspicious financial activity reporting', International Journal of Computers and Applications, 35.
7. Khanuja, Harmeet Kaur, and Dattatraya S Adane. 2014. 'Forensic Analysis for Monitoring Database Transactions'. in, Security in Computing and Communications (Springer).
8. S. Haider, "Clustering based anomalous transaction reporting," Procedia CS, vol. 3, pp. 606–610, 12 2011.
9. L.-T. Lv, N. Ji, and J.-L. Zhang, "A rbf neural network model for anti-money laundering," in 2008 International Conference on Wavelet Analysis and Pattern Recognition, vol. 1. IEEE, 2008, pp. 209–215.
10. Chen, Z., Van Khoa, L. D., Teoh, E. N., Nazir, A., Karuppiah, E. K., & Lam, K. S. (2018). Machine learning techniques for anti-money laundering (AML) solutions in suspicious transaction detection: a review. Knowledge and Information Systems, 57, 245–285.
11. Canhoto, A. I. (2021). Leveraging machine learning in the global fight against money laundering and terrorism financing: An affordances perspective. Journal of business research, 131, 441-452.
12. Zand, A., Orwell, J., & Pfluegel, E. (2020, June). A secure framework for anti-money-laundering using machine learning and secret sharing. In 2020 International Conference on Cyber Security and Protection of Digital Services (Cyber Security) (pp. 1–7). IEEE.
13. Pettersson Ruiz, Eric, and Jannis Angelis. "Combating money laundering with machine learning–applicability of supervised-learning algorithms at cryptocurrency exchanges." Journal of Money Laundering Control 25.4 (2022): 766–778.
14. Jullum, Martin, Anders Løland, Ragnar Bang Huseby, GeirÅnonsen, and Johannes Lorentzen. "Detecting money laundering transactions with machine learning." Journal of Money Laundering Control 23, no. 1 (2020): 173–186.
15. Alkhalili, Mohannad, Mahmoud H. Qutqut, and FadiAlmasalha. "Investigation of applying machine learning for watch-list filtering in anti-money laundering." IEEE Access 9 (2021): 18481–18496.

Accessible Chatbot Interface for Price Negotiation System

P. Muralidhar, K. Nagakeerthana, B. Sai Manvith Reddy, and P. Shivani

Abstract Globally, E-commerce has been evolving as the most popular business platform. In recent years, several e-commerce organizations have evolved such as Amazon, Flipkart, eBay, and Walmart, etc., with robust search engines and beginners-friendly interfaces. The benefits of this e-commerce marketing strategy may be made available at any moment in accordance with the interests of the consumer, which may also be the goal of making a profit in the market. Compared to physical brick-and-mortar stores, e-commerce has become more successful because of quick access and availability of various products at any instant but fails to provide negotiation in purchasing (bargaining approach), which most consumers prefer. With the development of machine learning and deep learning technologies, a variety of chatbot applications were developed that can be used to calculate pricing so that consumers can view them and make more informed and effective decisions. In this article, we intend to provide a scalable user-friendly interface for the chatbot e-commerce platform to enhance the relationship between two parties(the consumer and the merchant). As a result, we observed that enhancing negotiation with chatbot provides a win-win strategy for both consumers as well as merchants. Through this win-win model, consumers gain from purchasing the goods at the best price for the large quality and Merchants can improve their business.

Keywords Machine learning · Negotiation system · Bargain approach chatbot · E-commerce platform

1 Introduction

Over the last few decades e-commerce and shopping online became more familiar to people. E-commerce websites and online retailers are expanding daily. Many people

P. Muralidhar (✉) · K. Nagakeerthana · B. Sai Manvith Reddy · P. Shivani
CSE, Vignan Institute of Technology and Sciences, Hyderabad, Telangana, India

A. Patel et al. (eds.), *Advances in Machine Learning and Big Data Analytics II*,
Springer Proceedings in Mathematics & Statistics 442,
https://doi.org/10.1007/978-3-031-51342-8_25

make a disproportionate number of their everyday purchases online rather than in brick-and-mortar establishments or shopping centers. This change in e-commerce had a great effect on static and dynamic trade. Most people seem to be busy and are unwilling to travel out for window shopping as well and many do not prefer to purchase in the crowd because of fear of recent COVID-19 waves all over the world.

Instead, they opt for e-commerce platforms, virtual online stores where customers can select their products and choose the product of choice. The major pitfall of the e-commerce platform was the lack of a negotiation system for purchasing the products. The optimal decision, trust, and sharing of ideas are the key potentials of any negotiation system. Consumers can resolve their disputes in making the deal, and tailor the decisions to their needs. Merchants can learn the interests of the consumers and maximize business outcomes. The Chatbot e-commerce system enables users to communicate freely with the program, upload product-related questions and budgets, and discover the answer to the query. This article provides a user-friendly scalable interface for negotiation in an e-commerce system.The major objectives of this article are:

- Develop a negotiation system for the e-commerce website so that customers interact with the system.
- Provide an accessible interface to the negotiation system.
- Record the survey of customers and merchants on this negotiation system.

Section 2 deals with the literature review of several chatbot systems, and the methodology of the proposed work discussed in Sect. 3. Section 4 has the working screen of the negotiation system. The survey of the negotiation system with both merchant and customer is shown in Sect. 5. Section 6 deals with the conclusion followed by future enhancements.

2 Literature Survey

To develop the negotiation system, we created using chatbot. The first ever well-known chatbot system was ELIZA developed by Joseph Weizenbaum. These earlier chatbots recognize the keywords in the phrases and pre-programmed responses are produced as the output in a meaningful way [1]. With the help of natural language processing, the responses of the chatbot systems are more accurate and help the customers in many applications such as university websites, banking websites Insurance websites, etc. [4].

With the development of machine learning and deep learning techniques, a major breakthrough has come to chatbot applications [2, 3]. The development of price Negotiation system was developed with the help of chatbot systems [5–11]. Even though a large number of applications are developed unable to reach user satisfaction. In this work, we developed a chatbot negotiation system using sentiment analysis as a part of the work so the responses of the chatbots are developed based on the customer sentiments and the website is more accessible to the customer to deal with the negotiation system.

3 Methodology

In this proposed work the negotiation system was developed using a chatbot. The customer selects a product and wants to enquire about the relevant information about the product. The customer selects a query to raise the option of negotiator with the developed chatbot. The negotiator system reads the query and processes the query without the involvement of the merchant.

An agent-based automatic negotiation will be created to aid e-commerce websites in obtaining the finest service. This makes it possible for eCommerce sites to keep a variable price rather than a settled one. The rewards for buyers and sellers will be higher because of this flexibility and availability of prices. Thus it serves as a useful mechanism for business-related areas. Before negotiating, a customer can study the positive and negative reviews of a product they wish to purchase.

Figure 1 explains the customer interaction with negotiation systems. The proposed system processes the customer request, it performs three tasks Dialogue rollout, Sentiment analysis, and natural language processing as shown in the system architecture presented in Fig. 1. The dialogue rollout takes care of the complete conversation between the bot system and the customer or merchant.

The Sentiment analyzer component explores the sentiment in the sentence query raised by the customer. The System identifies the intention of the customer from the broken words assigned with tags which are selected by the negotiation system to estimate the sentiment.

The admin gathers relevant information about the product from the merchant and stores the knowledge related to the product in the database, which helps the natural language system retrieve the response for the query.

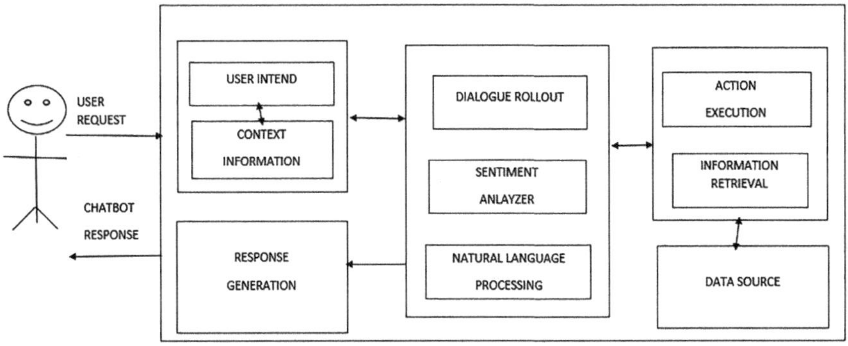

Fig. 1 Negotiation system architecture

4 E-Commerce System

Negotiation is an important part of an e-commerce platform. There is a need for major updating of existing e-commerce platforms by including the chatbot based-negotiation system architecture in the e-commerce

sites. For the test of the negotiation system, we developed an accessible e-commerce platform the screenshots of the negotiation system were presented in this section.

Figures 2, 3, 4, 5, 6a, and 6b are screens of the e-commerce environment developed to test the negotiation system. The Home page, Customer Registration page, Product selection page, and View product page are most common for any negotiation system. Figure 6b is the Negotiation screen page through which the customer negotiates with the query with our system and the negotiation system responds back based on the knowledge stored by the merchant of each and every product. The customer and merchant satisfaction survey was mentioned in the next section.

5 Survey on Negotiation System

The developed negotiated system was tested with the merchant and customers to check the usability and accessibility of the negotiation system. The survey summary is mentioned in the detailed graph presented in Fig. 7, 8, 9, 10, 11, 12, 13 and 14.

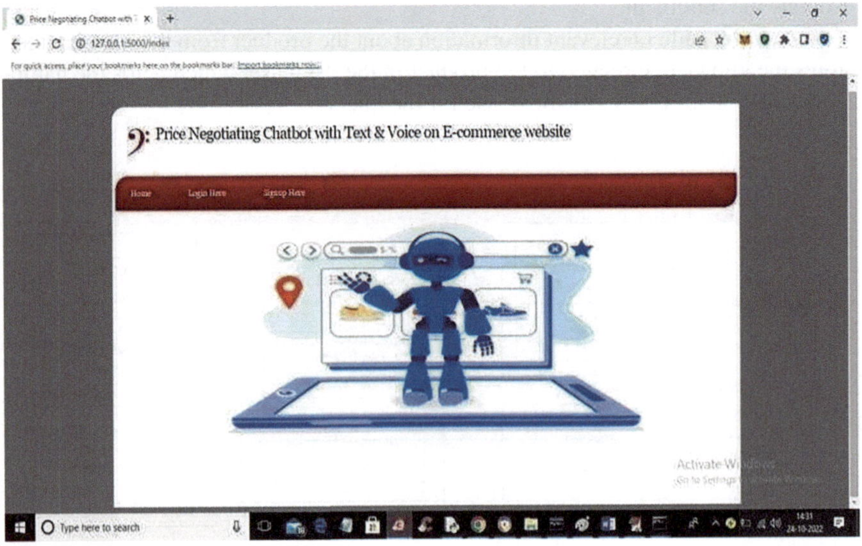

Fig. 2 Home Screen

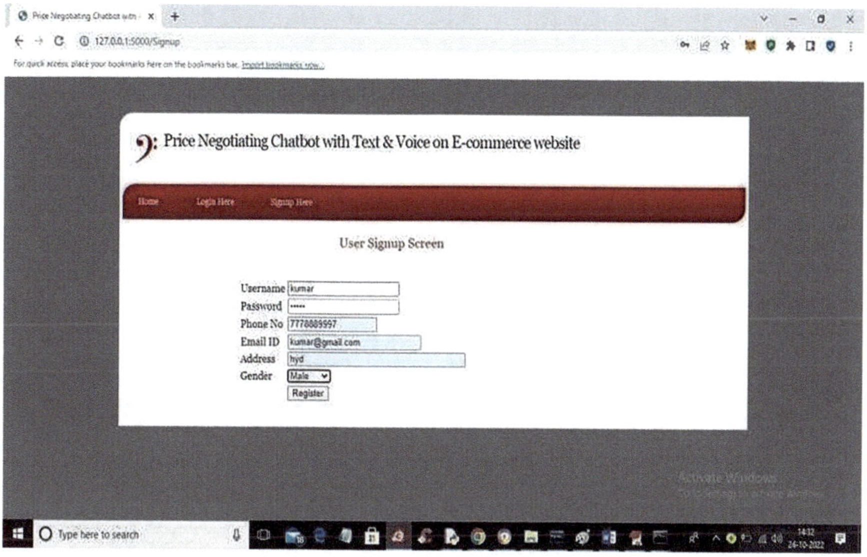

Fig. 3 Customer Registration Page

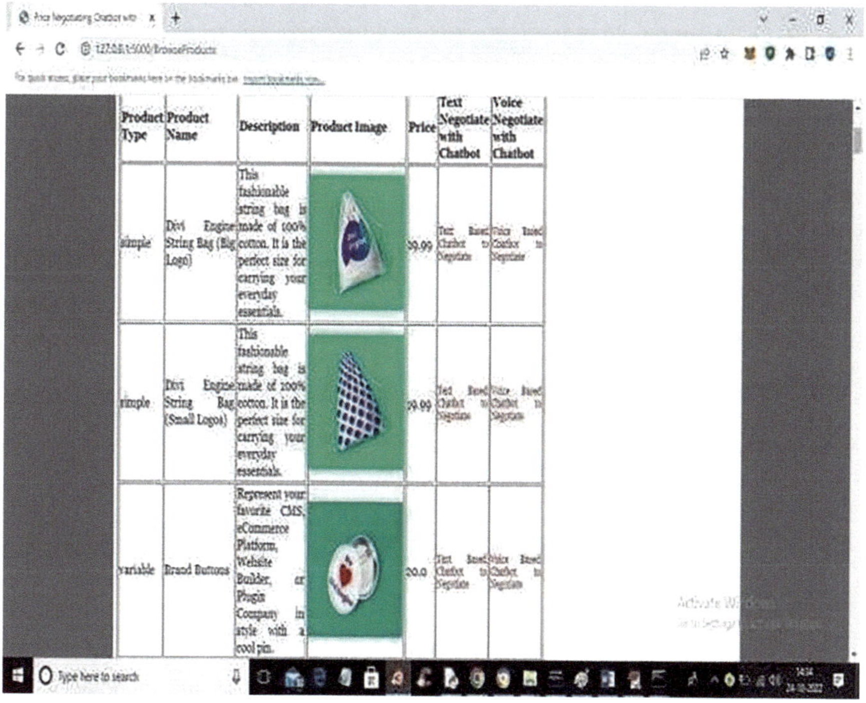

Fig. 4 Product Selection Page

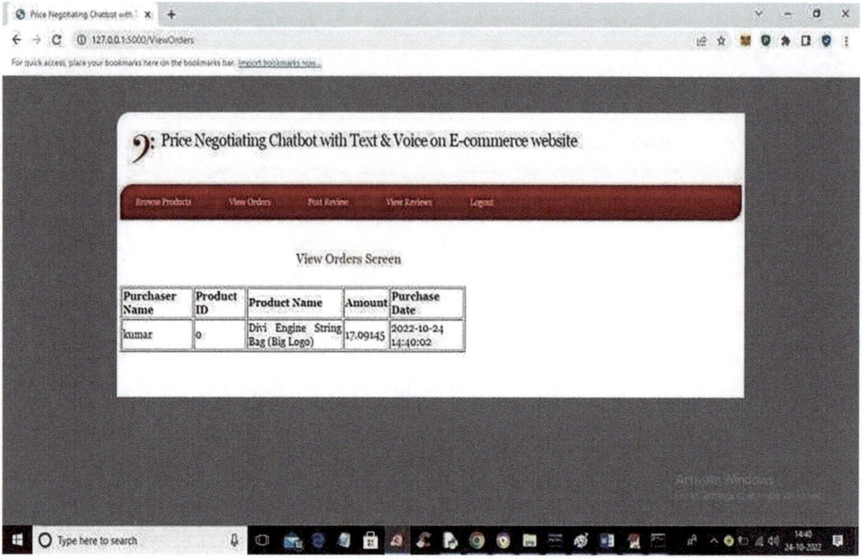

Fig. 5 View Product Page

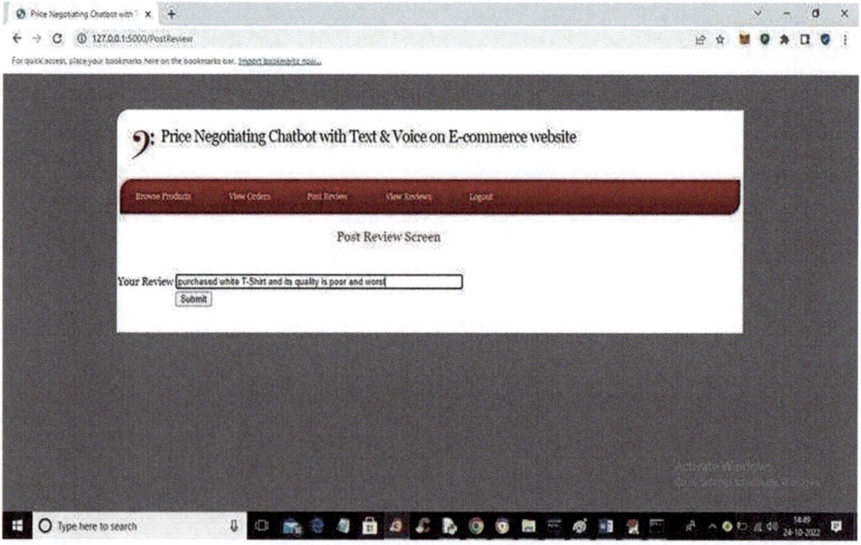

Fig. 6a Review Screen

The survey was done both by the customers as well as the merchants. The merchants are the one who places the product on the e-commerce site and provides relevant information to the chatbot about the product. The customer can interact query with the negotiation system and the negotiation system finds the sentiment of

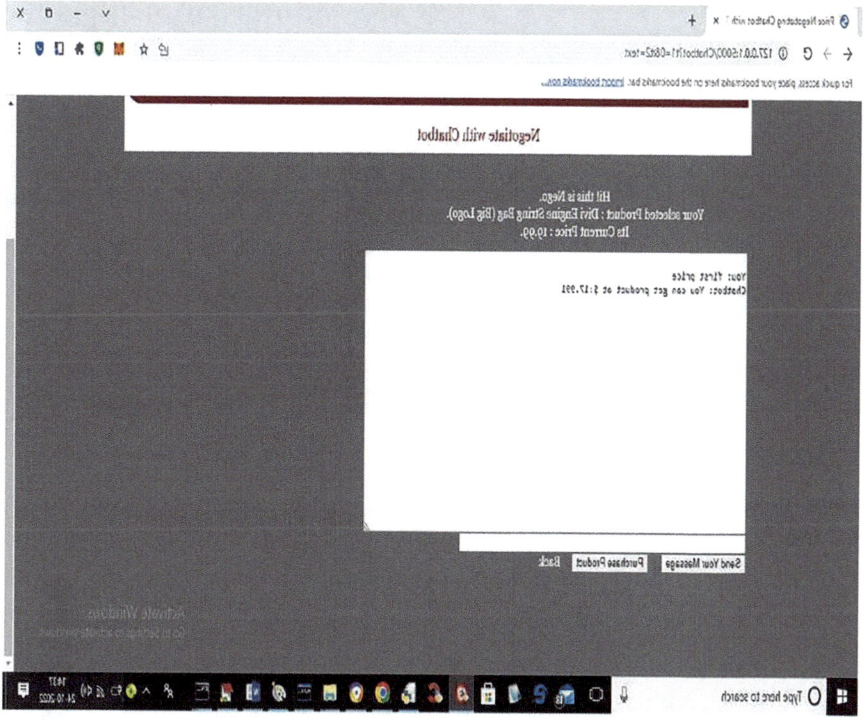

Fig. 6b Negotiation Screen with Chatbot

Fig. 7 Responses to Question 1

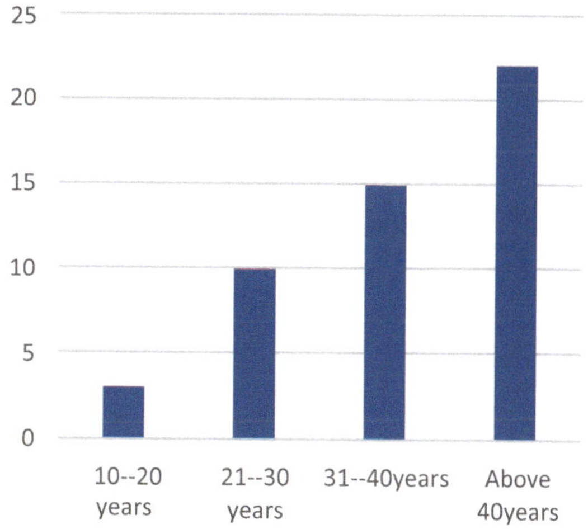

the query raised by the customer and from the knowledge stored in the database with the help of Natural language processing, the responses are generated to the user.

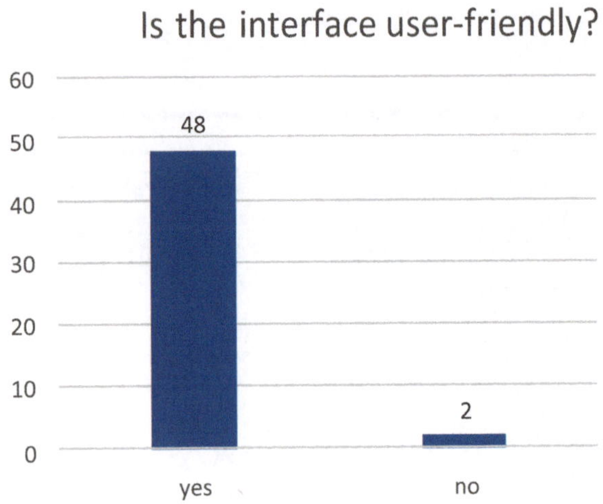

Fig. 8 Responses to Question 2

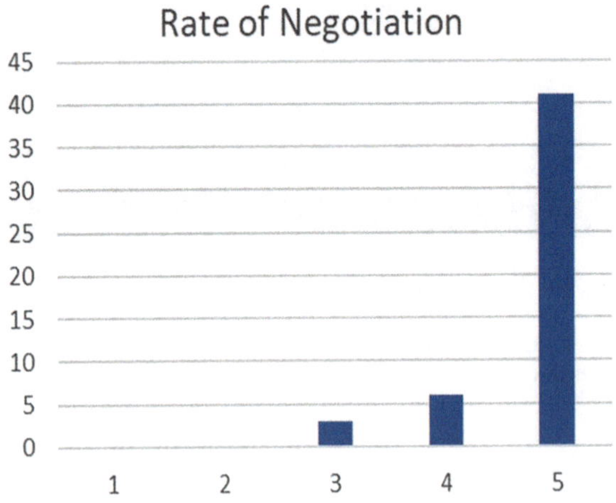

Fig. 9 Responses to Question 3

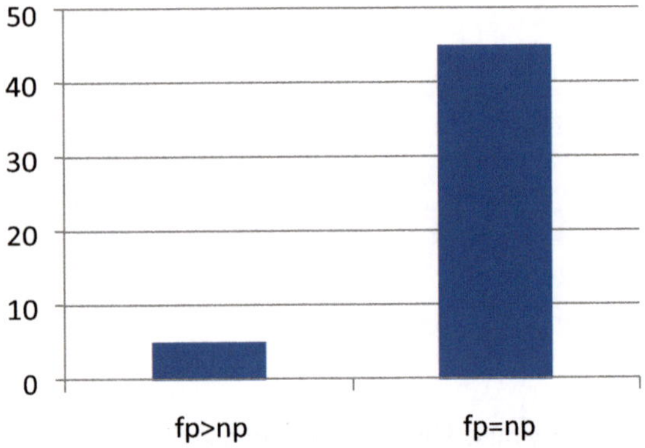

Fig. 10 Responses to Question 4

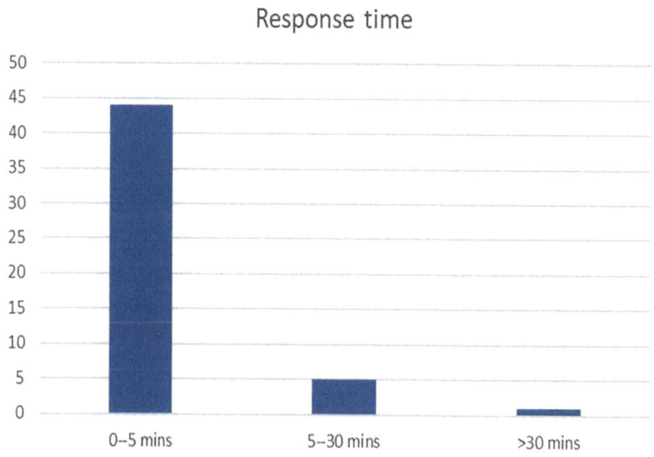

Fig. 11 Responses to Question 5

Fig. 12 Responses to Question 6

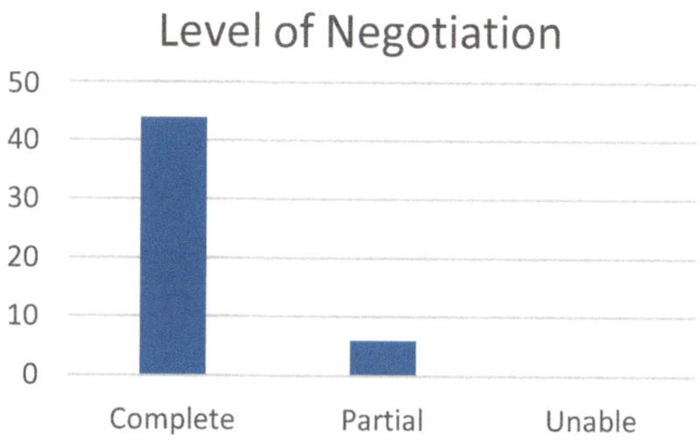

The customer and the merchant were raised with a number of questions after they worked with the system. At first, the merchants were asked to update the product and the relevant information on the e-commerce website and then asked to choose the product and interact with the negotiation system. The negotiation system responds back with the appropriate response based on the knowledge of the information in the database system.

If any questions are unaware by the negotiation system, the merchant will keep on updating the information with the negotiation system. The negotiation system then again responds to the customers with updated information.The following questions have been raised to the customer mentioned in Table 1.

Figures 6, 7, 8, 9, 10, 11 and 12 convey customer satisfaction with the negotiation system, most customers are of age above 40 years and accepted that the interface was user-friendly, the negotiation system above gives a response in less than 5 min when it is knowledgeable with the relevant information. The customers are satisfied with the negotiation system since the final price can be purchased at the rate of the

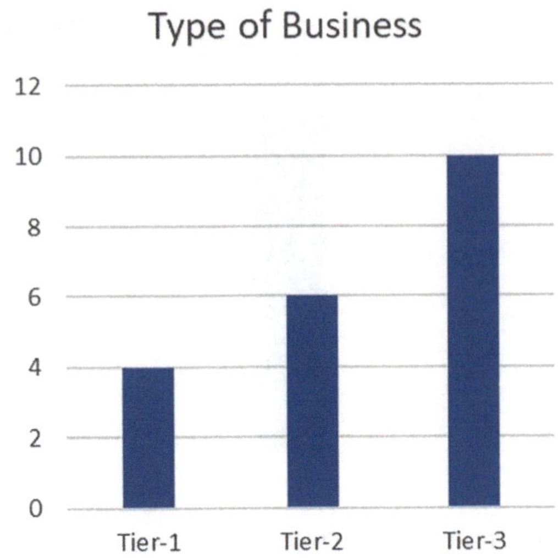

Fig. 13 Analysis of the type of business

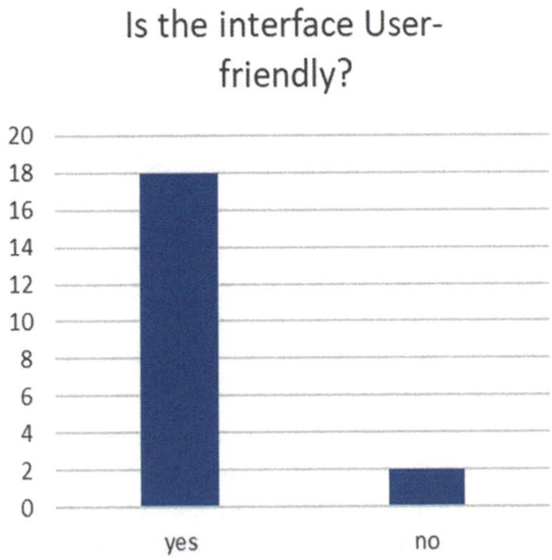

Fig. 14 Responses for Question 2

negotiation price in most cases and the customer specified that the conversion was complete with the negotiation system.

Apart from the customers, we have also surveyed merchants, the survey questions are presented in Table 2.

The merchants will provide the list of products and the information related to the products to the database, and the negotiation system accesses the information from the data store and provides relevant information to

the customer. For the provision of information, updating the e-commerce is accessible. To identify difficulties in the merchant, we conducted a survey at the merchant end.

Table 1 Caption

S. No.	Question	Answer
1.	Age group	Radio Button choices 10–20 21–30 31–40 Above 40
2.	Is the interface user-friendly?	Radio Button Yes No
3.	Rate the level of negotiation	Likert Scale 1 (Low) 2 3 4 5 (High)
4.	Is negotiation satisfiable Final price>negotiated price (Or) final price = negotiated price	Radio Button Fp > Np FP=Np
5.	Is the response quick?	Radio Button 0–5 min 5–30 min >30 min
6.	Mention the level of negotiation with a chatbot.	Radio Button Complete Partial Unable

Figures 13 and 14 are the responses from the merchant regarding the negotiation system, the merchant responses are recorded after 2 months of practice with a set of customers on the particular product. The responses to the survey questions acknowledge that the negotiation system is completely satisfactory on both the customer side and the merchant side.

Figure 11 shows that most of the merchants belong to tier -3 whose income is greater than 30,000 and Some of them whose income is Rs 10,000 to 30,000. A very few of them were of tier-1 whose income was less than 10,000. From the graph, it can be concluded that our chatbot is more profitable to tier -2 and tier- 3.

Figure 14 specifies that our interface was user-friendly. Most of them answered yes and only two answered no, when we ask for the reason, they claimed that they are using the system for the first time.

Figure 15 points that the rate of increase in sales after employing this negotiation method, most of them said that their sales increased and Fig. 16 responses claim that the merchants are very much happy with the negotiation system developed and able to increase sales with the help of the negotiation system

Table 2 Survey Questions for Merchant

S. No.	Question	Answer type
1.	Type of business	Radio button Tier 1 (income<10,000) Tier 2 (Income between 10,000–30,000) Tier 3 (Income >30,000)
2.	Is the interface user-friendly?	Radio button Yes No
3.	The rate of customers increased using this method	Radio button Fallen Moderate Increased
4.	Is the chatbot system adjustable to your requirements?	Radio button Yes Satisfactory Not up to the mark Dissatisfactory

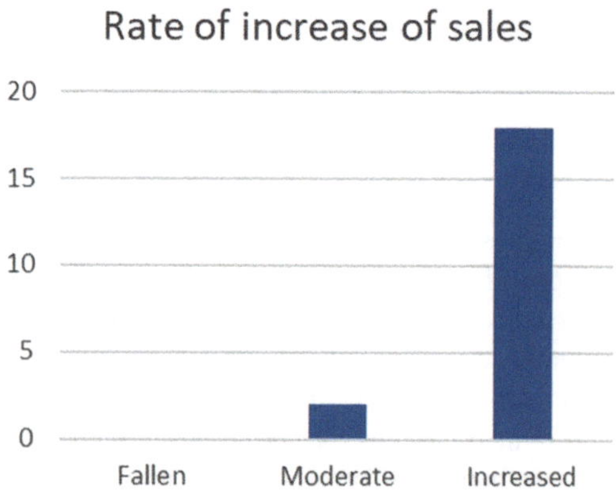

Fig. 15 Responses for Question 3

6 Conclusion

When it comes to e-commerce systems, negotiating products is a difficult issue. We attempted a primary chatbot that covers a wide range of topics and scenarios for discussion, we estimated that the negotiation system would produce the best outcomes.

The chatbot we developed occasionally accepts clients' preferred prices, even though they are always more than the minimum price. If this happens frequently, however, the seller may incur losses. Such circumstances need to be managed.

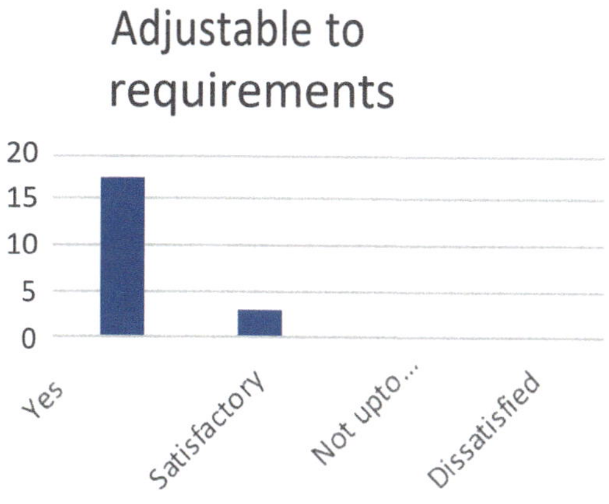

Fig. 16 Responses for Question 4

Despite the disadvantages, the basic product of the negotiation system gives much satisfaction to both customers and merchants who use the product.

In the future, we will extend this negotiation system to more language interpretations and will decrease the merchant's involvement.

Declaration

1. All authors do not have any conflict of interest.
2. This article does not contain any studies with human participants or animals performed by any of the authors.

References

1. J. Weizenbaum, "ELIZA – A Computer Program for the study of Natural Language Communication between Man and Machine," Communications of the ACM, vol. 9, no. 1, pp. 53–62, January 1966
2. A. Porselvi, Pradeep Kumar A.V Hemakumar S. Manoj Kumar, "Artificial Intelligence Based Price Negotiating E-commerce Chat Bot System," International Research Journal of Engineering and Technology (IRJET).
3. Tingwei Liu, Zheng Zheng, "Negotiation Assistant Bot of Pricing Prediction Based on Machine Learning," International Journal of Intelligence Science > Vol. 10 No. 2, April 2020.
4. Eleni Adamopoulou and Lefteris Moussiades, "An Overview of Chatbot Technology," Springer.
5. How to Negotiate with a Chatbot –and win! https://online.hbs.edu/blog/post/how-to-negotiate-with-a-chatbot-and-win.
6. Rohan Asrani, Paras Nandwani, Sahil Sidhani, Vidya Pujari "Price Negotiating Chatbot on E-commerce website" (IRJET).
7. David Traum, Stacy Marsella, Jonathan Gratch, Jina Lee, and Arno Hartholt, "Multi-party, Multi-issue, Multi-strategy Negotiation for Multi-modal Virtual Agents."

8. RAZ LIN 1, SARIT KRAUS, TIM BAARSLAG, DMYTRO TYKHONOV, "GENIUS: An Integrated Environment for Supporting the Design of Generic Automated Negotiators," Journal Compilation C Wiley Periodicals, Inc.
9. Towards Automated Negotiation Agents that use Chat Interface Inon Zuckerman, Erel Segal-Halevi, Sarit Kraus, Avi Rosenfeld, "Towards Automated Negotiation Agents that use Chat Interface," 12th International Conference on Autonomous Agents and Multiagent Systems (AAMAS 2013).
10. Amir Reza Asadi, RezaHemadi, "Design and Implementation of a chatbot for e-commerce." R. E. Sorace, V. S. Reinhardt, and S. A. Vaughn, "Highspeed digitalto-RF converter," U.S. Patent 5,668,842, Sept. 16, 1997.
11. Yinon Oshrat, Sarit Kraus, RazLin, "Facing the challenge of human-agent negotiations via effective general opponent modeling", 8th International Joint Conference on Autonomous Agents and Multiagent Systems (AAMAS 2009), Budapest, Hungary, May 10–15, 2009, Volume.

Recognizing and Logging Vehicles by Scanning License Plates and Driver's Face

P. Uma Sankar, G. Sri Devi, Praveen Kumar Karri, P. LaxmiKanth, Sana Saraswathi, and K. Ganga Parvathi

Abstract Over the years, logging traffic at checkpoints has been a manual job. Vehicles are needed to be logged for security and validation purposes. This can be automated by using the technology of computer vision and neural networks to recognize the license plates of the vehicles and the driver's face. Our project is to develop software that uses a camera to recognize the license plate and driver of the vehicle. The developed system initially detects the vehicle and captures the vehicle image. From the image, the portion of the license plate and driver's face is extracted using image segmentation and then they are searched for recognizable patterns in the database. The details are logged for future reference. By conducting various experiments on our proposed work by taking some sample images collected from the Kaggle dataset and then training the model using these images to identify and detect the vehicle's number plate accurately and corresponding we try to use the LBPH model to identify the human faces who are inside that vehicle. Our simulation results clearly state that our proposed approach is very accurate in identifying and logging the vehicle and driver's details into the database.

P. Uma Sankar (✉) · P. Kumar Karri
Department of CSE, Sri Vasavi Engineering College(A), Tadepalligudem, Andhra Pradesh, India
e-mail: umasankar.cse@srivasaviengg.ac.in; praveenkumar.cse@srivasaviengg.ac.in

G. Sri Devi
Department of ME, Vignan's Institute of Information Technology, Visakhapatnam, Andhra Pradesh, India

P. LaxmiKanth
Department of Information Technology, Anil Neerukonda Institute of Technology & Sciences (ANITS), Visakhapatnam, Andhra Pradesh, India

S. Saraswathi · K. Ganga Parvathi
Department of CSE, Vignan's Institute of Engineering for Women, Visakhapatnam, Andhra Pradesh, India

A. Patel et al. (eds.), *Advances in Machine Learning and Big Data Analytics II*,
Springer Proceedings in Mathematics & Statistics 442,
https://doi.org/10.1007/978-3-031-51342-8_26

Keywords Deep learning · OCR (Optical character recognition) · LBPH (Local Binary Pattern Histogram) · ANPR (Automatic License/Number Plate Recognition) · Face recognition

1 Introduction

One of the effective methods for vehicle monitoring in recent years has been automatic license plate recognition. It can be used in a variety of public settings to meet goals including parking management, traffic safety enforcement, and automatic transportation log systems. The last 20 years have seen a significant increase in face recognition as a result of increased research activity. It entails the computer's identification of a person based on geometric or statistical information extracted from face photographs [1–4]. Every country now faces serious issues with traffic management and car ownership identification. The scale of our project allows us to meet this need. Our current goal is to implement the project for our college's transport buses.

Automatic Number Plate Recognition (ANPR), commonly referred to as "vehicle number plate recognition" [5], is a technology that employs optical character recognition (OCR) and image processing methods to automatically scan and decipher vehicle number plates. It is frequently utilized for many different things, such as traffic control, parking management, toll collecting, and law enforcement. The following steps are usually involved in the vehicle number plate recognition process:

Image Acquisition ANPR systems take pictures of vehicles and their license plates using cameras that are often placed close to highways or entry/exit points. These cameras might be gantry-mounted, mobile, or stationary.

Image Pre-processing To improve the quality and clarity of the number plate area, the collected photos may go through pre-processing. Tasks like image scaling, noise reduction, and contrast are examples of this.

Number Plate Localization The ANPR system examines the previously processed image to locate the number plate—containing region of interest (ROI). The number plate area is separated from the remainder of the image using a variety of approaches, including edge detection and morphological procedures.

Character Segmentation The ANPR system divides up the characters on the number plate after locating the number plate region. To separate each character for later processing and recognition, it is essential to take this step.

Optical Character Recognition (OCR) In this step, the segmented characters are supplied into an OCR engine, which recognizes and transforms the characters into machine-readable text using pattern recognition techniques. To accomplish accurate OCR, the OCR engine can use a variety of methods, such as template matching, neural networks, or machine learning algorithms.

Post-processing and Interpretation The ANPR system may use extra algorithms after character recognition to confirm and interpret the detected characters. To produce the desired result, methods including character grouping, error checking, and text formatting may be used.

Application and Data Integration For particular applications, the information from the recognized number plate can be further processed and combined with databases or other systems. This may entail running the number plate number through a database of stolen automobiles, getting vehicle owner data from a registration database, or launching particular operations in accordance with predetermined regulations.

It is important to remember that ANPR systems' implementation and precision can differ based on elements including camera caliber, lighting, image-processing algorithms, and database integration. Additionally, the implementation of ANPR technology is subject to legal and regulatory considerations in various jurisdictions, and it presents privacy issues.

2 Literature Survey

The most important step in the software development process is the literature review. This will describe some preliminary research that was carried out by several authors on this appropriate work and we are going to take some important articles into consideration and further extend our work.

Tahir Muhammad Qadri and Muhammad Asif [6] proposed a research article, "Automatic Number Plate Recognition System for Vehicle Identification," in 2009. The conclusion is that using a high-resolution camera can boost the system's speed and robustness. The affine transformation can be utilized to improve the OCR recognition from diverse sizes and angles. The OCR methods employed in this project for the recognition will not be successful if there is a misalignment and varied sizes.

A Kashyap [7] proposed a research article, "Automatic Number Plate Recognition System: Using Character Recognition Is Done by the Existing Optical Character Recognition (OCR) Tool," which was designed by Dipti Shah Atul Patel. Numerous software programs are available for OCR processing. Tesseract, an open-source OCR program maintained by Google, supports multiple languages. The conclusion is that it is pretty obvious that ANPR is a challenging system due to the variety of phases, and that at this time it is not possible to reach 100% overall accuracy because each step depends on the phase before it.

Using the dataset and approach, **Vijay Kumar Sharma** [8] created the research paper "Face Recognition" in 2019. Using OpenCV's LBPH algorithm for:

1. Face recognition.
2. Feature extraction.
3. Classification.

The technique has an efficiency of roughly 80% and has been verified by more than 150 people. It has been tried with many cameras under various environmental and lighting situations, and the results are essentially the same.

Y. Lei [9], X. Jiang, Z. Shi, D. Chen, and Q. Li, "Face Recognition Method Based on SURF Feature." SURF is used to identify faces in the Reference and Test photos, after which the SURF features are extracted and descriptors are constructed for each image. After that, the Descriptors are compared to identify any similarities between the Images. The SURF algorithm, in our opinion, has superior performance and accuracy. SIFT, on the other hand, proved to take more time.

A. Choudhury, J. Mukherjee, and S. Roy [10]. They split the numbers to identify each number separately and developed a mechanism for the localization of number plates, primarily for automobiles in West Bengal (India). The method presented in this study is based on the Sobel edge detection method and a straightforward and effective morphological operation. He also offers a straightforward method for dividing up the number plate's letters and numbers. We attempt 3 to improve the contrast of the binarized image using histogram equalization after removing noise from the input image. We mainly focus on two steps: finding the license plate and segmenting the letters and numbers to identify each number separately.

M. Shehata, W. Badawy, and S. Du [11]. By classifying them based on the features utilized in each stage, provide a thorough analysis of the current (Automatic License Plate Recognition) ALPR Techniques. It was discussed to compare them in terms of advantages, disadvantages, recognition outcomes, and processing speeds. Also provided at the conclusion was ALPR's potential future. Future ALPR research should focus on ambiguous-character recognition, multi-style plate recognition, video-based ALPR employing temporal information, processing multiple plates, and high-definition plate image processing.

3 Existing System

ANPR (Automatic Number Plate Recognition) is a technique that reads car registration plates and generates vehicle position data from photos using optical character recognition. It may employ already-in-use cameras for enforcing traffic laws or cameras created especially for the job. A popular technique for face detection is OpenCV. By removing the facial Haar features from the image, it first extracts the feature photos into a sizable sample set. The following are the primary drawbacks of the current system.

1. The system that is currently being used by the traffic control is outdated and they are being run on the LBPH algorithm for face recognition.
2. The system used at some universities or colleges in India is manual work.
3. Manual work is prone to human errors and maintenance of logs can be hard.

4 Proposed System

In this stage, we are going to explain the proposed system and its model by using some model diagrams and now we can clearly identify the step-by-step procedure of our proposed system (Fig. 1).

The proposed system is an integrated software solution for recognizing the license plate of the vehicle and the face of the driver. The image of the vehicle is captured by a camera and the points of interest (license plate and driver's face) in the image are segmented for further processing by neural networks. The license plate is recognized by the OCR algorithm using Tesseract Neural Networks in OpenCV. The driver's face is recognized by the SURF algorithm [7–12] in OpenCV which is the fastest of the recent developments. The logs are then maintained in the Database for future reference. This model is proposed to fulfill the following purposes:

1. ANPR and face detection is used for security purposes.
2. It reduces manual work.

4.1 Proposed Algorithms

The proposed system contains several deep-learning algorithms to capture the vehicle image, process the image, separate the license plate and driver's face, and later extract data from the images.

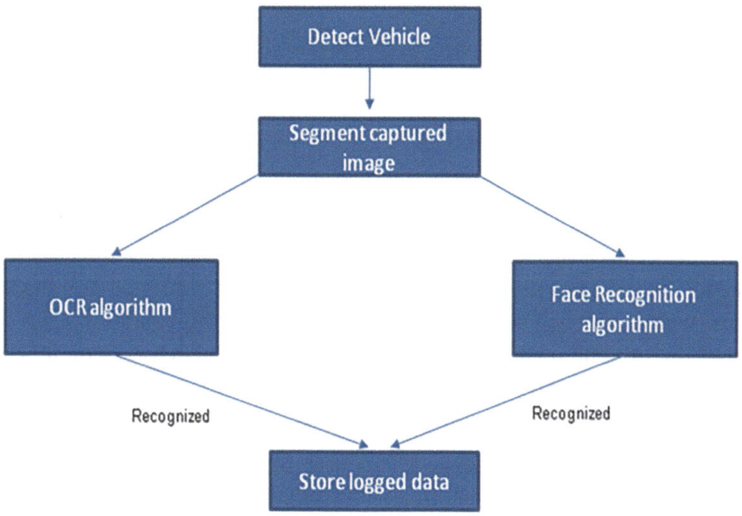

Fig. 1 Representation of the proposed system architecture

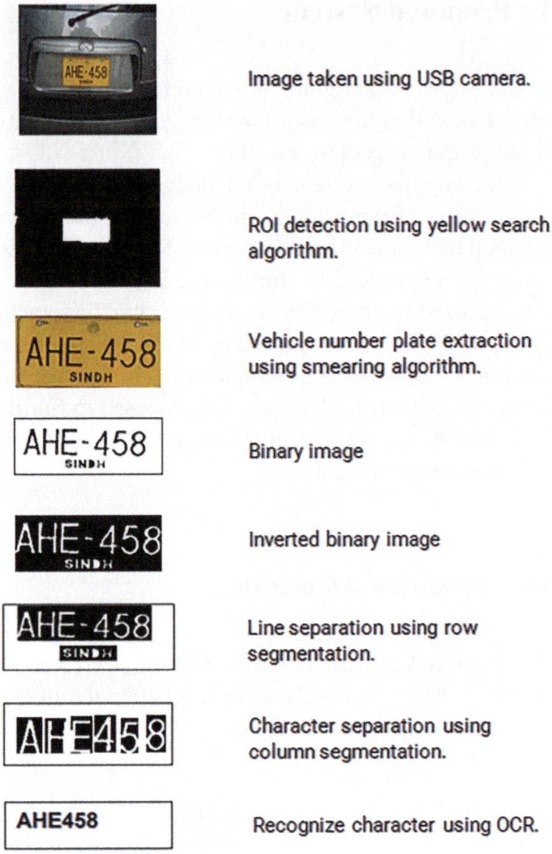

Fig. 2 Representation of the Proposed OCR algorithm for number plate recognition

4.1.1 OCR Algorithm

OCR (Optical character recognition) [12] works by dividing up the image of a text character into sections and distinguishing between empty and non-empty regions. An OCR program extracts and repurposes data from, camera images, scanned documents, and image-only pdf. OCR software extracts letters on the image, puts them into words, and then puts the words into sentences, thus enabling access to and editing of the original content. OCR algorithms and approaches come in a variety of forms, each with a unique approach and degree of complexity. Here are a few frequently employed OCR formulas (Fig. 2):

Template Matching To recognize characters, this fundamental OCR technique compares the input image or text region with predefined templates or patterns. Based on similarity measures, it finds the best match between the input and template photos [13].

Feature Extraction To build a character representation, this OCR method extracts important character properties like lines, curves, or corners. The characters in the input image are then recognized by comparing these features to a training dataset or a reference [14].

Convolutional Neural Networks (CNNs) in particular, have demonstrated outstanding success in optical character recognition (OCR). Accurate character recognition is made possible by CNN-based OCR algorithms, which employ many layers of convolutional and pooling operations to automatically learn and recognize patterns in the input images [15].

Hidden Markov Models (HMMs) HMM-based OCR algorithms employ statistical techniques to identify the most likely character sequence for a given input. They model characters as sequences of states. To increase accuracy, HMMs are frequently integrated with other methods like feature extraction or neural networks.

Support Vector Machines (SVMs) Using a supervised learning method, SVM-based OCR algorithms categorize characters based on their attributes. SVMs determine the ideal decision boundary to divide several character classes, facilitating precise recognition.

Tesseract OCR Tesseract is a Google-developed open-source OCR engine. To achieve precise character recognition, it integrates a number of OCR approaches, including neural networks and HMMs. Tesseract has received a lot of attention and is renowned for its performance and adaptability. These are but a few illustrations of OCR algorithms; several more configurations and strategies are employed in various OCR systems. The complexity of the input images, the level of precision desired, available computing power, and the application-specific requirements all play a role in the algorithm selection process [16].

LBPH Algorithm LBPH (Local Binary Pattern Histogram) [17] is a face recognition algorithm that is used to recognize the face of a person. It is known for its performance and how it is able to recognize the face of a person from both the front face and the side face.

Ojala et al. first presented the LBP technique as a texture descriptor in 1994. With encouraging outcomes, it was then expanded and used for facial recognition tasks. The technique is based on the study of local patterns in the image and works with grayscale images. Here is a quick explanation of how the LBP algorithm for face recognition functions:

Image Preparation The input facial image is frequently improved through preprocessing to normalize elements like lighting and posture fluctuations.

LBP Computation The LBP method separates the facial picture into discrete cells, or small local regions. The technique calculates a binary value (0 or 1) for each pixel in a cell based on whether the surrounding pixels are higher or less intense than the pixel in question. Each pixel in the cell receives a binary pattern as a result of this operation.

LBP Histogram Computation A histogram is created by counting the occurrences of various LBP patterns in the image after computing the LBP patterns for each pixel in the image. The distribution of small patterns in the face image is shown by the histogram.

Feature Extraction and Representation The local texture information of the face is captured by the LBP histogram, which acts as a feature vector. Face recognition algorithms frequently make use of this feature vector to depict faces.

Face Matching Using similarity metrics like Euclidean distance or cosine similarity, the LBP histogram of the input face image is compared with the histograms of a database of recognized faces during the recognition phase. Potential matches are those whose faces are most similar.

While LBP Histogram is a popular method for face recognition, there are other algorithms and methods utilized in the field as well, including Eigenfaces, Fisherfaces, and deep learning-based techniques like Convolutional Neural Networks (CNNs). Each approach has advantages and disadvantages, and the selection of an algorithm is influenced by a number of variables, including the particular application needs, the available processing resources, and the characteristics of the dataset.

5 Experimental Results

In this proposed work, we try to use Python as a programming language and implement the above two algorithms for identifying the driver's face and vehicle number plate recognition accurately and try to store all that information in a log file.

Main Window

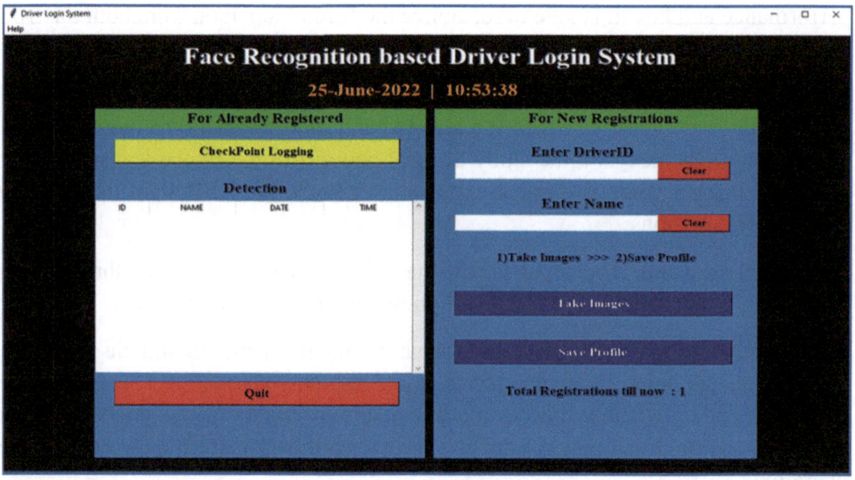

Explanation From the above window we can there are two frames, one is to check already registered driver details and another frame is to register a new driver.

Driver Registration

Explanation From the above window, we can see a new driver is registered and for that driver, one unique ID and name are taken as input along with his face.

Logging Driver's Face and Number Plate

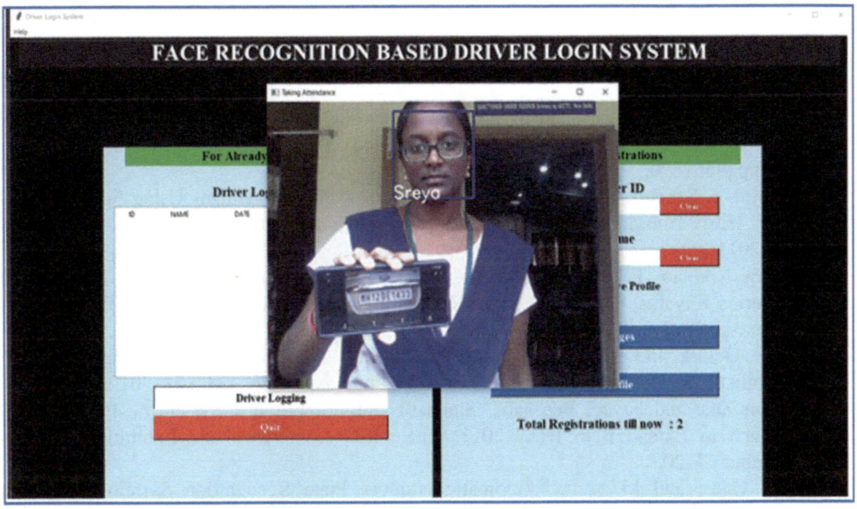

Explanation From the above window, we can see one person is holding a car image on her mobile and the web camera captures the picture and tries to extract the driver's face and also identify the vehicle number.

6 Conclusion

By using ANPR and OCR recognition algorithms, we can able to identify the driver's face accurately despite light or other factors that affect existing Eigenface-recognizing algorithms. Also by using OCR in OpenCV we can able to extract the vehicle number plate easily and can able to store that vehicle number and the corresponding driver's face in the log file. If this type of model is implemented in all large-scale companies and colleges a lot of manual work is reduced and there will be no errors in log information. As a future work, we want to extend these points: The system's robustness can be increased if a bright and sharp camera is used. The system detection can be increased in night vision. The system detection can be increased in infrared rays.

Declaration

1. All authors do not have any conflict of interest.
2. This article does not contain any studies with human participants or animals performed by any of the authors.

References

1. Anton Satria Prabuwono and Ariff Idris, "A Study of Car Park Control System Using Optical Character Recognition," in International Conference on Computer and Electrical Engineering, 2008, pp. 866–870.
2. A Albiol, L Sanchis, and J.M Mossi, "Detection of Parked Vehicles Using Spatiotemporal Maps," IEEE Transactions on Intelligent Transportation Systems, vol. 12, no. 4, pp. 1277–1291, 2011.
3. Christos Nikolaos E. Anagnostopoulos, Ioannis E. Anagnostopoulos, Vassili Loumos, and Eleftherios Kayafas, "A License Plate-Recognition Algorithm for Intelligent Transportation System Applications," pp. 377–392, 2006.
4. Fikriye Öztürk and Figen Özen, "A New License Plate Recognition System Based on Probabilistic Neural Networks," Procedia Technology, vol. 1, pp. 124–128, 2012.
5. Xiaojun Zhai and Faycal Bensaali, "Standard Definition ANPR System on FPGA and an Approach to Extend it to HD," 2013 IEEE GCC Conference and Exhibition, pp. 214, November 17–20.
6. M. T. Qadri and M. Asif, "Automatic Number Plate Recognition System for Vehicle Identification Using Optical Character Recognition," *2009 International Conference on Education Technology and Computer*, Singapore, 2009, pp. 335–338. https://doi.org/10.1109/ICETC.2009.54.

7. A. Kashyap, B. Suresh, A. Patil, S. Sharma and A. Jaiswal, "Automatic Number Plate Recognition," *2018 International Conference on Advances in Computing, Communication Control and Networking (ICACCCN)*, Greater Noida, India, 2018, pp. 838–843. https://doi.org/10.1109/ICACCCN.2018.8748287.
8. Thavani, S., Sharma, S. & Kumar, V. Pose invariant non-frontal 2D, 2.5D face detection and recognition technique. *Int. j. inf. tecnol.* (2023). https://doi.org/10.1007/s41870-023-01335-2
9. Y. Lei, X. Jiang, Z. Shi, D. Chen and Q. Li, "Face Recognition Method Based on SURF Feature," *2009 International Symposium on Computer Network and Multimedia Technology*, Wuhan, China, 2009, pp. 1–4. https://doi.org/10.1109/CNMT.2009.5374808.
10. International Journal of Innovative Technology and Exploring Engineering (IJITEE), vol. 2, no. 4, March 2013, Retrieval Number: D0572032413/13©BEIESP, Blue Eyes Intelligence Engineering and Sciences Publication (BEIESP), ISSN: 2278-3075.
11. S. Du, M. Ibrahim, M. Shehata and W. Badawy, "Automatic License Plate Recognition (ALPR): A State-of-the-Art Review," in *IEEE Transactions on Circuits and Systems for Video Technology*, vol. 23, no. 2, pp. 311–325, Feb. 2013. https://doi.org/10.1109/TCSVT.2012.2203741.
12. Zhigang Zhang and Cong Wang, "The Research of Vehicle Plate Recognition Technical Based on BP Neural Network," AASRI Procedia, vol. 1, pp. 74–81, 2012.
13. Xiaojun Zhai, Faycal Bensaali and Reza Sotudeh, "OCR-Based Neural Network for ANPR," IEEE, pp. 1, 2012. 34
14. P. Viola and M. J. Jones, "Robust real-time face detection," Int. J. Comput. Vis., vol. 57, no. 2, pp. 137–154, May 2004.
15. X. Tan and B. Triggs, "Enhanced local texture feature sets for face recognition under difficult lighting conditions," IEEE Trans. Image Process., vol. 19, no. 6, pp. 1635–1650, Jun. 2010.
16. O. Barkan, J. Weill, L. Wolf, and H. Aronowitz, "Fast high dimensional vector multiplication face recognition," Proc. IEEE Int. Conf. Comput. Vis. (ICCV), pp. 1960–1967, Dec. 2013.
17. S. R. Arashloo and J. Kittler, "Energy normalization for pose-invariant face recognition based on MRF model image matching," IEEE Trans. Pattern Anal. Mach. Intell., vol. 33, no. 6, pp. 1274–1280, Jun. 2011.

Machine Learning-Based Classification of X-Ray Images Using Convolutional Neural Networks

Manisha Uttam Waghmare, Vipul V. Bag, and Mithun B. Patil

Abstract In many low-income countries, a significant portion of the population faces challenges in accessing quality healthcare. This study proposes an automated machine-learning approach to segment X-ray images, aiding medical personnel in identifying fractures. Medical facilities typically have their own image databases and X-ray machines. The objective is to classify digital X-ray images into five categories: elbow, leg, spinal cord, toes, and other items. This report presents the findings from assessing X-ray image identification for the Imaging CLEF-2019 challenge. The study utilizes advanced feature extraction and classification algorithms to categorize images, focusing on features that highlight bone size and shape. Techniques include edge detection, classification algorithms, and convolutional neural networks to enhance bone fracture detection based on bone morphology.

Keywords X-ray images · Classification · CNN · Fracture detection · Machine learning

1 Introduction

The rapid accumulation of diagnostically relevant data stems from the increasing volume of medical imaging in clinical practice. This study focuses on using Convolutional Neural Networks (CNNs) for classifying and detecting fractures in X-ray images. Content-Based Image Retrieval (CBIR) is gaining traction in image-based content retrieval, enabling efficient search capabilities within large medical image databases. X-ray imaging, widely used in healthcare facilities, offers valuable insights. Various methods exist for acquiring and classifying medical image data. Local Binary Patterns (LBPs) are employed for classification, while Yijie Luo utilizes gray-scale co-occurrence matrices for character extraction. Cunzhe Lu

M. U. Waghmare (✉) · V. V. Bag · M. B. Patil
Department of CSE, N. K. Orchid College of Engineering & Technology, Solapur, Maharashtra, India

A. Patel et al. (eds.), *Advances in Machine Learning and Big Data Analytics II*,
Springer Proceedings in Mathematics & Statistics 442,
https://doi.org/10.1007/978-3-031-51342-8_27

combines local and global features using pixel values and Canny edge detection. IRMA code categorizes images based on modality, body position, anatomical location, and biological system complexity. The Bag of Visual Words (BoVW) paradigm assesses X-ray images by extracting feature descriptors from regional patches. Cosimo Lacinato employs fuzzy set theory, using the Canny Edge Detector and Discrete Gabor Transform for feature extraction. Zare et al. propose photo annotation methods using Probabilistic Latent Semantic Analysis (PLSA).

In this study, we designed a unique CNN architecture tailored for segmentation by extracting detailed features from X-ray images. Our dataset includes 525 images of 19 body parts, imbalanced without spread adjustment, deemed adequate for clinical applications. Optimized for the ImageCLEF-2019 competition, our approach emphasizes features indicative of bone size and shape. We outline our architecture and thoroughly evaluate its performance, incorporating post-processing techniques to reduce false positives and enhance the F1 score. Whereas our dataset aligns with standard AI applications, our method demonstrates superior accuracy and generalizability compared to traditional image processing techniques. Our architecture also outperforms leading image segmentation networks across diverse applications, offering insights into X-ray image categorization for the ImageCLEF-2019 competition, particularly optimizing traits related to bone morphology.

2 Related Work

Please keep in mind that the purpose of this section is not to conduct a comprehensive writing review, but rather to establish the location of this work's procedure and outcomes within the flow of research seen. Because of its role in picture handling and other examination-based tasks, methods for fragmenting X-Beam images have long been a topic of discussion. Utilizing conventional picture-handling methods has been the focus of a lot of previous work. A common strategy is to group pixels according to similarity within specific boundaries. Kubilay Pakin et al. [1] use bunching in their division calculation, which yields high exactness scores but necessitates hyperparameter tuning for each class of body parts and makes it difficult to produce smooth limits. Essentially, Wu and Mahfouz achieved excellent results by employing ghostly grouping techniques and producing many smoother limits; however, this application was specifically tuned for knee picture examination. Bandyopadhyay et al. utilized entropy-based methods, which have enabled clean limit ID but suffer from the negative effects of unnecessary edges like bone breaks or image twists [2]. Kazeminia and others expand on this work by examining the force variances in pixel columns and utilizing existing edge location calculations like Sobel and Shrewd to select the bone limit more precisely. Although this work is less susceptible to noise, some limited progression is lost [3]. Essentially, chart book models were created for clinical picture division, and Candemir et al. provide an application for rib confine division. The creators can generate complete divisions using these methods, but the limits remain tense and the area under the ROC curve (AUC) is highly dependent on the dataset breakdown. The improvement of computer-based intelligence applications in the clinical benefits region has been noteworthy of late. There is no doubt that the vast amount of data encoded in X-Beam images has prompted extensive focused research into their investigation [4]. Many AI methods for finding oddities like pneumonia, pneumonic tuberculosis, and thoracic illness have been developed using such datasets. These methods also use chest X-Beams to diagnose various illnesses [5]. Additionally, Qin et al. give connections of an extent of mind networks applied to perceiving eccentricities in chest X-Pillars. These recognition technologies are meticulously tuned to these applications and typically perform image segmentation to identify the location of the anomaly or of a specific bone structure (most frequently the rib confine) [6]. Neural networks have been at the forefront of this investigation and have adopted various architectures including the encoder-decoder architecture like that presented in this paper, and fully connected networks. The image augmentation technique is also used by some of these networks to support network speculation and reduce the risk of overfitting [7]. In a similar vein, Badrinarayanan, Handa, and Cipolla present a refined version of the widely used SegNet engineering, demonstrating improved performance on smaller datasets. Undoubtedly, it is against this deal with the network, known as SegNet-Fundamental, that we benchmark our association execution. Although these applications have typically been restricted to the division of cell structures, it has been observed that some networks were specifically designed for the complete division of clinical images. There have been instances where businesses [8].

3 Proposed Methodology

Our strategy divides digital images of X-rays into five groups: bones, limbs, spinal cord, elbow, and miscellaneous objects (see Fig. 1). With an emphasis on traits signaling bone size and structure, our objective is to apply cutting-edge classifiers and feature selection techniques to optimize them for this difficult task. A collection of (1) tagged images of X-rays from five groups, (2) feature extraction strategy, and (3) classifiers are the inputs used by our procedure. With the provided five-group classification challenge, our method generates a set of ten feature-classifier pairs that have the highest accuracy classification.

- Extraction of color: The color distribution of the image is represented by a gray-scale histogram. From the picture, we choose seven 7×7 patches as well as compute an 8-bit histogram for each of them. The method's color feature is then created by merging every histogram based on the patch into the individual vector.
- Texture extraction: The Local Binary Pattern (LBP) by Khalid M. Hosny [16], which gives hugely discriminative texture data and is utilized to give reliable.

Figure 2 includes classifiers, feature extraction techniques, and tagged X-ray pictures [26]. There are two stages to our process. To choose the optimal features-

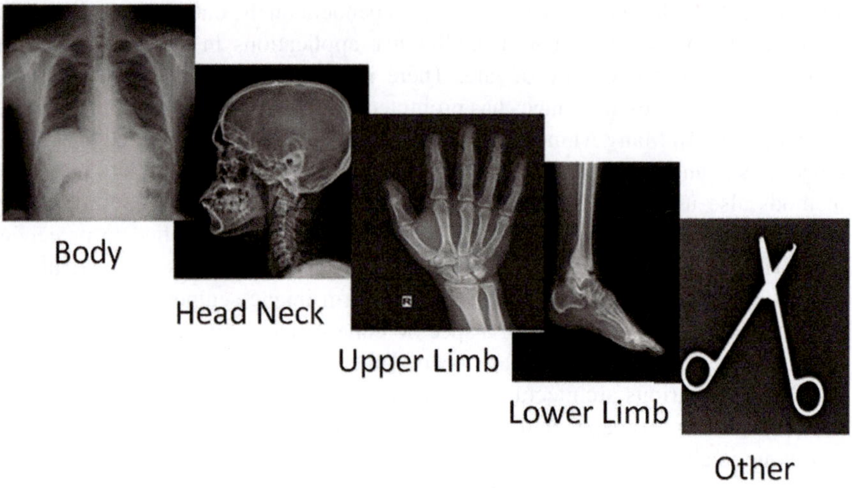

Fig. 1 Classify the X-ray body parts by using the KNN algorithm [26]

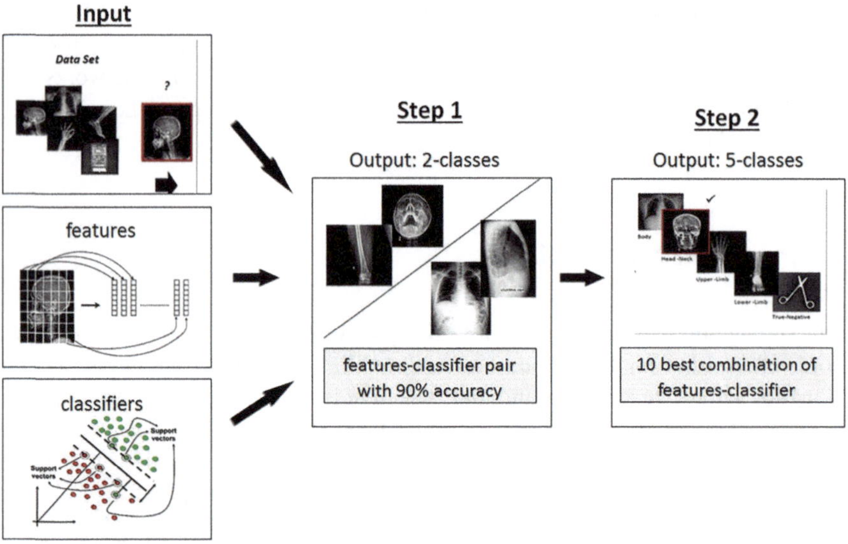

Fig. 2 Components of our system input

classifier pair as the initial step, a two-class experiment was employed. Distinguish between images of X-rays showing long, huge bones as well as those showing short, tiny bones. Phase 2: To choose the top 10 features-classifier combinations, the characteristics-classifier pairs that produced accuracy levels better than 90% in the earlier stage were trained on the five sets of X-ray pictures. The top 10 feature-classifier combinations were shown for the challenge evaluation.

- HoG: A histogram of neighboring pixels that are ordered by gradient direction and weighted by gradient magnitude. Every 10 × 10 patch in the image has its HoG values obtained. These values are then used to create the individual vector that represents the HoG characteristic of our approach [17].
- BoVW [18]. An approach that may be used to develop a visual vocabulary. The methods listed below were employed to create method descriptions from the identified key points
- Scale Invariant Feature Transform (SIFT) [19].
- Speeded-Up Robust Features (SURF) [20]
- Binary Robust Independent Elementary Features (BRIEF): BRIEF (ORB) quickly rotated while alighted.

We investigated the following classification methods:

1. *KNN* uses the labels shared by an object's *k-nearest neighbors* to identify it in space [22].
2. The SVM divides the spatial points into two different groups [23].
3. The predictor variables in the LR probability model result in a binary outcome.
4. Starting with a visual representation of the training data, DBN creates deep hierarchical layers. The DBN initially performs unsupervised pre-training in order to effectively apply supervised learning for classification, and then generates the network weights [24].

   k-NN calculations: Picture Classifier

   The *k-Nearest Neighbor (KNN)* classifier is, without a doubt, the most fundamental AI and image classification algorithm. In our case, the distance between feature vectors, which are the raw RGB pixel intensities of the images, is the only thing that matters in this calculation. The fundamental assumption that images with comparable visual elements lie close to one another in an n-dimensional space is necessary for the k-NN calculation to function. Here, dogs, cats, and pandas are each represented as a separate category of images. In this envisioned model, we have plotted the "fluffiness" of the animal's coat along the x-axis and the "delicacy" of the coat along the y-axis. In our n-dimensional space, each of the information focuses on creatures assembled somewhat close together. This suggests that the distance between two images of felines is much smaller than the distance between canines. However, before we can use the k-NN classifier, we must select a distance metric or similarity function. A typical choice integrates the Euclidean distance (commonly called the L2 distance):

   (a) Nevertheless, other distance metrics like the Manhattan/city block (commonly called the L1-distance), can be used as well:
   (b) In general, you can use any distance metric or similarity function that works best with your data and produces the best classification results. Nevertheless, the most well-known distance metric will be used until the end of this example: the Euclidean distance.
   (c) The *Support Vector Machine (SVM)* is a supervised artificial intelligence algorithm that can be used for both classification and relapse issues. How-

ever, most of the time, it is used in arrangement issues. We plot every data point as a point in n-dimensional space for this SVM algorithm, where n is the number of elements you have, and the value of each feature is equal to the value of a specific coordinate. Then, we perform classification by finding the hyperplane that separates the two classes very well. Some key parameters of SVM are:

- Gamma: Determines the extent to which biased results from the impact of single preparation models on values.
- C: Controls the cost of misclassification.
- Minimal C: Sets the cost of misclassification LOW.
- Immense C: Sets the cost of misclassification HIGH.
- Kernel: A collection of numerical capabilities that are utilized in SVM algorithms. Kernels are categorized as Direct, Polynomial, and RBF (Radial Basis Function).

4 Experimental Setup

Our objective is to identify the ten feature-classifier combinations that achieve the highest accuracy for classifying five sets of X-ray images. We start by simplifying the task with a two-class approach to distinguish between X-ray images of long bones (such as those in the arm, head, and leg) and those of short or small bones (such as those in the chest and belly). This step helps reduce the complexity of the classification task and highlights the features that differentiate various bone types. We then select and train different feature-classifier pairs on the five groups of X-ray images and compare their performance. During our experiments, these pairs consistently achieved a mean accuracy of over 90% in the initial two-class classification task. The top ten feature-classifier combinations were further tested on a set of unlabeled test images provided by the ImageCLEF group. For feature extraction and classification, we used the OpenCV-Python library along with the machine-learning module from Python's scikit-learn. The experimental setup involved the following steps:

- Data Preprocessing: Classify the X-ray images into two categories based on bone size (long vs. short bones).
- Feature Extraction: Apply various methods to extract features that distinguish between different bone types.
- Classifier Selection: Combine the extracted features with different classifiers.
- Training: Train each feature-classifier pair on the five sets of X-ray images.
- Performance Evaluation: Evaluate and compare the performance of each pair to identify the top ten combinations based on accuracy.
- Testing: Test the top ten feature-classifier combinations on a set of unlabeled test images provided by ImageCLEF.

This approach allows us to pinpoint the most effective feature-classifier combinations for accurately identifying bone types in X-ray images.

4.1 Identification of Fractures

To locate the fracture site and set boundaries, a deep convolutional neural network (DCNN) technique was used. The DCNN is a type of nonlinear regression model. These capabilities integrate elements from radiographs as inputs to generate at least one output (pathologies identified in the radiographs). There are no restrictions whatsoever on the arcs that represent input/output links. Fitting these flexible bounds to a dataset that contains model information and mixed outcomes (the *training set*) is essential in order for building a model. The fracture localization model we developed frames this task as a problem of dependent semantic segmentation and binary classification. Or, to put it another way, one result is a single probability that supports a *yes* or *no* decision, and the other is a dense, detailed likelihood map that performs like a heatmap—a single probability.

A standard radiograph present to the model is shown in Fig. 3a left, right X-ray image A with heatmap overlay [27]. The intensity chart, with yellow and blue denoting high and low confidence, represents the model's confidence that the specific region is indicative of the fracture if the model confirms the existence of a fracture. (b) An enlarged heatmap with four additional inputs for illustration.

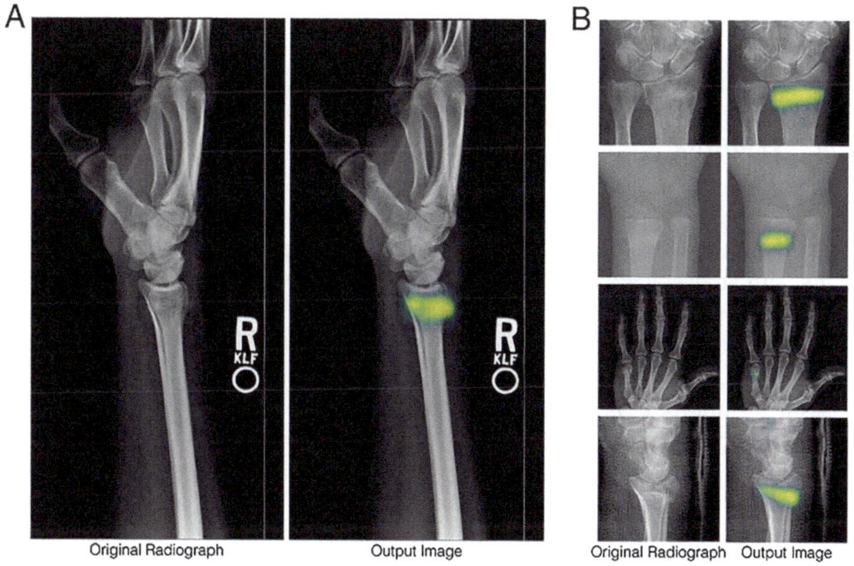

Fig. 3 Left, right X-beam image

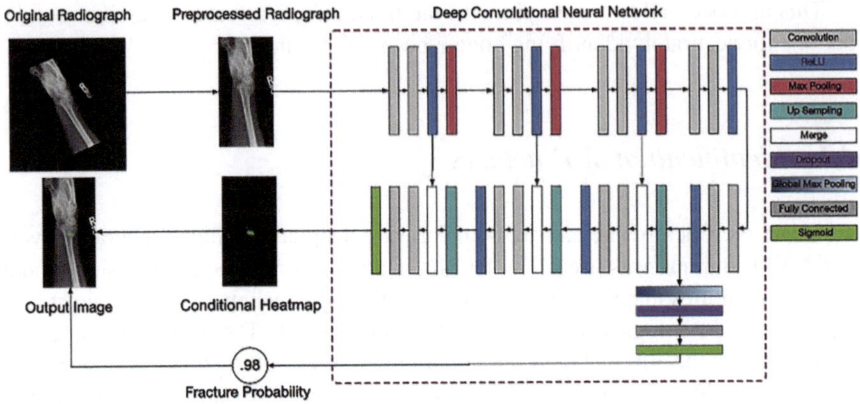

Fig. 4 How to identify and avoid radiography malfunctions

Figure 4 depicts that by rotating, cutting, and applying an aspect ratio–preserving rescaling effort [27], an input radiograph is first preprocessed to provide a nice object of 512 × 512 pixels. A DCNN also handles the inverted image. This DCNN's design now includes U-Net architecture [18]. The DCNN produces two results: (i) the probability that any region of the radiograph has a visible fracture, and (ii) an intensity chart based on the presence of a fracture, illustrating the likelihood that the fracture spans every region of the image. When a fracture's probability is high enough to warrant a clinical decision, action is taken. As a result, we used KNN to categorize human bones and subsequently DCNN to identify body fractures. This will aid doctors in making clinical decisions.

4.2 Training

In the two-class experiment, we combine 36 feature classifiers by employing all four classifiers and all nine feature sets. In the first step, we select feature-classifier pairs with an average accuracy of greater than 90%. With this choice, the number of possible combinations was reduced from 36 to 16. The 36 feature classifiers. The nine sets of characteristics were used to differentiate between long and short bones. The accuracy of the BoVW and HoG features is poor regardless of the classifier employed. When used together, color and texture yield an impressive 89–93% accuracy. The combination of color, texture, and HoG features yields the highest average classification accuracy for the classifiers. The second step's five-class approach was used to classify all five categories. We select 16 feature-classifier combinations that provide results with greater than 90% accuracy. The second stage involves assessing the efficiency of the strategies and narrowing the pool of feature classifiers down to the top ten.

5 Results

Given a small dataset and a break discovery strategy, we develop a completely programmed method for dividing clinical X-Beam images. We present a design with two encoder-decoder modules to bridge the gap between a comprehensive network that can separate in incontrovertible coarse-level features and fine-grained details and one that avoids overfitting small datasets. This network is trained using a dataset of 525 images that have been misleadingly increased to create 7000 preparation images. We discovered an overall accuracy of 92%, an F1 score of 0.92, and an AUC of 0.98 when evaluating the network's performance. We benchmarked our architecture against the popular SegNet plan. Beginning with the Segnet-Essential architecture [13] and carrying out a hyperparameter search similar to the one we use for XNet, the best result provides a 2% increase in bone classification precision compared with XNet. However, this comes at the expense of only having a true positive (TP) rate of 75% and a false positive rate of 25% in the delicate tissue class—in contrast to our network, which achieves a TP rate of 82%. Additionally, when it comes to distinguishing the open shaft area, we fundamentally outperform SegNet-Basic. For the diaphysis, soft tissue, and bone regions separately, SegNet-Basic achieves F1 scores of 0.96, 0.83, and 0.90, with an overall weighted F1 score of 0.89. This results in a lower F1 score in each classification when compared with XNet. There are many boundaries in the full implementation of SegNet [23] that prevent successful fitting to such a small dataset. Additionally, its size necessitates substantially more computational power, rendering it unsuitable for training in many medical facilities. When compared with cutting-edge traditional image-processing procedures, we also observe a significant improvement. The XNet segmentation separates well between bone and soft tissue regions and creates well-defined boundaries around bone locations. It is important to note that the X-ray scanner provided a number of high-resolution TIF images for our algorithm, but this analysis was conducted on substantially lower-quality JPEG images, so our network was unable to perform to its full potential on this image. In addition, our preparation set does not include any foot views from the point shown in Fig. 5, demonstrating our network's generalizability. The ROC curve of the model built from the two test sets is depicted in Fig. 3. In run 1, the AUC of the model was 0.967 ($n = 3500$; 95% CI, 0.960–0.973). For a subset of images from test set 2, the model reached an AUC of 0.994 ($n = 1243$; 95% CI, 0.989–0.996); however, there was no ambiguity regarding the reference standard (no interexpert conflict) case.

The radiograph within each bin's median unaided response time in seconds is shown by a point's horizontal location. The vertical location of a point indicates the average across-clinician diagnostic accuracy of the radiographs in the bin. After fracture detection in hand X-rays, you will get the above graph as a result. The accuracy disparity between aided and unaided reading conditions expands as unassisted reading time increases. The dashed-black horizontal line reflects the accuracy that a doctor would have attained if they had reported "no fracture" on each radiograph. Image CLEF [15] has made accessible a dataset of 750 X-ray

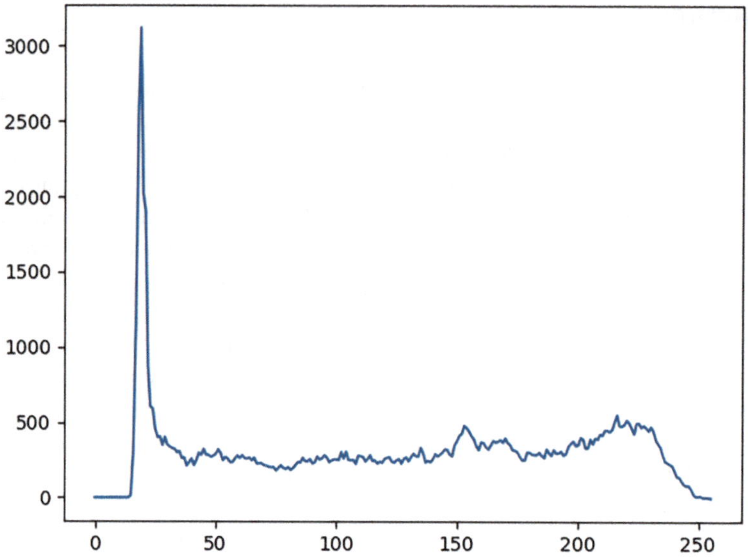

Fig. 5 Train image prediction accuracy

images: (a) 500 labeled photographs (100 images from each group) were made accessible for training purposes. The five image groupings are X-rays of toes, limbs, spines, elbows, limbs, and miscellaneous body parts, and 250 labeled photographs for benchmarking and difficulty evaluation. We begin by providing the results of our two-step methodology's training set of 500 labeled X-ray images. The outcomes of unlabeled images of 250 images are then displayed, as confirmed by the organizers of the challenge.

6 Conclusion

We evaluated various feature extraction approaches and X-ray image classification methods, utilizing classifiers such as SVM, KNN, LR, and DBN. These classifiers employed features extracted from the images, including color, texture, HoG, and BoVW. Our training dataset consisted of 500 X-ray images from the ImageCLEF-2019 body component X-ray challenge, and we tested the classifiers on 250 images. The combination of the KNN classifier with intensity, texture, and HoG features achieved the highest classification accuracy, with 86% accuracy on the training set and 73% on the test set. In future work, we plan to explore additional classifiers and further refine our approach to improve accuracy by examining how the original clusters can be effectively partitioned into subclusters.

References

1. Kubilay Pakin: "Clustering approach to bone and soft tissue segmentation of digital radiographic images of extremities," Journal of Electronic Imaging 12, 12–10 (2003).
2. Bandyopadhyay: "Entropy-based automatic segmentation of bones in digital x-ray images," Pattern Recognition and Machine Intelligence, 122–129 (2011).
3. Kazeminia: "Bone extraction in x-ray images by analysis of line fluctuations," in [2015 IEEE International Conference on Image Processing (ICIP)] (2015)
4. Sema Candemir: Atlas-based Rib-Bone Detection in Chest X-rays April 2016
5. Wang, X: "Chestx-ray8: Hospital-scale chest x-ray database and benchmarks on weakly- supervised classification and localization of common thorax diseases," in [2017 IEEE Conference on Computer Vision and Pattern Recognition (CVPR)], 3462–3471 2017
6. Islam: "Abnormality detection and localization in chest x-rays using deep convolutional neural networks," arXiv preprint arXiv:1705.09850 (2017).
7. Dung Nguyen: Memory-Augmented Theory of Mind Network (March 2023).
8. Badrinarayanan: "Segnet: A deep convolutional encoder-decoder architecture for image segmentation," IEEE Transactions on Pattern Analysis and Machine Intelligence 39(12), 2481–2495 (2017).
9. Xiaoyuan Guoa: An Outlier-Sensitive Content-Based Radiography Retrieval System (6 April 2022).
10. Yijie Luo: Local Binary Pattern Based on Magnitude Ranking for TextureClassification Date Added to IEEE Xplore (10 Jan 2022).
11. Cunzhe Lu: Xiaogang An Improved FAST Algorithm Based on Image Edges for Complex Environment Published online (20 Sep 2022).
12. Lehmann: The Irma code for unique classification of medical images. In: Medical Imaging 2003. pp. 440–451. International Society for Optics and Photonics. 2003
13. K France: Classification and retrieval of thoracic diseases using patch-based visual words: a study on chest x-rays Published (10 march 2020).
14. Cosimo Ieracitano: A fuzzy-enhanced deep learning approach for early detection of Covid- 19 pneumonia from portable chest X-ray images (7 April 2022).
15. Amin Overview of the ImageCLEF 2019 medical clustering task. In: CLEF2019 Working Notes. CEUR Workshop Proceedings, CEUR- WS.org, Toulouse, France (8 September 2021).
16. Khalid M. Hosny: Refined Color Texture Classification Using CNN and Local Binary Pattern Published (16 Nov 2021).
17. Adarsh Joshi: A Review on Human Activity Recognition using HOG Feature Issue (2016).
18. Manisha Saini: Bag-of-Visual-Words codebook generation using deep features for effective classification of imbalanced multi-class image datasets (2021).
19. Tao He: Content-Based Image Retrieval Method Based on SIFT Feature (January 2018).
20. Bay: Speeded-up robust features (surf). Computer vision and image understanding 110(3), 346–359 (2008).
21. Sumit Lakhera: Binary Robust Independent Elementary Features (March 23 2022).
22. Padraig Cunningham: k-Nearest Neighbour Classifiers: 2nd Edition (with Python examples) (9 April 2020).
23. Sai yeshwanth chapati: Image Classification using SVM and CNN (March 2020).
24. Matteo Zambra: A Developmental Approach for Training Deep Belief Networks (22 Dec 2022)
25. Amin: Overview of the ImageCLEF 2019 medical clustering task. In: CLEF2019 Working Notes. CEUR Workshop Proceedings, CEUR- WS.org, Toulouse, France (September 8-11 2019)
26. Automatic Classification of Body Parts X-ray Images Moshe Aboud, Assaf B. Spanier, and Leo. Joskowicz published in conference 2015
27. Deep neural network improves the fracture detection by clinicias Robert Lindsay 22 October 2023.

A Comparative Study of Inception Models for Bone X-Ray Classification and Pathology Detection

K. Anusha, Karanam Sai Surya, Jadhav Prashanth, and Chaluvadi Kartheekeya

Abstract This work presents a novel two-stage technique for categorising and locating bone abnormalities in X-ray images. The suggested approach concentrates on the upper extremities and takes into account the bones in the humerus, forearm, wrist, elbow, shoulder, hand, and finger. To achieve optimal performance with the least amount of computing work, it combines eight original models that were created from scratch and were influenced by the InceptionV3 model. In the initial step, the area of the bone X-ray image is classified into one of seven groups. Following that, seven classifiers were trained to recognise irregularities in bones sent in the image. Eight models are therefore used in the classification step: one for categorisation and seven for abnormality recognition. The study evaluated the efficacy of the proposed method using the largest publicly accessible dataset of bone X-ray scans, the MURA database. The outcomes showed that, while retaining excellent accuracy levels, the suggested approach significantly cuts down on computation and processing time. Furthermore, the hierarchical structure of the system enables the simultaneous examination of bone categorisation and anomaly detection problems. This characteristic sets the suggested method apart from earlier research. Furthermore, this work is the first to incorporate all seven bone sections into the same system, providing a thorough way for classifying and identifying abnormalities in bones in X-ray pictures. Overall, the suggested methodology provides good levels of accuracy and reliability for classifying and identifying abnormalities in bone X-rays.

Keywords Deep learning · InceptionV3 · MURA database · X-rays · bones classification · Abnormality detection · Accuracy

K. Anusha (✉)
Department of Computer Science and Engineering, Vignan Institute of Technology and Science, Yadadri-Bhuvanagiri District, Telangana, India

K. S. Surya · J. Prashanth · C. Kartheekeya
Department of Computer Science and Engineering, Vignan Institute of Technology and Science, Yadadri-Bhuvanagiri District, Telangana, India

A. Patel et al. (eds.), *Advances in Machine Learning and Big Data Analytics II*, Springer Proceedings in Mathematics & Statistics 442,
https://doi.org/10.1007/978-3-031-51342-8_28

1 Introduction

A huge section of the world's population suffers from musculoskeletal problems, which are a serious public health concern. The wrong diagnosis can lead to unnecessary diagnostic procedures, patient injury, and long-term suffering and impairment from these illnesses. The usefulness of radiographic investigations as a diagnostic tool depends on how well the pictures are interpreted. Sadly, physical and psychological conditions can have an impact on a clinician's diagnostic accuracy. Hence, to improve patient outcomes and lessen the strain on healthcare systems, it is essential to develop methods for precisely detecting anomalies in radiographic scans. Clinicians can benefit from computer-aided diagnosis (CAD), which offers quick, accurate diagnoses while minimising time, effort, and human error. Doctors frequently use X-rays to identify abnormalities in the bones, but this procedure can be time-consuming and error-prone. X-ray-based computer-aided diagnosis (CAD) systems have been created as a result of recent developments in image processing and machine learning, which can help physicians make more accurate diagnoses of musculoskeletal diseases. Unfortunately, due to variances in bone forms and different types of abnormalities, Most studies exclusively focus on a single kind of bone, which is unsuitable for therapeutic application. Deep learning models have demonstrated tremendous promise in this work, yet it remains a tough task to develop a reliable CAD system that can reliably detect defects in many types of bones. Deep learning models are being utilised more frequently in medical diagnosis, bioinformatics, and computer vision despite their computational complexity since they have shown encouraging results in the detection of bone abnormalities. Nevertheless, creating a real-time CAD system employing deep learning models necessitates a significant amount of computing time and resources, which poses practical implementation difficulties.

The classification and anomaly identification in bone X-rays using a two-stage process in this paper combines eight models drawn from the InceptionV3 architecture. The suggested approach concentrates on seven upper extremity regions and delivers optimal performance with little computational overhead. In the first stage, the X-ray picture is classified into a single of seven bone categories based on the region, and in the second stage, the image is transmitted to a single of seven classifiers trained to discover abnormalities in the bones. The MURA dataset, the biggest collection of publicly available X-ray scans of bones, is used to assess the suggested method. The outcomes show how the suggested approach can shorten computation and processing times while retaining excellent accuracy. The suggested method's hierarchical structure enables simultaneous consideration of bone classification and abnormality detection, providing a thorough solution to the issue. This work differs from earlier research in that it makes use of InceptionV3-inspired models, which also help to attain cutting-edge performance. It has been demonstrated that InceptionV3 achieves greater accuracy and is built to handle input photos of various sizes. InceptionV3 uses regularisation approaches, such as batch normalisation and dropout, to enhance the model's generalisability and reduce overfitting. Due to the use of convolutional layers with various kernel sizes, which enables the network to collect characteristics at various scales, InceptionV3 is more computationally efficient. The model becomes quicker and more efficient as a result of the decrease in the number of parameters. Overall, the suggested strategy is a trustworthy and efficient method that provides excellent accuracy with little computational effort.

2 Related Works

Many people in recent years have had research projects that concentrated on the classification and anomaly detection of bone X-rays. We will highlight some of the most important results and contributions of these studies in this literature review.

One of the early studies in this field was conducted by Rajpurkar et al. (2017), who classified bone X-ray pictures into several groups using a deep convolutional neural network (CNN), such as normal, abnormal, and pathological. They used a large dataset of over 100,000 images and attained a precision of 80%.

Another study by Wang et al. (2018) used a similar approach, but focused specifically on detecting vertebral fractures in bone X-ray images. They used a two-stage framework that first detected the presence of fractures and then classified them according to their severity. They achieved an accuracy of 84.2% in detecting vertebral fractures.

Kim et al. (2018) proposed a deep learning-based approach for osteoporosis diagnosis using bone X-ray images. They used a CNN model to extract features from bone X-ray images and trained a classifier to predict the risk of osteoporosis based on the extracted features. They achieved an accuracy of 75.2% in their experiments.

Rahimzadeh et al. (2018) developed a CNN model for bone X-ray classification using transfer learning. On a dataset of bone X-ray images, they optimised a pre-trained Convolutional Neural Network model. They achieved an accuracy of 92.9% in classifying bone X-ray images into normal and abnormal categories.

Gondara (2018) proposed a bone fracture detection method using CNNs. They used a two-stage approach that first detected the presence of fractures and then classified them according to their type. They achieved an accuracy of 92.8% in detecting bone fractures.

More recently, Zhou et al. (2020) proposed a framework that combined CNNs and recurrent neural networks (RNNs) for bone X-ray classification and abnormalities detection. They used the CNNs to extract features from the bone X-ray images and fed them into an RNN model to model the temporal dependencies between the images. They achieved an accuracy of 91.3% in their experiments.

Another research, A Hybrid Two-Stage GNG—Modified VGG Method for Bone X-Rays Classification and Abnormality Detection (2021) Developed using the VGG model which has achieved an accuracy of 85%. A total of eight VGG models are built from scratch.

Inferring from prior research investigations, it can be said that creating precise models for categorising bone anomalies in X-ray pictures has a number of difficulties. These difficulties include a dearth of publicly accessible information, the diversity of bone forms, the difficulty of classifying different types of abnormalities, and the limitations of conventional feature extraction and classification techniques. Deep learning models also need a lot of computer power and training time, despite their promising results. Instead of only one bone as in most previous investigations, we suggest a unique technique in this study to classify anomalies in seven bones in the upper limbs: the elbow, wrist, shoulder, hand, finger, forearm, and humerus. This technique is based on a two-stage classification method based on the InceptionV3 model.

Overall, this research shows the potential of deep learning-based approaches for X-ray categorisation of bones and anomaly identification. They have proven to be highly accurate at spotting various anomalies and may help radiologists diagnose patients more precisely. For these strategies to be validated on bigger and more varied datasets and to be incorporated into clinical practise, more study is necessary.

3 Proposed Methodology

3.1 Data Collection

The collection of datasets is done using the GitHub repository. For testing and training, the **musculoskeletal radiographs** are used. The largest publicly accessible radiographic imaging databases overall are listed below, along with the largest public dataset for bone anomalies. It involves a variety of musculoskeletal analysis of medical images of the seven bones that make up the upper limbs, including the finger, elbow, wrist, forearm, humerus, shoulder, and hand. The bone is seen from different angles in each investigation. Consequently included are 40,561 multi-view X-ray pictures from the MURA (Musculoskeletal Radiographs) database. This study utilised both the training and testing sets (Fig. 1).

3.2 Dataset Pre-processing

The pipeline for the classification and abnormality detection of bone X-rays includes pre-processing techniques that serve to enhance the quality of the input data, reduce noise, and eliminate artefacts that could impair the effectiveness of machine learning models. In this section, several pre-processing techniques will be discussed. Methods that are frequently employed in this industry.

Image resizing: Bone X-ray pictures can vary in size and resolution, which can have an impact on how well machine learning models perform. To maintain

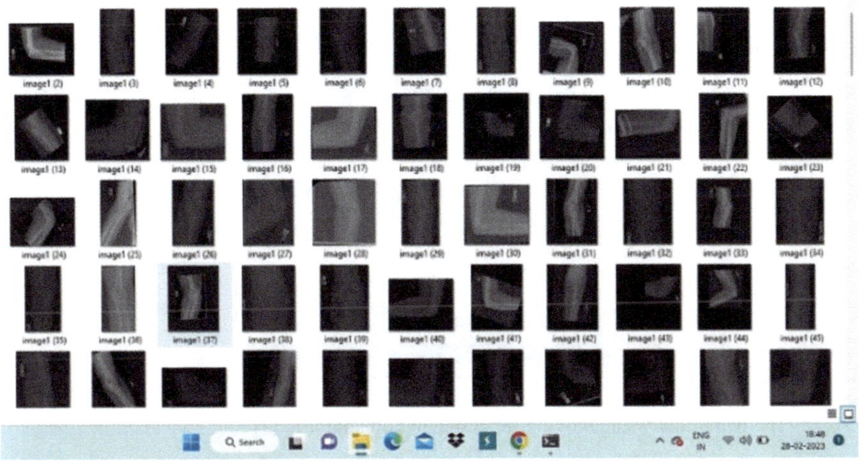

Fig. 1 Dataset sample [5]

uniformity in the input data, image resizing is a standard pre-processing step that entails scaling the photos to a predefined size, often 256 × 256 or 512 × 512 pixels.

Image normalisation: Image normalisation is a pre-processing procedure that involves converting the image's intensity values to a standard scale. To do this, divide the result by the standard deviation after subtracting each pixel's mean pixel value. Normalisation enhances visual contrast and helps to lessen the impact of varying lighting. Image enhancement: The quality of bone X-ray images can be increased by applying image enhancement techniques to certain elements, such as edges or textures. Histogram equalisation, contrast stretching, and adaptive filtering are examples of common improvement methods.

Image denoising: Many factors, including patient mobility, equipment artefacts, and radiation dispersion, can cause noise in bone X-ray images. The precision of machine learning models can be increased by using image denoising techniques such as median filtering, wavelet denoising, and Gaussian filtering to make the photos noise-free.

Image augmentation: Using transformations such as rotation, flipping, and zooming, picture augmentation algorithms generate new images from the original dataset. This can help to broaden the dataset's diversity and strengthen the stability of machine learning models.

Region of interest (ROI) extraction: Sometimes unrelated data, such as patient information or images of other body parts, may be present in the bone X-ray images. Techniques for ROI extraction can be used to separate the important portions of the image—like the bone or the area of abnormality—from the background.

Overall, pre-processing techniques are essential to the pipeline for classifying and detecting abnormalities in bone X-rays and can help to enhance the quality

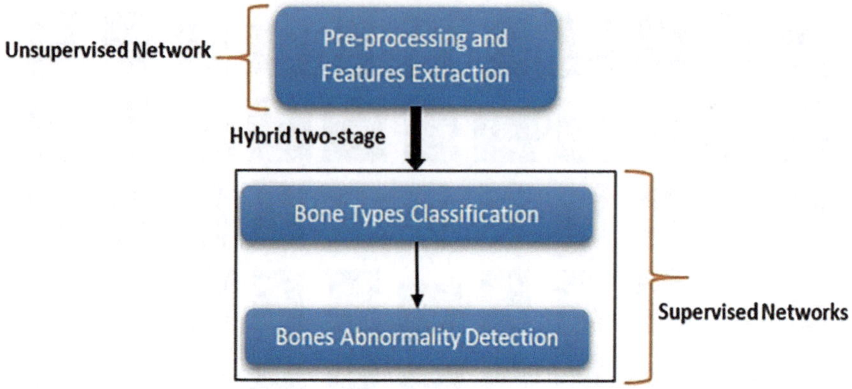

Fig. 2 The two primary stages of the suggested hybrid approach [7]

and consistency of the input data, which in turn enhances the precision of machine learning models.

The collected features are processed using two classification phases. To begin, each type of bone must be categorised into one of the seven groups (elbow, wrist, forearm, shoulder, humerus, finger, and hand). Identifying whether the categorised bone is diseased or normal is the next step. At the second stage of anomaly identification, every bone has its own classifier. Eight creative models created from nothing using the InceptionV3 model as inspiration are used to find the anomalies of each of the seven bones (Fig. 2).

3.3 Feature Extraction

Two main types of machine learning models, supervised and unsupervised networks, are employed in various applications.

When a model is trained on a labelled dataset, which means that the input data is accompanied by corresponding output or target values, the process is known as supervised learning. The objective of supervised learning is to develop a mapping between the input and the intended output so that the model can make predictions on future data that has not yet been observed. Logistic regression, neural networks, linear regression, and decision trees are supervised learning techniques examples.

Contrarily, in unsupervised learning, a model is trained using data that has not been assigned a label, meaning that the input data does not have a matching goal value or output. Unsupervised learning seeks to uncover patterns or structures in data without having any prior understanding of their meaning. Autoencoders, principal component analysis (PCA), and clustering are a few examples of unsupervised learning techniques. The decision between supervised and unsupervised learning relies on the issue at hand as both offer advantages and disadvantages. When the

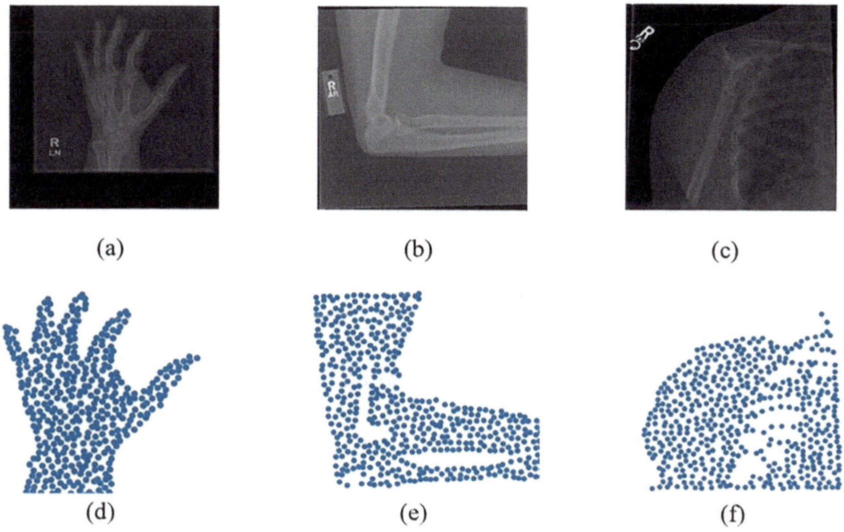

Fig. 3 (**a**, **b**, and **c**) are the original photos; (**d**, **e**, and **f**) are the images following the image pre-processing [8]

objective of a classification or regression problem is to make predictions based on labelled data, supervised learning is frequently used. In exploratory data analysis and anomaly detection, when the objective is to find patterns or anomalies in the data without knowing the values of the outputs beforehand, unsupervised learning is frequently utilised (Fig. 3).

The suggested method makes use of models drawn from the InceptionV3 architecture in a two-stage classification approach. Building a classification model to classify the various types of bones is required in the first stage, and the second stage comprises developing seven different models to find anomalies in each bone type. Eight models total—for stage 1, one, and stage 2, seven—are created from scratch using these models. To account for differences in the morphology of distinct bone types, different complexity, and layering levels were used to generate the models for stage 2. Yet, they all adhere to the Inception model's basic structure. For instance, the Humerus bone requires more layers than other bones to work optimally since it has a significant degree of variability in the views that the Mura database captures. As a result, compared to the models for other bone types, the model for stage 2 Humerus bone identification has more layers. The suggested method uses a multi-stage classification strategy with specific models for each type of bone to address the problem of identifying medical images with high structural and visual variability. The method can better account for the distinctive features and variances of each bone type by creating different models for each bone type, which increases the accuracy in spotting problems.

3.4 InceptionV3 Model

The primary reason to utilise InceptionV3 is because of its many benefits over other convolutional neural networks. Following are some of InceptionV3's main benefits:

Improved computational efficiency: Convolutional layers with various filter widths are combined in InceptionV3 to make better use of the available computer power. InceptionV3 can achieve equivalent or greater accuracy than other models with less computational cost by lowering the number of parameters needed for training.

Reduction of overfitting: InceptionV3 employs a method known as "dropout" to avoid overfitting. Randomly removing certain neurons during training causes dropout, which forces the network to acquire more robust features that are more transferable to fresh data.

Improved accuracy: In various image classification benchmarks, including ImageNet and CIFAR-10, InceptionV3 has produced results that are at the cutting edge of technology. This is because it can accurately forecast outcomes by capturing both local and global information in the input image.

Flexibility: By altering the layers' number and size, InceptionV3 is easily adaptable to various image classification applications. It is a flexible design as a result, suitable for a variety of applications (Fig. 4).

InceptionV3 supports larger convolution matrices known as inception modules (1×1, 3×3, 5×5, 7×7). The larger the module, the more relevant features are picked as in the case of our research, the bone images are converted into linear convolutions and then maximum relevant feature values from each sub-matrix are selected and form a separate convolution matrix. This is done by the Maxpooling layer. The obtained outputs are given as input to the flatten layer which arranges them in a vertical unit to give input to the fully connected layer. When the model is trained with an excess amount of data, i.e. the data more than required, it tries to recognise unnecessary or irrelevant patterns. This is known as overfitting. To avoid such issues, InceptionV3 offers a special layer called the dropout layer. A dropout

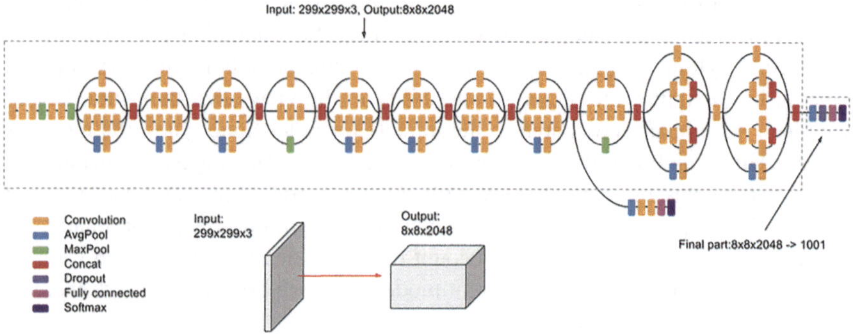

Fig. 4 InceptionV3 architecture [10]

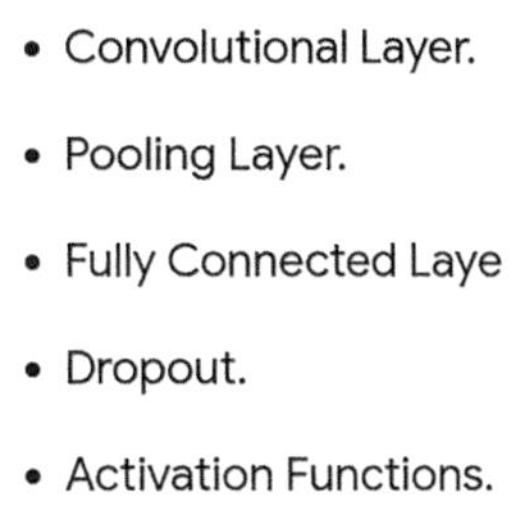

Fig. 5 Basic layers of InceptionV3 [11]

layer assigns zero to random units in a network which helps to reduce the overfitting. The Fully Connected (FC) layer, which links the neurons between two layers, is made up of the weights and biases as well as the neurons. The final few layers of a CNN Architecture are often placed before the output layer. This process flattens the input image from the preceding layers and feeds it to the FC layer. The flattened vector is subsequently subjected to a few more FC levels, which are often where mathematical function operations happen. At this point, the classification process gets started.

Last but not least, the activation function is one of the most crucial variables in the CNN model. They are used to identify and estimate any kind of complicated and continuous link between variables in the network. Simply described, it decides which model information should be sent forward to the network end and which information should not. It gives the network more nonlinearity. Many activation functions are often used, including the ReLU, Softmax, tanH, and sigmoid functions. These functions each have a particular use. While sigmoid functions are chosen for binary classification in CNN models, Softmax is often utilised for multi-class classification.

Concluding the fact that the InceptionV3 model has a wide range of unique functionalities that help to maintain a balance between computational time and accuracy (Fig. 5).

4 Experimental Results

As previously noted, eight unique models drawn from InceptionV3 are fed into a two-stage categorisation procedure. The first step in the classification process is to assign the X-ray of each of the seven different types of bones, and the second stage determines if the bone type is normal or pathological.

4.1 Datasets

Images that are employed from the musculoskeletal radiographs dataset are used to test and train. Among the most populous datasets for radiographic pictures overall, it is recognised as the greatest publicly available database of bone abnormalities.

There are 9067 normal and 5915 pathological upper extremity musculoskeletal radiography scans in it, covering the elbow, wrist, shoulder, finger, hand, forearm, and humerus. Several images of the bone are shown in each study. The MURA database's 40,561 multi-view X-ray images are a result. Broadly speaking, it is separated into two sets: training (37,111 images), validation (3225 images) and testing (559 images). As the testing set is not published, experiments are done on both training and validation sets.

4.2 Evaluation Metrics

Sensitivity, Specificity, and Average Accuracy are used to assess the success of the suggested two-stage approach. The normalised confusion matrices' TN, TP, FN, and FP values were used to calculate the metrics. Following are the metrics' formulas:

$$\text{Senistivity} = \frac{\text{TP}}{\text{TP} + \text{FN}}^{*} 100$$

$$\text{Specificity} = \frac{\text{TP}}{\text{TP} + \text{FP}}^{*} 100$$

$$\text{Average Accuracy} = \frac{\sum_{1}^{\text{NC}} \text{Accuracy of Each Bone}}{\text{NC}}$$

where, TP (True Positive) indicates the number of predicted values that match the actual values; FP(False Positive) indicates the number of predicted values that were falsely predicted; TN(True Negative) indicates the predicted value that matches the actual value; FN(False Negative) indicates the predicted value that was falsely predicted and NC is the number of classes taken into account (seven).

4.3 Results

The outcomes of the suggested Inception approach are shown in this section. For all of the tests, a GIGABYTE GeForce GTX 1080 Ti overclocked 11G graphics card and an Intel(R) Core(TM) i5 CPU M 460 @ 2.53GHz 2.53 GHz series were both utilised. To avoid the overfitting issue, the parameters of the suggested approach are experimentally altered using only the training data (Figs. 6, 7, 8, 9, 10, 11, 12, 13 and Tables 1, 2, 3).

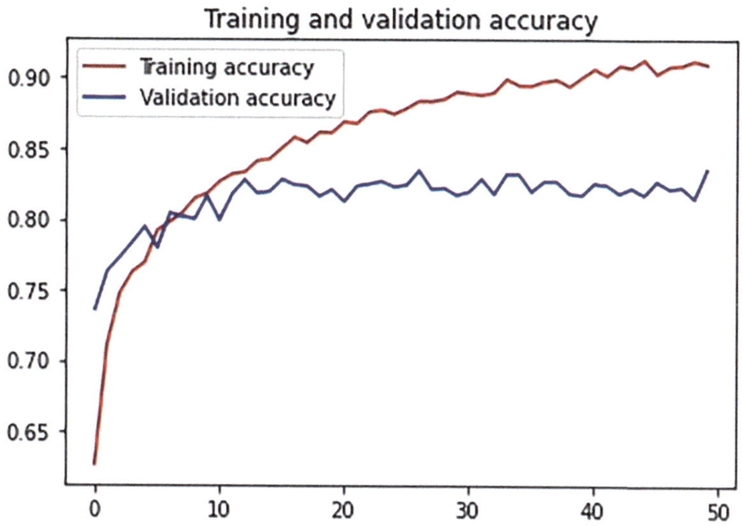

Fig. 6 Stage 1 classification accuracy graph [15]

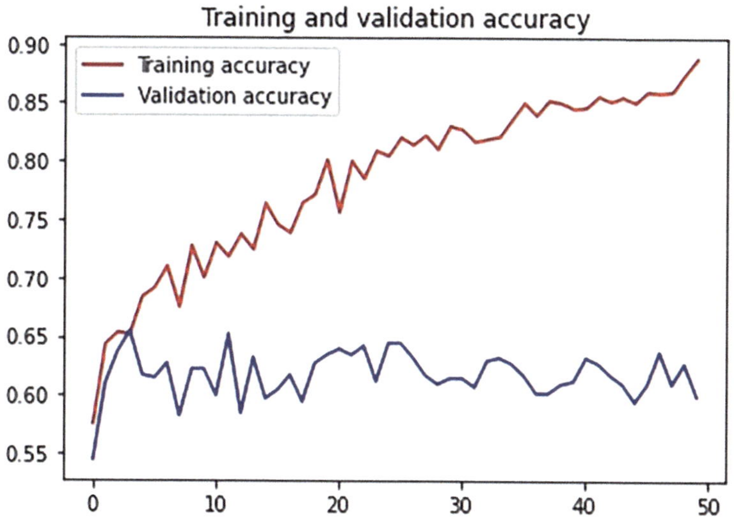

Fig. 7 Stage 2 detection of Elbow type [15]

5 Conclusion and Future Work

A hybrid hierarchy technique with two stages that is automated is created for classifying and diagnosing anomalies in the bones of the upper body i.e., the elbow, shoulder, wrist, finger, hand, forearm, and humerus. A two-stage hybrid

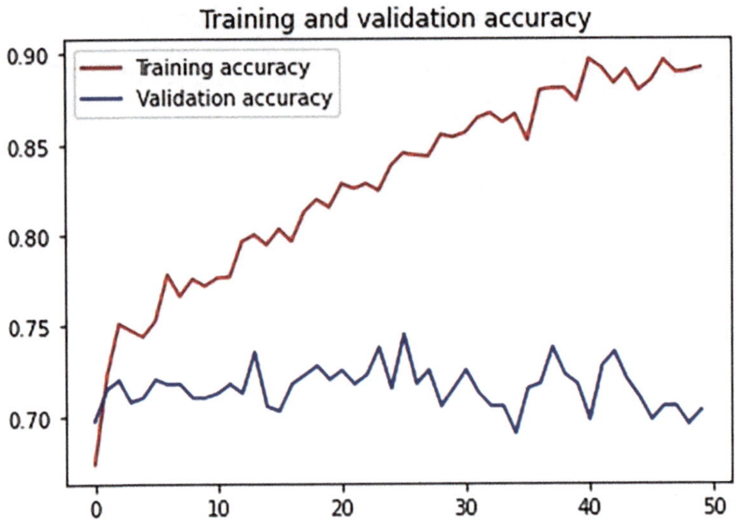

Fig. 8 Stage 2 detection of Finger type [15]

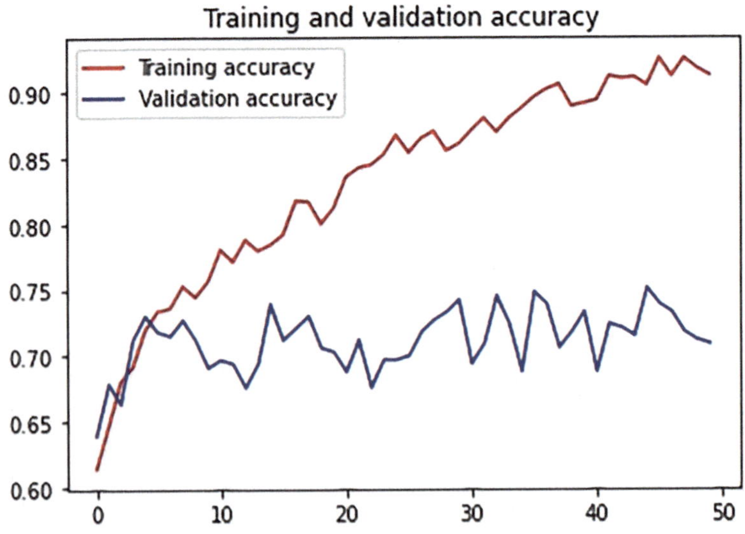

Fig. 9 Stage 2 detection of Forearm type [15]

classification phase using the extracted features employs eight alternative models that were built from the ground up and inspired by the InceptionV3 model. One model is used in stage one to categorise bones, and seven models are used in stage two, one for each kind of bone. To find abnormalities, one of seven models that successfully classified a bone X-ray from the first stage is provided in the

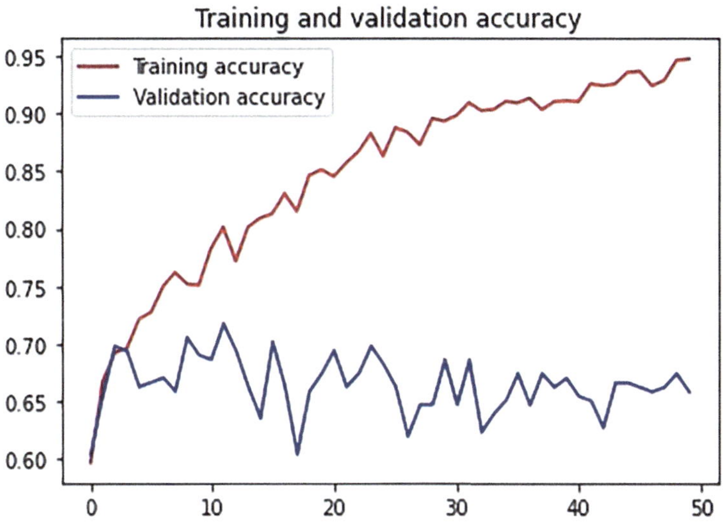

Fig. 10 Stage 2 detection of Humerus type [15]

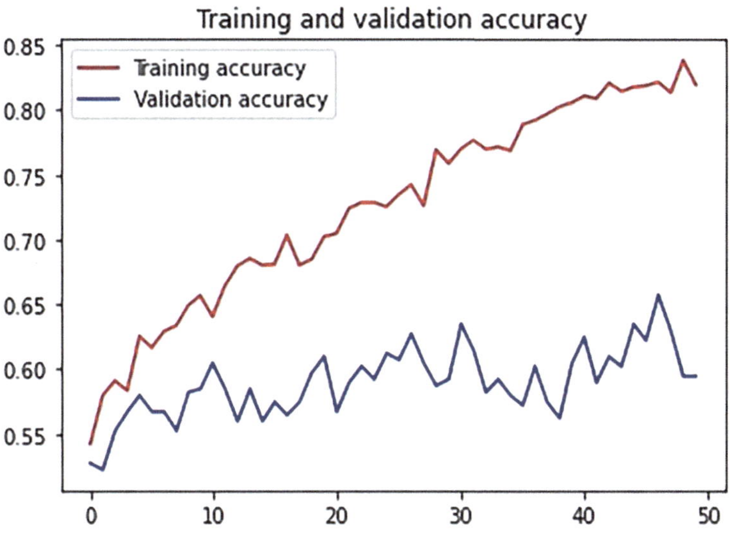

Fig. 11 Stage 2 detection of Hand type [15]

second stage. The MURA database was used in each study. The results show how trustworthy the proposed InceptionV3 models are. In the future, we anticipate (1) having a database with additional bone kinds to take advantage of the scalability of the suggested structure, and (2) taking into account other bone deformity categories.

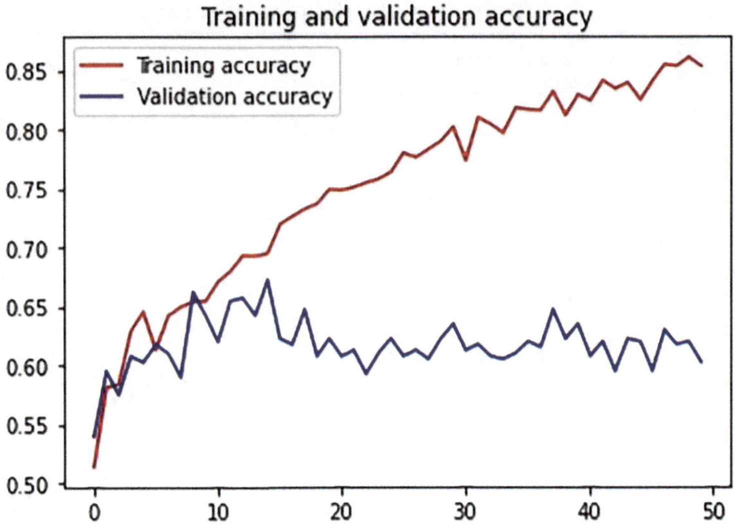

Fig. 12 Stage 2 detection of Shoulder type [15]

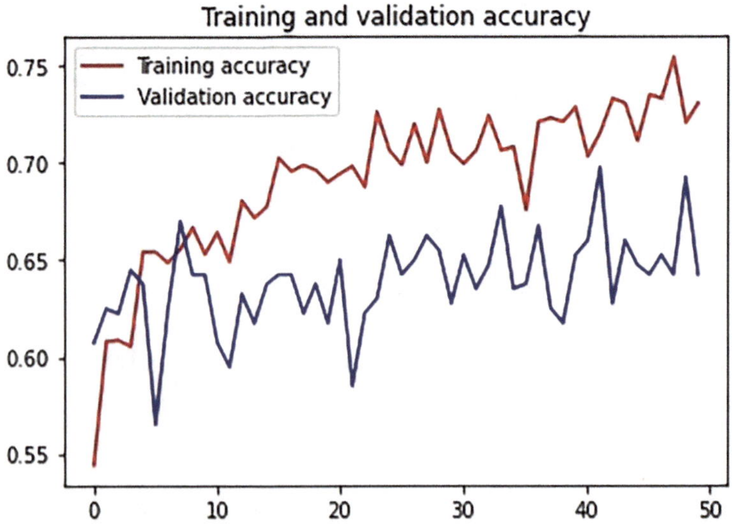

Fig. 13 Stage 2 detection of Wrist type [15]

As there is no such information in the MURA database, they are not taken into account in this study.

Table 1 The number of samples utilised in the studies for training and testing [15]

Bone structure	Wrist	Forearm	Humerus	Shoulder	Finger	Elbow	Hand
Quantity of training samples	4011	1825	1272	4009	3969	4011	3492
Quantity of testing samples	401	1125	575	1009	2609	3025	2120

Table 2 Stag using pre-trained existing and proposed models, stage 1 results are obtained [15]

Bone structure	Shoulder	Humerus	Finger	Elbow	Wrist	Forearm	Hand	Avg. accuracy
Inception	99.82%	90.63%	96.53%	97.63%	98.79%	82.39%	95.22%	94.43%
VGG16	99.82%	90.28%	96.31%	94.84%	97.12%	78.07%	98.04%	93.50%
VGG19	99.82%	79.51%	95.01%	92.47%	97.12%	67.77%	97.17%	89.84%

Table 3 Results from abnormality detection stage 2 utilising trained current and proposed models [15]

Bone structure	Shoulder	Elbow	Finger	Forearm	Hand	Humerus	Wrist	Avg. accuracy
Inception	79.00%	82.97%	60.32%	65.61%	88.16%	80.38%	71.09%	75.36%
VGG16	50.00%	65.00%	70.00%	89.00%	81.36%	56.94%	73.93%	69.46%
VGG19	50.00%	60.00%	75.00%	87.00%	80.70%	56.94%	84.00%	70.52%

References

1. Urinbayev, Kudaibergen, et al. "End-to-end deep diagnosis of x-ray images." 2020 42nd Annual International Conference of the IEEE Engineering in Medicine & Biology Society (EMBC). IEEE, 2020.
2. Caron, Rodrigue, et al. "Segmentation of trabecular bone microdamage in Xray microCT images using a two-step deep learning method." Journal of the Mechanical Behavior of Biomedical Materials 137 (2023): 105540.
3. Candemir, Sema, and Sameer Antani. "A review on lung boundary detection in chest X-rays." International journal of computer assisted radiology and surgery 14 (2019): 563–576.
4. Qin, Chunli, et al. "Computer-aided detection in chest radiography based on artificial intelligence: a survey." Biomedical engineering online 17.1 (2018): 1–23.
5. Joshi, Deepa, and Thipendra P. Singh. "A survey of fracture detection techniques in bone X-ray images." Artificial Intelligence Review 53.6 (2020): 4475–4517.
6. Rajaraman, Sivaramakrishnan, et al. "Chest x-ray bone suppression for improving classification of tuberculosis-consistent findings." Diagnostics 11.5 (2021): 840.
7. Amiya, Gautam, et al. "Assertion of Low Bone Mass in Osteoporotic X-ray images using Deep Learning Technique." 2022 4th International Conference on Advances in Computing, Communication Control and Networking (ICAC3N). IEEE, 2022.
8. El-Saadawy, H., Tantawi, M., Shedeed, H. A., & Tolba, M. F. (2021). A Hybrid Two-Stage GNG–Modified VGG Method for Bone X-Rays Classification and Abnormality Detection. IEEE Access, 9, 76649–76661.
9. Mery, Domingo, et al. "Automated fish bone detection using X-ray imaging." Journal of Food Engineering 105.3 (2011): 485–492.
10. Christian Szegedy, et al. "Rethinking the Inception Architecture for Computer Vision," 2016 IEEE Conference on Computer Vision and Pattern Recognition, (2016): 2818–2816.
11. Anu, T. C., and R. Raman. "Detection of bone fracture using image processing methods." Int J Comput Appl 975 (2015): 8887.

12. Sarki, Rubina, et al. "Automated detection of COVID-19 through convolutional neural network using chest x-ray images." Plos one 17.1 (2022): e0262052.
13. "Bone X-ray Classification Using a Deep Convolutional Neural Network" by Dong Yang et al. (2019)
14. Li, Shuo, et al. "An automatic variational level set segmentation framework for computer aided dental X-rays analysis in clinical environments." Computerized Medical Imaging and Graphics 30.2 (2006): 65–74.
15. "Automated Classification of Bone X-ray images using Deep Learning" by Muhmmad Arsian et al. (2019)
16. "A Deep Learning Approach for Bone Age Assessment and Abnormality Detection in Pediatric X-ray images" by Emanuele Marini et al. (2018).
17. "Abnormality Detection in Chest and Bone X-ray images using Deep Convolutional Neural Networks" by Tushar M, Pachori et al. (2019).

Dietary Assessment Report Generation Using RCNN

V. Sesha Bhargavi, M. Aishwarya, Mounika Jadi, Sara Fathima, P. Soumya, and K. Ramya

Abstract Food is an important part of human life and we must keep track of dietary details before consuming them. It is necessary to know what sort of food we are consuming and have an idea about the content present in it. This paper majorly focuses on the idea behind the project, methodology, results and conclusions thus obtained. The basic aim of the project is to detect single or multiple foods from an image using a faster RCNN algorithm and to provide a detailed report about the dietary details present in it. This would provide a comprehensive understanding of a healthy diet and direct their daily recipe to enhance well-being. It would help people to know their consumption and make changes in their lifestyle using the given report to ensure they are going on the healthy track. It also saves us the doctors' consultancy fee and reduces the gym fee since eating healthy can definitely enhance their health conditions.

Keywords Region-based convolution neural network · Faster RCNN · Hypertext markup language · Machine learning · Deep learning · Image detection · Image segmentation · Classification · Bounding boxes

1 Introduction

Dietary assessment is a necessary component of the nutritional status assessment of an individual and is useful for other purposes such as reducing weight and maintaining health. Nowadays, we can see people show more interest in the food and the calories they intake, and having a tracker that allows them to calculate would make their lives easier. This project uses methodologies to detect the content of food that we are consuming. It could accurately help us detect the levels of food content and give us insights on how to improve our health through the report. An

V. S. Bhargavi · M. Aishwarya (✉) · M. Jadi · S. Fathima · P. Soumya · K. Ramya
Department of Information Technology, GNITS, Hyderabad, India
e-mail: lncs@springer.com

A. Patel et al. (eds.), *Advances in Machine Learning and Big Data Analytics II*,
Springer Proceedings in Mathematics & Statistics 442,
https://doi.org/10.1007/978-3-031-51342-8_29

algorithm we found very efficient from the literature survey was Faster RCNN. We have performed object detection for this project using the Faster RCNN algorithm. First, the user is required to upload the picture. Then the system checks whether the uploaded image contains food or not. If there is food present in the image, the trained model will create bounding boxes and then calculate the quantity of proteins, calories and carbohydrates present in the image. The name of the food present in the uploaded image will be displayed along with the calories, proteins and carbohydrates present in the uploaded image.

2 Literature Survey

In Zhu et al.'s article [1], the authors have provided a method called Progressive Self Distillation, also called PSD which is more effective than other methods which use multiple regions in a picture to detect a food. This refers to the suggested technique for gradually enhancing the network's capacity to mine additional information for food recognition. It talks about the dataset used to set up this network and steps in the whole process. It describes in detail the self-supervised learning, knowledge distillation, self-distillation, progressive training and implementation. It portrayed effective results in the end.

In Mohanty et al.'s study [2], the authors have come up with a huge dataset of 24,119 images. They wanted to create a huge database of food images so that everyone could use it for further food analysis if they wanted to carry out research. The dataset is called MyFoodRepo-273. It showed a result of 39,325 segmented polygons divided into 273 distinct classes. They processed the data using a variety of segmentation and annotation strategies, using IoU and mAP as assessment metrics.

In Salim et al.'s article [3], they talk about how evaluation techniques play a major role in dietary appraisal. It explains about deep learning and the procedure involved in deep learning. It talks about layers and assessment techniques. It explains how deep learning is used in the processing of foods such as fruits, vegetables, oil, fish and many more. It also includes a brief explanation of multiple food datasets and algorithms that can help us classify the food and assess them accordingly.

In Adhira et al.'s study [4], they explore a range of methods for identifying foods, some of them are traditional and use SVM while others are more advanced and, as a result, faster and more precise. Deep Learning and neural networks are used most frequently in these. Its accuracy is around 79%. It is a very time-consuming process and only detects the food items and gives the name but does not give the intake of calories.

In Jiang et al.'s article [5], the authors talk about a three-step technique that uses a deep convolutional neural network (CNN) for object classification and candidate region detection to recognise multi-item (food) photos. They have used UEC-FOOD256 and UECFOOD100 datasets in the project. Various assessment Metrics are used to evaluate the model. We have taken our major inspiration from this paper and used Faster RCNN as it gave good results.

In Mohammed et al.'s article [6], they mentioned the use of the K Nearest Neighbor algorithm, and Random Forest algorithm for food segment and feature extraction. SVM and certain Deep Learning models are used for food recognition. The model has more time complexity. It uses many algorithms which makes it a little complex and accuracy also decreases. The overall accuracy is 76%.

In Aguilar-Torres et al.'s study [7], they talk about food segmentation and food detection. They have used SVM for food segmentation. So, with the help of food segmentation, we can detect if the food is present on the plate or not. They have also used YOLO for food detection through which they can detection the food and give the results. It makes it more complex as it uses two algorithms. It has an accuracy of up to 90 percent only. If the dataset has more noise like target classes are overlapping them SVM does not work efficiently.

In Shimoda et al.'s article [8], the authors talk about a CNN-based approach to food image segmentation that does not call for pixel-by-pixel annotation. The suggested approach entails the identification of food regions through bounding box clustering and selective search, the estimation of saliency maps using the CNN model trained on the UEC-FOOD100 dataset, GrabCut guided by the maps, and region integration using non-maximum suppression. In the experiments, both the PASCAL VOC identification task and the food region detection task performed better using the suggested technique than RR-CNN.

3 Methodology

3.1 Work Flow

The landing page of the website is the login/ signup page. The user can log in or sign up as per requirement. The password is stored in the database. They can use it again when he logs in for the second time. After logging in, the user selects an image and specifies the quantity of food in grams for which they want the calories, fats, proteins and carbohydrates. Then, they could click on the Food Detect button to know the results. The food is detected accurately and the dietary details are displayed in a tabular format beneath the image. If there is more than one image in the picture then the dietary details of all the food items are displayed respectively. The carbohydrates, fats, proteins and calories are displayed in quantity. After having obtained the results, the user can log out of the portal.

Figure 1 shows the diagram flow of our proposed system. There are two stakeholders respectively, the user and the system database. The user can perform functions such as login, signup, giving the image, performing calculations and logging out of the application. When the food is analysed the fats, carbohydrates, proteins and calories are provided by the system database on the screen in a tabular format. The details as per quantity are fetched from the database where details of dietary content are stored in categorical format.

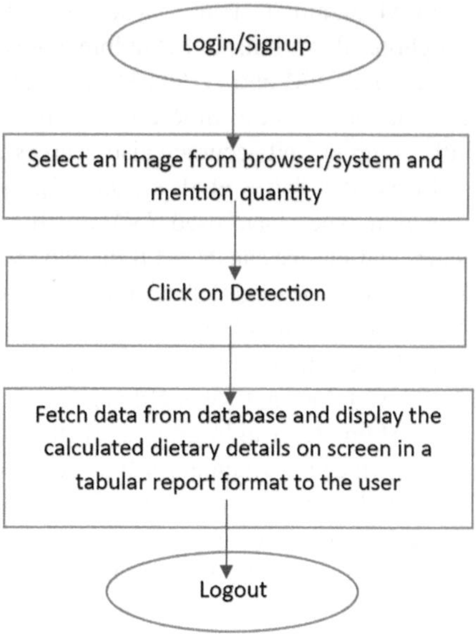

Fig. 1 Diagram flow of our proposed system

3.2 Architecture

The models were trained using TensorFlow. Our models were trained using the Faster RCNN Resnet152 v1 640 × 640 as the pretraining model.

There are 152 layers in the ResNet-152 model, including convolutional, pooling, fully connected, and shortcut connections. It is a deep convolutional neural network design that has been widely used for a range of computer vision tasks, including segmentation, object recognition, and image classification. Faster R-CNN is a well-liked approach for reliable object detection that combines region proposal generation and region-based convolutional neural network modules.

Feature Extraction In the Faster R-CNN model, the features are extracted using the ResNet152 v1 architecture. The 640 × 640 pixel input pictures are processed, and many convolutional layers are used to extract useful and differentiating information from the images. The architecture's hierarchical structure enables the extraction of characteristics at various degrees of abstraction.

Region Proposal Network (RPN) The RPN module is in charge of producing potential bounding box regions or region proposals in the input picture. Based on anchor boxes and objectness scores, it offers probable item positions using the feature maps acquired by the ResNet152 v1 feature extractor.

Region-Based Convolutional Neural Network (RCNN) The RCNN module classifies and improves the bounding box predictions by using the suggested regions of interest from the RPN. For object categorization and bounding box regression, it

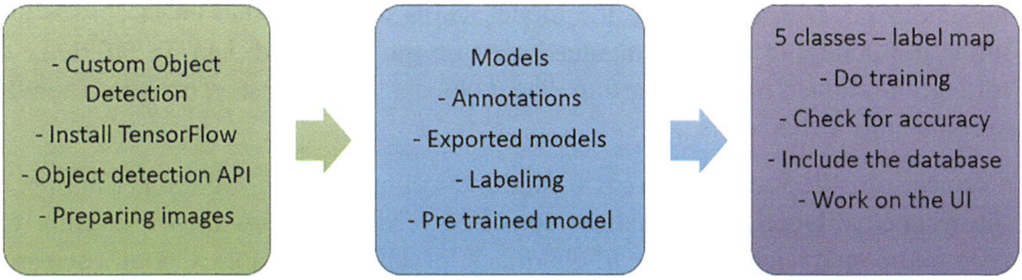

Fig. 2 Sequence of implementation

uses region pooling to align the suggested areas to a predetermined size and a string of completely linked layers.

Input Image Size The Faster R-CNN ResNet152 v1 640 × 640 variation only handles input pictures with a 640 × 640 pixel size. This size influences the model's object identification ability and computing needs by determining the resolution of the photos the model uses.

3.3 *Implementation*

Figure 2 depicts the sequence of steps followed to implement this project. We have used TensorFlow to train the models. We used the Faster RCNN Resnet152 v1 640 × 640 as the pretraining model and trained our models based on that. Several pre-trained object detection models are provided in TensorFlow 2 and may be used for a variety of applications, including keypoint recognition, instance segmentation, and object detection. These models are created using deep learning architectures like Faster R-CNN and are a component of the TensorFlow Object Detection API. These models may be found in the TensorFlow Model Zoo as pre-trained models. Each model may be used for a variety of object identification tasks and is trained on large-scale datasets like COCO (Common Objects in Context). The TensorFlow Object Detection API also offers tools for training and optimising these models on unique datasets. The variable classifier receives a Sequential model instance that has been generated for it. The classifier is given additional layers (Conv2D, MaxPooling2D). The Flatten layer is used to flatten the model. A loss function, metrics, and optimizer are built into the model.

Training and Inference The Faster R-CNN model using a ResNet-based feature extractor is trained using annotated datasets that contain item bounding boxes and the class labels that correspond to each one. To train them we have generated the annotations of images by considering the x1, x2, y1 and y2 coordinates of each image. After the regions of proposals are created, the images are labelled as per the category cause all the images taken as input should be categorised into five categories. The labelling of the images is done according to the five categories of images taken. During training, the model learns to predict the object classes and refine the bounding box coordinates based on the annotated data. In inference or testing, the trained model is used to detect objects in new, unseen images. Then the pretrained model is loaded.

Connection We have prepared a database containing the values pertaining to the respective labels. The database consisted of dietary assessment per gram for each compound such as fats, carbohydrates, proteins and calories. And a separate database for user credentials was prepared and as soon as a user registers himself the details are stored in it. This database prepared in sqlyog using MySQL connector is linked to the application made in PyCharm.

Interface The UI is made for login, signup, food detection and logout pages. Routing is created and a button is created to take input of quantity in grams. The user can enter a quantity and click on the Detect button to get the report of the dietary details of the input image.

4 Results and Discussion

4.1 Dataset

We have used the UEC FOOD-100 dataset which is publicly available for research purposes. Each class in the dataset consists of 100 images, which have been split into 70–30 for training and testing. Under that, we have used five different categories of food. The training set contains 50 images of each category. In total 250 images. So, by using the above images we are training the RCNN algorithm.

There are five categories respectively Caesar salad, cheese cake, chicken wings, cupcakes and hamburger. The calories, fats, proteins and carbohydrates are tallied in the tables and will be displayed when requested. After making the schemas and tables, the tables are connected to the UI respectively.

4.2 Database Tables

We have manually entered the data into the tables corresponding to each class. Each class is allotted a certain amount of carbohydrates, fats, proteins and calories. A database is created and linked to the application.

Table 1 shows the database created for the class labels and the quantity of each dietary detail for 1 gram of that nutrient. When we give the quantity manually in grams to generate a report then the quantity specified is multiplied with these dietary components and given out in the form of a table.

Table 1 Database for the food class labels

Food_name	Calories	Protein	Carbohydrates	Fats
caesar_salad	1	1	0.1	0
cheese_cake	3	1	0.2	1
chicken_wings	3	1	0.3	1
cup_cakes	3	1	1	1
hamburger	3	1	2	1
(NULL)	(NULL)	(NULL)	(NULL)	(NULL)

Table 2 Database for storing the user credentials

Name	Userid	Passwrd	Email	Mno
aishwarya	aishwarya	aishwarya	aiahu.m2552@gmail.com	7386635052
sara	Sara	Sara12345	sara@gmail.com	1234567390
(NULL)	(NULL)	(NULL)	(NULL)	(NULL)

Table 2 shows the database which stores the details of the users and their credentials so whenever the user logins again, it is auto-populated in the interface using this table.

4.3 *Output*

Figure 3 depicts the landing page of the website. The user is allowed to click on 'User' if they have already registered else click on User Signup if they have not registered earlier. A registration page is opened by clicking on User Signup as shown in Fig. 4.

Figure 4 depicts the interface where the user can register themself. They could enter a few personal details such as name, email and contact number.

Figure 5 is the major Food Detection page which takes the input of the image and the quantity for which it needs to display the dietary details. The user needs to upload the image, enter the number of grams and then click on the detection button so the output page is viewed.

Figure 6 shows the output displayed to the users. The food item is recognized accurately and the name of the category it belongs to is displayed at the top. It has great accuracy of more than 90% in detection of the food and it displays its accuracy at the top every time it detects a food item. At the bottom of the page, a table is displayed which contains the number of proteins, carbohydrates, fats and calories as per the specified quantity. It calculates the details for the quantity specified and displays that content on the page.

Figure 7 depicts a case that shows the result when the user gives an input image containing multiple objects and it would detect the dietary details for all the food items detected in the image.

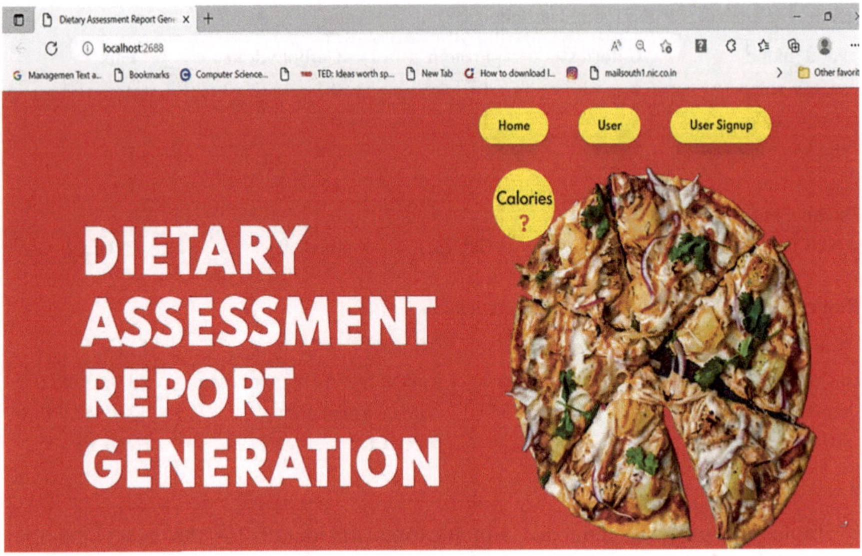

Fig. 3 Home page of the website

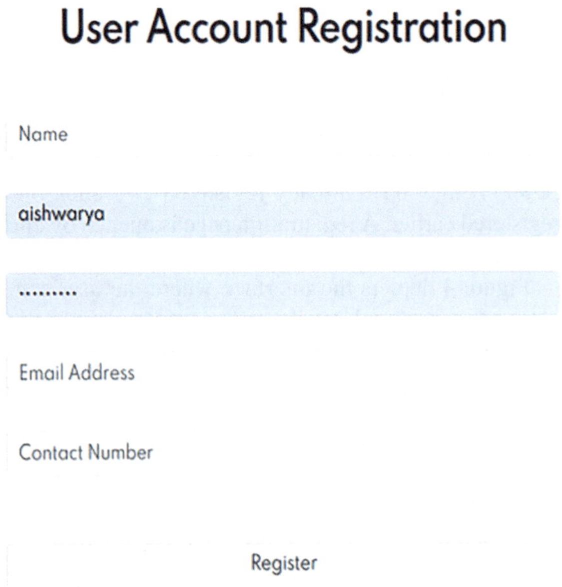

Fig. 4 Registration page

5 Results and Discussion

In recent trends, the identification of food objects from the images is a popular and trending research topic because food identification would be helpful for different

Fig. 5 Upload the image and enter the calories

FOOD CALORIE'S DETAILS

Food Name	Calorie's	Proteins (grams)	Carbohydrates (grams)	Fats (grams)
caesar_salad	100.0	100	100	0

Fig. 6 Food detection result

purposes such as food recognition, calorie estimation, identification of contents in the food, etc. In this project, we focused on the Identification of the Dietary details using the faster R-CNN algorithm. We have performed the training in 2000 steps and the output at every step gave a lossy value of 7–8%. It gave us an accuracy of 97%–100%. We could detect one or more images present in the picture and give calories, fats, proteins and carbohydrates estimation of the food. Based on the food's name and its appearance, we can determine the item's specifics. Compared to other algorithms such as VGG, K Nearest Neighbour, and Random Forest we have produced better results. Compared to Jiang et al.'s study [5], we have chosen fewer categories of only five foods because we trained using 3000 steps which could be a major disadvantage of our project. However, the project was able to detect the food items accurately.

Food Detection Results

FOOD CALORIE'S DETAILS

Food Name	Calorie's	Proteins (grams)	Carbohydrates (grams)	Fats (grams)
cup_cakes	3000.0	1000	1000	1000
hamburger	3000.0	1000	2000	1000

Fig. 7 Food detection result—Scenario 2

The future scope of the project would be to provide user recommendations on the intake of food items so that they can cut down on the excess intake of calories or fats. It could give them a detailed plan of what to take and create a timetable for them with respect to intake. Apart from that, this project could be extended to many more food items, which could give a wider approach to the users. It could be extended such that it could send the data to the user's respective doctors so they can track their patient's intake.

References

1. Zhu, Yaohui., Liu, Linhu., Tian, Jiang.: "Learn More for Food Recognition via Progressive Self-Distillation." Published under subject Computer Vision and Pattern Recognition (cs.CV), https://doi.org/10.48550/arXiv.2303.05073 (2023).
2. Mohanty, Sharada Prasanna., Singhal, Gaurav., Scuccimarra, Eric Antoine., Kebaili, Djilani., Héritier, Harris., Boulanger, Victor., Salathé, Marcel.: "The Food Recognition Benchmark: Using Deep Learning to Recognize Food in Images." In Frontiers in Nutrition Journal, https://doi.org/10.3389/fnut.2022.875143 (2022).
3. O. M. Salim, Nareen., Zeebaree, Subhi., M. Sadeeq, Mohammed., Radie, A., Shukur, Hanan., Najat, Zryan.: "Study for Food Recognition System Using Deep Learning." In Journal of Physics: Conference Series. 1963. 012014. https://doi.org/10.1088/1742-6596/1963/1/012014 (2021).
4. Adhira, Gupta., Sanjay, Sharma.: A Review: Food Recognition for Dietary Assessment/Calorie Measurement using Machine Learning Techniques. In (IJERT) ICACT – 2021 (Volume 09 – Issue 08) (2021).

5. L, Jiang., B, Qiu., X, Liu., C, Huang., K, Lin.: "DeepFood: Food Image Analysis and Dietary Assessment via Deep Model." In IEEE Access, vol. 8, pp. 47477–47489, 2020, https://doi.org/10.1109/ACCESS.2020.2973625 (2020).
6. Mohammed, Ahmed, Subhi.; Sawal, Hamid, Ali., Mohammed, Abulameer, Mohammed.: "Vision Based Approaches for Automatic Food Recognition and Dietary Assessment: Survey." In IEEE Access, vol. 7, pp. 35370–35381, 2019, https://doi.org/10.1109/ACCESS.2019.2904519 (2019).
7. Aguilar-Torres, Eduardo., Remeseiro, Beatriz., Bolaños, Marc., Radeva, Petia.: Grab, Pay and Eat: Semantic Food Detection for Smart Restaurants. In IEEE Transactions on Multimedia. 20. https://doi.org/10.1109/TMM.2018.2831627 (2017).
8. Shimoda, W., Yanai, K.: CNN-based Food Image Segmentation Without Pixel-Wise Annotation. In: Proc. of IAPR International Conference on Image Analysis and Processing 2015,https://doi.org/10.1007/978-3-319-23222-5_55(2015).

A Blockchain-Based Networking Approach for Advanced HealthCare Services

Bithi Mukhopadhyay, Subham Misra, Satyam Vatsa, Sovan Bhattacharya, Dola Sinha, Udit Narayana Kar, and Chandan Bandyopadhyay

Abstract In the recent pandemic we have witnessed several issues regarding not getting the bed available in hospital, and even if they got it, it might be with respect to a huge cost or after a hard way of searching for a bed. Recently we have also witnessed that the electronic health records are sold out or hacked by some organizations. Due to limitations in privacy, security, and full ecosystem interoperability, current healthcare technologies cannot adequately meet these requirements. Several studies have reported the availability of blockchain technology has led to the most secure transaction possible till date. Here we prepare a blockchain model for management of healthCare where the hospital beds are made available irrespective

Authors Subham Misra, Satyam Vatsa, Sovan Bhattacharya, Dola Sinha, Udit Narayana Kar, and Chandan Bandyopadhyay have contributed equally to this work

B. Mukhopadhyay (✉) · S. Misra · S. Vatsa
Department of CSE, Dr. B. C. Roy Engineering College, Durgapur, West Bengal, India
e-mail: dola.sinha@bcrec.ac.in

S. Bhattacharya
Department of CSE (Data Science), Dr. B. C. Roy Engineering College, Durgapur, West Bengal, India

Department of CSE, National Institute of Technology, Durgapur, West Bengal, India

D. Sinha
Department of Electrical Engineering, Dr. B. C. Roy Engineering College, Durgapur, West Bengal, India

U. N. Kar
Vellore Institute of Technology, Vellore, Andhra Pradesh, India
e-mail: uditnarayana.k@vitap.ac.in

C. Bandyopadhyay
Department of CSE (Data Science), Dr. B. C. Roy Engineering College, Durgapur, West Bengal, India

Department of CSE, University of Bremen, Bremen, Germany

A. Patel et al. (eds.), *Advances in Machine Learning and Big Data Analytics II*,
Springer Proceedings in Mathematics & Statistics 442,
https://doi.org/10.1007/978-3-031-51342-8_30

of the patient's wealth. Moreover, the blockchain here acts as a decentralized network which prevents manipulation of the data of hospital Beds and medical records available which will be helpful for every section of society. This chapter discusses the challenges and implementation of blockchain in the healthcare field. In this way, our research extends and complements existing blockchain research in healthcare by finding affordable hospital or medical packages with respect to the financial details of the patient.

Keywords Blockchain technology · HealthCare · Security · Data privacy · Shortest Path Graph (G)

1 Introduction

Blockchain is a decentralized and public digital ledger that records transactions on many computers so that no record involved can be altered retroactively without altering any blocks afterward. It provides an honest deal of accountability, and data is maintained on networks rather than a central database, improving stability and showing its proneness to be hacked. It can also be a distributed ledger network that adds and never deletes or modifies records without a common consensus, making it accessible and accountable to all network users. It facilitates better control of health records and patient care by minimizing the data redundancy, saving both practitioners and patients time and resources. The bed allotment algorithm is an efficient and time-efficient way to determine which hospital is best for a patient's ENT or plastic surgery needs.

Patient data will be stored in a medical data bank and shared easily, with an Electronic Health Record System used to store past hospital records, laboratory data, radiology reports, vital signs, and past physicians and consulted doctors. This data can be used for research and to understand the disease. In Kumar et al. [1], Blockchain technology can be used to create trustless and transparent healthcare systems. In Girija Moona et al. [2], measurement is essential for quality control in industrial manufacturing, supporting the objective of "make it right in the first time." In Mohamad Kassab et al. [3] Blockchain technology can support and challenge the healthcare domain for all stakeholders and assets. In Usharani Chelladurai & Seethalakshmi Pandian et al. [4], Blockchain smart contracts provide a regulated solution to the need of patients, physicians, and health service providers to exchange health information on a blockchain platform. In Rajesh Gupta et al. [5] HaBiTs framework provides security, privacy, and interoperability for telesurgery, addressing security, privacy, and interoperability issues. In Gautam Srivastava et al. [6] IoT and Blockchain Technology are bringing new applications and efficiency to Healthcare, with potential applications in other sectors. In Rim Ben Fekih et al. [7] Blockchain technology is being used in healthcare to share medical records. In Ahmed Afif Monrat et al. [8] Blockchain technology provides decentralization, immutability, transparency, and auditability, making transactions more secure. In Alaa A. Abd-alrazaq et al. [9] this study explores proposed or implemented blockchain technologies used to mitigate COVID-19 challenges, focusing on contact tracing and immunity passports. Blockchain technologies are expected to play an integral role in the fight against COVID-19, but further studies are needed to assess their effectiveness. In Abid Haleem et al. [10] Blockchain technology can improve data efficiency and security in healthcare by providing versatility, interconnection, accountability, and authentication for data access. In Suveen Angraal et al. [11] Blockchain technology offers secure, distributed database to improve healthcare data authenticity and transparency. In Anshuman Kalla et al. [12] Blockchain technology can play a key role in safeguarding humanity during the COVID-19 pandemic.

2 Key Concepts on Blockchain

Blockchain is a decentralized peer-to-peer (P2P) network of personal computers called nodes that stores, stores, and records historical or transaction data. It was introduced in 2008 as a peer-to-peer network over the Internet. Blockchains are public ledgers composed of blocks, which hold a complete history of transactions in a network. Smart contracts are self-contained code on the blockchain foundation that allows for direct show-through clarification and becomes permanently tamper-proof. Each member of the community can interact with the blockchain with a generated address that does not reveal the real identity of the user.

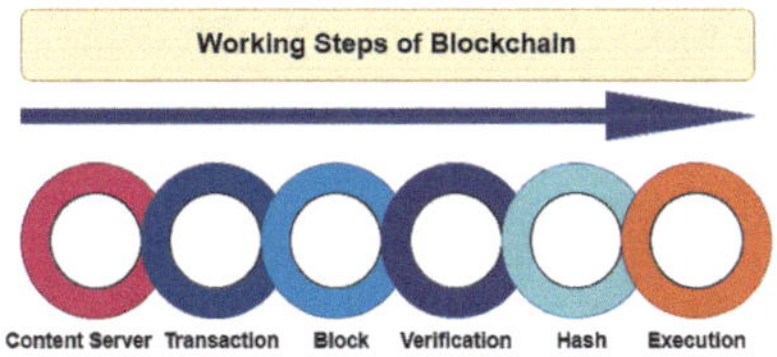

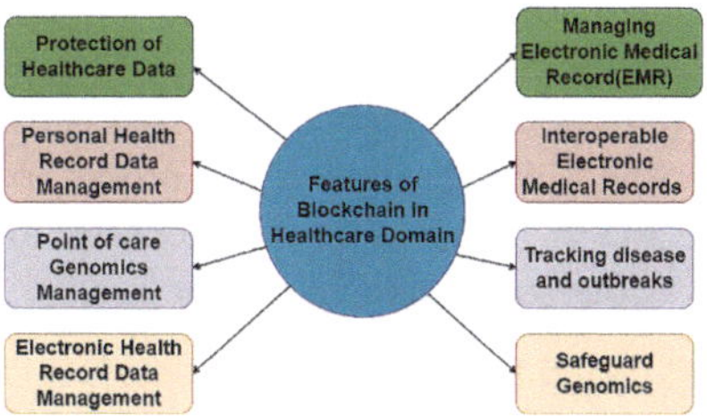

2.1 *Taxonomy of Blockchain Systems*

Current blockchain systems are classified into four types: public, private, consortium, and hybrid blockchain:

- **Public Blockchain:** Public blockchain provides a fully open network, allowing anyone to read and write within the blockchain.

- **Private Blockchain:** A private blockchain is a distributed ledger secured with cryptographic concepts, but only those with authorization can run a full node, negotiate, or validate changes.
- **Consortium Blockchain:** A consortium blockchain is a permission network used to securely store data for a privileged group, making it auditable and reliable.
- **Hybrid Blockchain:** Blockchain technology is playing a critical role in healthcare, enhancing patient care and high-quality health facilities. It also facilitates health information exchange, allowing citizens to participate in health study programs and better research.

2.2 Some Particular Concepts Used in the Paper

- **Hashing function**—A unique hash code created using user input data, fingerprint, and aadhar card Id to login in to the user account in the network.
- **Spanning tree**—According to the Kruskal algorithm, a spanning tree is a linked and undirected graph subgraph that connects all vertices. Many spanning trees can exist in a graph. The minimum spanning tree (MST) on each graph is the same weight or less than all other spanning trees. As V is the number of vertices in the graph, the minimum spanning tree has edges of $(V - 1)$, where V is the number of edges.

3 Proposed Technique

We have created a decentralized network using blockchain to store health data and hospital details, with strict orders to maintain the rules. Hospital beds are made available irrespective of patient wealth.

Four stages to achieve our goal:

1. Graph formation
2. Status calculation of hospitals
3. Improved Kruskal's algorithm
4. Bed allotment

3.1 Finding Graph from the Hospital Location

To find the shortest path from the user location to the desired hospital location, we have to connect all the hospitals location and form a graph. Connect all hospitals to

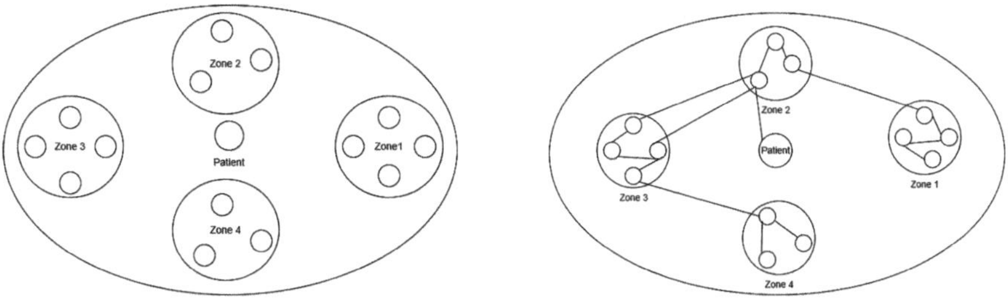

(a) Map of hospitals in the area (b) The source node from the patient

Fig. 1 Finding graph from the hospital location

form a graph to find the shortest path to the desired hospital. If needed, the graph will direct the user to the next nearest hospital. Let $h1, h2, h3 \ldots\ldots h7$ be the vertices of the graph and $p1, p2, p3 \ldots,$ $p5$ be the paths between the two vertices as you can see in Fig. 1b.

3.2 Status Calculation of Hospitals

We are going to calculate the status of the nodes/hospitals before going to the shortest path algorithm. The weights of the node will vary according to the details of the patients and parameters of the hospital.

$$\text{Status of the hospital} = \sum_{i=1}^{n} P_i W_i \Big/ \sum_{i=1}^{n} W_i \tag{1}$$

where P_i = Parameter status, W_i = Weight factor of patient and n = number of parameters. Here we have done the status of the parameter using NLP. Homa Alemzadeh et al. [13] NLP-based cognitive decision support system automatically identifies disease control status from clinical notes.

Status is used to calculate credibility of hospitals, and graph containing nodes and weight V is used to sort them according to their status. The parameters include . . .:

1. Doctors' availability
2. Specialization
3. Bed available
4. Cost factor
5. Distance

The next step includes choosing the m number of hospital based on credibility from n no. Of hospitals and then the sorting algorithm is to be used. The weight of the graph is to be assigned differently for the poor, middle, and rich sections of society so that the poor gets the higher priority to get their bed allotted.

Example 1

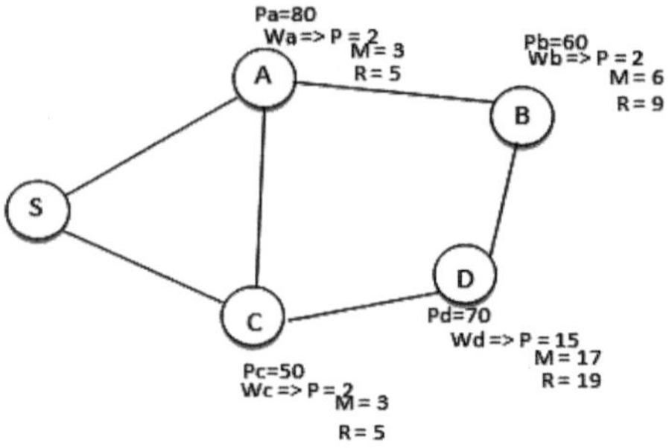

Table 1 Weight and parameter for the graph will be like

Weight	Wa	Wb	Wc	Wd	P1	55
Poor	2	6	2	15	P2	74
Middle	3	7	6	17	P3	78
Rich	5	8	9	19	P4	60

Calculating status of graph.
Using Eq. 1.

$$\text{Poor} = (50 \times 2 + 80 \times 6 + 60 \times 2 + 70 \times 15)/25 = 70$$
$$\text{Middle} = (50 \times 3 + 80 \times 7 + 60 \times 6 + 70 \times 17)/33 = 68.36$$
$$\text{Rich} = (50 \times 5 + 80 \times 8 + 60 \times 9 + 70 \times 19)/41 = 67.317$$

Here in example Table 1 a graph is given, and its respective table is also given. With the given data status of the hospital is calculated from Eq. 1. Each node contains parameter value and weight which is further classified into rich, middle, and poor. Each of these classifications contains different data for a particular node so that when the status is calculated the poor gets the higher priority.

3.3 *Implementation of Improved Kruskal's Algorithm*

To find the graph of the minimum cost spanning tree we have used Improved Kruskal's algorithm. Here:

The most important details are that the patient hospitals need to be connected in such a way that it shows the best route covering all the hospitals in the area. The minimum spanning tree graph should have (vertices—1) to avoid the formation of a cycle, which can lead to a loop hole in the design.

It is clearly shown how a map for the hospital route would look like and how it will be functional. The algorithm gives the time complexity of less than $O(n)$ and space complexity of $O(1)$. The shortest path from the graph is $h5$->$h6$->$h7$->$h4$->$h3$->$h2$. For example you can see Fig. 2b.

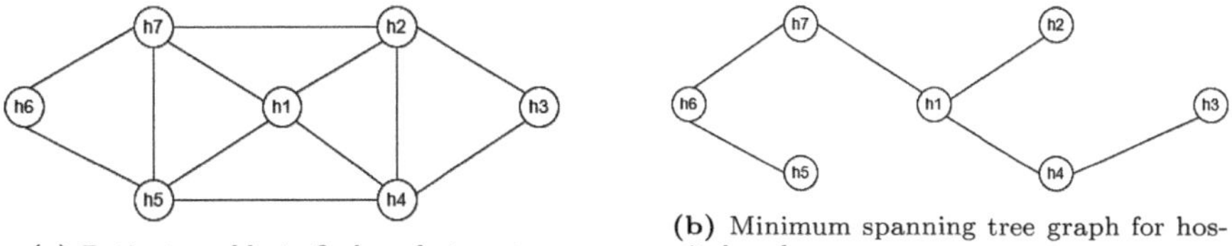

(a) Patient unable to find perfect route

(b) Minimum spanning tree graph for hospital nodes

Fig. 2 Finding graph from the hospital location

Algorithm 1 Improved Kruskal's algorithm

1. Initialization.
2. A set of primary inputs of hospital locations Hi= i1, i2,. . . ,in and patient location Pi=i1,i2,...,in.
3. For V is not completed and for each node Vi belong V.
4. Begin.
5. Find the distance dist from the patient's position to the hospital position.
6. If the dist is minimum then.
7. Return the index of the node.
8. END

Procedure: IM Kruskal

1. Sorting the graph weights in increasing order.
2. For G is not completed and for each node Gi belong V.
3. Find the set of x hospital and y hospital.
4. If x hospital is not equal to y hospital then.
5. Add the edge to the minimum spanning tree.
6. Call union set function for x hospital and y hospital.

Union Set

1. Parent of x hospital is equal to parent of y hospital.
2. Increment the rank of y hospital by one.

Procedure: Find Set

1. If the node n is the parent of itself, return the node.
2. Else if the node is not the parent of itself, we will recursively call the find set function for the node n.

3.4 Allotting the Hospital Bed to Patient

We will ask for the details of the beds from the hospitals. The government hospitals have to provide 100% of the beds details, and private hospitals will give 60% of

Table 2 Hospital index with their bed available

Hospital Id	Total beds	Beds for poor	Beds for middle	Beds for rich
Hospital 1	85	38	28	19
Hospital 2	40	18	13	9
Hospital 3	77	34	26	17
Hospital 4	85	38	28	19
Hospital 5	49	22	16	11
Hospital 6	68	30	23	15

the beds details. For example, beds available in a hospital for each section of the society:

For poor = 40% of total beds available
For lower middle = 30% of total beds available
For middle = 20% of total beds available
For rich = 10% of total beds available

The private hospital will have 40% of beds available for the rich and middle class (Table 2).

Algorithm 2 Bed allotment

1. INPUT = User financial details and the Graph G.
2. OUTPUT = Shortest distance hospital with all the facilities from the user location.
3. If finance details are satisfying the rich section criteria, then return rich.
4. Else if finance details are satisfying the middle section criteria, then return middle.
5. Else if finance details are satisfying the lower section criteria, then return lower.
6. END

Procedure: Bed_allotment

1. For $i = 0$ to total bed count.
2. If finance amount satisfying the lower section, then return the total bed available for the poor section with respect to the hospital Id.
3. Else if finance amount satisfying the middle section, then return the total bed available for the middle section with respect to the hospital Id.
4. Else if finance amount satisfying the rich section, then return the total bed available for the rich section with respect to the hospital Id

Procedure: Specialization

1. INPUT: User specification and 2D array of hospitals and their specialization.
2. OUTPUT: Return the hospital Id which is satisfying the specialization.
3. For $i = 1$ to number of hospitals.
4. If the specialization is satisfying the user specialization, then return the hospital Id.

Table 3 Average response time of the algorithm for different regions

Region name	State	Area (sq. km)	No. of hospitals	Total beds	Population (in crore)	Average response time (ms)
Kolkata	West Bengal	206.1	530	25,698	1.49	2,811,590
Delhi	Delhi	1483	972	57,194	3.18	3,549,270
Mumbai	Maharashtra	603.4	6504	45,000	2.06	4,451,920
Chennai	Tamil Nadu	426	557	33,527	1.12	2,931,200

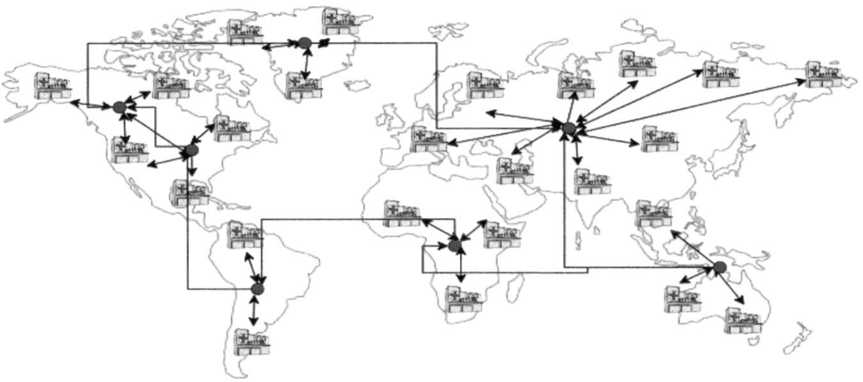

Fig. 3 Decentralized global network

The algorithm first asks the patient about their financial condition. As per the input, the algorithm filters the beds from the hospital and then shows them to the patient. And then the patient is asked to select the specialization they are seeking for. According to the specialization selected, the hospital graph is then formed which will be shown in a map. The hospital with the shortest distance from the patient is then recommended to them.

The algorithm shows the time complexity of less than $O(n)$ and the space complexity of $O(1)$ Table 3.

4 Experimental Evaluation of Global Network

The network system is working globally, requiring countries to accept protocols and sign contracts to store medical data in a smart contract. This will lead to benefits for international patients, as countries can fix healthcare packages according to GDP growth and provide medical insurance for lower backgrounds (Fig. 3).

5 Conclusion and Future Scope

The current study presents an overview of the application of blockchain in healthcare. It has been applied in several use cases, with the main focus being on sharing Computerized Health Records. However, there is still a need for more investigation to better recognize, effectively and securely develop and access this technology. Ongoing efforts have been conducted to overcome limitations in extensible, security and privacy to improve stakeholders' confidence in using this technology and increase its adoption in healthcare.

References

1. Kumar, T., Ramani, V., Ahmad, I., Braeken, A., Harjula, E., Ylianttila, M.: Blockchain utilization in healthcare: Key requirements and challenges. In: 2018 IEEE 20th International Conference on E-health Networking, Applications and Services (Healthcom), pp. 1–7 (2018). IEEE
2. Moona, G., Jewariya, M., Sharma, R.: Relevance of dimensional metrology in manufacturing industries. MAPAN **34**, 97–104 (2019)
3. Kassab, M., DeFranco, J., Malas, T., Laplante, P., Destefanis, G., Neto, V.V.G.: Exploring research in blockchain for healthcare and a roadmap for the future. IEEE Transactions on Emerging Topics in Computing **9**(4), 1835–1852 (2019)
4. Chelladurai, U., Pandian, S.: A novel blockchain based electronic health record automation system for healthcare. Journal of Ambient Intelligence and Humanized Computing, 1–11 (2022)
5. Gupta, R., Tanwar, S., Tyagi, S., Kumar, N., Obaidat, M.S., Sadoun, B.: Habits: Blockchain-based telesurgery framework for healthcare 4.0. In: 2019 International Conference on Computer, Information and Telecommunication Systems (CITS), pp. 1–5 (2019). IEEE
6. Srivastava, G., Parizi, R.M., Dehghantanha, A.: The future of blockchain technology in healthcare internet of things security. Blockchain cybersecurity, trust and privacy, 161–184 (2020)
7. Ben Fekih, R., Lahami, M.: Application of blockchain technology in healthcare: A comprehensive study. In: The Impact of Digital Technologies on Public Health in Developed and Developing Countries: 18th International Conference, ICOST 2020, Hammamet, Tunisia, June 24–26, 2020, Proceedings 18, pp. 268–276 (2020). Springer
8. Monrat, A.A., Schelén, O., Andersson, K.: A survey of blockchain from the perspectives of applications, challenges, and opportunities. IEEE Access **7**, 117134–117151 (2019)
9. Abd-Alrazaq, A.A., Alajlani, M., Alhuwail, D., Erbad, A., Giannicchi, A., Shah, Z., Hamdi, M., Househ, M.: Blockchain technologies to mitigate covid-19 challenges: A scoping review. Computer methods and programs in biomedicine update **1**, 100001 (2021)
10. Haleem, A., Javaid, M., Singh, R.P., Suman, R., Rab, S.: Blockchain technology applications in healthcare: An overview. International Journal of Intelligent Networks **2**, 130–139 (2021)
11. Angraal, S., Krumholz, H.M., Schulz, W.L.: Blockchain technology: applications in health care. Circulation: Cardiovascular quality and outcomes **10**(9), 003800 (2017)
12. Kalla, A., Hewa, T., Mishra, R.A., Ylianttila, M., Liyanage, M.: The role of blockchain to fight against covid-19. IEEE Engineering Management Review **48**(3), 85–96 (2020)
13. Alemzadeh, H., Devarakonda, M.: An NLP-based cognitive system for disease status identification in electronic health records. In: 2017 IEEE EMBS International Conference on Biomedical & Health Informatics (BHI), pp. 89–92 (2017). IEEE

Fine-Grained Sentiment Analysis on COVID-19 Tweets Using Deep Learning Techniques

P. Appalanaidu, K. Deepthi Krishna Yadav, and P. M. Manohar

Abstract With the outbreak of COVID-19, we understand how social media played a crucial role by providing a platform as a means of emotional outlet for peer support and relief in the event of a health crisis. Nowadays there has been an exponential growth in the number of complex documents, user-generated data, and texts on social media, mostly on Twitter and Facebook. Social media serves as a common platform, where people share their opinions, experiences, thoughts, and precautions on COVID-19. The main purpose of this work is to find the fine-grained sentiment analysis from COVID-19 tweets from Twitter by using NLP (Natural Language Processing) techniques that classify the feature sets. User sentiment about COVID-19 tweets is determined to be positive or negative based on extracting opinions at deeper levels, that is generating a sentiment based on Polarity and Subjectivity using Textblob, rather than assigning only sentiment polarity to the tweets. Feature Extraction is performed using TF-IDF and GLOVE Word Embedding techniques on Machine Learning and Deep Learning Classifiers such as Random Forest, SVC, LOG, XGBoost, DT, NB, LSTM, and BI-LSTM. Out of all the above ML and DL approaches, LSTM on GLOVE Word Embedding achieved the highest accuracy for binary-class sentiment analysis on the *COVID-19 tweets dataset.*

P. Appalanaidu
Department of Computer Science and Engineering at Raghu Engineering College, Vishakhapatnam, India
e-mail: appalanaidu.p@raghuenggcollege.in

K. D. K. Yadav
Department of Computer Science and Engineering at Neil Gogte Institute of Technology, Hyderabad, India

P. M. Manohar (✉)
Department of Computer Science and Engineering at GITAM School of Technology, GITAM Deemed to Be University, Visakhapatnam, India
e-mail: mpattisa@gitam.edu

A. Patel et al. (eds.), *Advances in Machine Learning and Big Data Analytics II*,
Springer Proceedings in Mathematics & Statistics 442,
https://doi.org/10.1007/978-3-031-51342-8_31

Keywords NLP · COVID-19 · Sentiment analysis · Text Blob · Long Short Term Memory (LSTM) · Support Vector Machine (SVM) · RF (Random Forest) · Bi-LSTM

1 Introduction

The majority of people used social media to express their sadness, hate, respect, anger, and grief, and some also spread their positivity and happiness during the COVID-19 pandemic. During this hard period, people also used social media to get help related to vaccines or hospitals in case of emergency. In this virtual world, understanding the user opinions behind the text is very crucial, especially in the business of decision-making. By using sentiment analysis, we can know the sentiment behind the text. This concept works by labeling and classifying the texts. Manual classification and labeling of the texts can be time-consuming and inefficient. Sentiment analysis helps to magnify this process of automatic separation. In addition to product reviews, many topics such as political topics, medicine, disasters, stock markets, and many other topics increase the use of emotional analysis. Other areas where this concept is ready for review, are ratings and recommendation programs. This research work aims at the Sentiment Analysis of COVID-19 Tweets. The infectious disease coronavirus (COVID-19) is responsible for numerous fatalities. In December 2019, COVID-19 was discovered for the first time in Wuhan, China. Since then, it has spread to almost every continent. During the COVID-19 epidemic, almost all countries took steps to isolate themselves. Therefore, people prefer to share information about COVID-19 through various social networks such as Facebook, WhatsApp, and Twitter. Most people use Twitter to share their thoughts on the epidemic. During pandemics on social media, like Twitter, the process of sharing information about COVID-19 was a trending and leading topic throughout the globe. To understand public opinion, responses, and discussions related to COVID-19, many opinions have been collected from Twitter API. A large number of responses suggests that the first tweet's message was successfully spread. Consequently, interpreting responses to social tweets has emerged as a crucial area of study in emotional analysis. As an illustration, various cultural differences in COVID-19 tweets were examined with the aid of particular emotional detection techniques. Few studies have examined how a tweet's sentiment polarity (SP) affects the responses to it up to this point. Our research aims to close this gap. Building systems that take input as a set of texts (such as reviews of social media texts, etc.) is the goal of feature/aspect-based sentiment analysis, a fine-grain sentiment analysis problem. The systems attempt to identify the key characteristics of the entity and estimate the typical sentiment based on the opinions expressed in the surrounding context. Text is classified using Sentiment Analysis into predetermined sentiment classes as a supervised machine learning problem. Positive and negative classes are conceivable in a coarse or binary classification. However, there are five classes in fine-grained sentiment analysis: Strongly-negative, Weakly-negative, Neutral, Weakly-positive, and Strongly-positive. In this research, tweets from Twitter messages created via Twitter are referred to as tweets in order to determine the participants' opinions on COVID-19. Tweets are subjected to sentiment analysis to determine the text's underlying attitude based on polarity. The algorithms Support Vector Machine (SVM), Random Forest, and Long Short-Term Memory (LSTM) are used to assess and enhance accuracy.

1.1 Literature Review

Sentiment analysis involves specific mathematical calculations to study people's emotions in a particular aspect. Analysis of subjectivity, concept mines, and Sentiment segregation are other related words in texts. Emotional analysis can be done by Sent WordNet, Affective Norms for English Words (NEW) lexicon-based methods, SentiStrength, Linguistic Inquiry Word Count (LIWC); Machine learning methods such as Naïve Bayes (NB), Multinomial Naïve Bayes (MNB), Multi-Layer Perceptron (MLP), Random Forest (RF), Support Vector Machine (SVM) and Maximum Entropy (ME) or Hybrid method using both learning methods machine-based and dictionary-based. Studies have shown that using the Hybrid method not only increases accuracy and stability but also provides better results than using the standard method.

Jongeling [1] used NLTK, SentiStrength, and Alchemy to do an emotional analysis in an article he wrote. Even while these tools are far from hand labeling, SentiStrength came in second after NLTK, he added. Ahmad [2] investigated the operations of the Support Vector Machine (SVM) in determining the sentiment polarity of text data using WEKA (a prominent data mining tool). According to his research, the SVM model uses a huge dataset and accuracy is largely dependent on the input data. Khan and Bahrudin [3] proposed a method for classifying words as negative, positive, and neutral phrases based on Sent WordNet. To design a scheme for opinion categorization, they applied every machine-learning approach and cosine similarity. They recommended the data which is preprocessed, to improve textual data structure. Based on the likelihood for classification, Han [4] assessed sentiments using latent semantic features and improved classification using SVM with a fisher kernel. In their study, B. Huang [5] used the Nave Bayes and SVM algorithms to analyze Twitter data using different emotional monitoring in their training data to predict emotions and concluded that SVM performed better. For sentiment analysis, Singh and Rani [6] employed SVM. They acquired Twitter data using the Twitter API, and the results revealed that Linear SVM outperformed kernel in terms of accuracy. The tree kernel model is used by Agarwal et al. to represent tweets as trees. These tweets are categorized as positive, negative, or neutral by using the subjectivity and polarity of tokens with associated part of speech (POS) tags. Turney [7] and Abd et al. [8] used unrestricted reading algorithms to determine the central semantic structure of texts. A product feature based on latent semantic analysis (LSA) is described in C.L. Liu et al.'s article [9]. Use mathematics to locate clauses as well [10]. expressed information using words with commentary. Collect adjective words from web pages, then write them out by hand and compare them to straight and negative sentences using a comparison sentence. R. Liu et al. [11] used basic and supporting vector machine (SVM) methods to determine sentence variability. By identifying the feelings of the dictionaries and categorizing them based on textual data knowledge, they were able to determine the variability of the entire pages of specific words [12]. It is advised to pre-process web data to enhance text formatting. Some researchers have used a variety of techniques to analyze emotions using neural network techniques [13, 14], Artificial Intelligence [15], and data mining [16]. There are many different studies done on the emotional analysis of tweets, but we suggest an advanced method that improves accuracy and identifies Emojis expressions.

2 Dataset and Features

The covid_19 tweets dataset contains 179,108 tweets and 13 variables associated with Twitter namely the user_account, user_location, user_description, user_created, user followers, user_ friends, user_ favourites, user verified, user_ date, text, user hashtags, and user accounts that retweeted. We explored and collected the dataset from the open-source platform Kaggle. Using Tweepy, an official Python Twitter API library, the tweets were collected. Our dataset does not provide us with labelled sentiment data (Table 1).

3 Pre-processing

Text Pre-processing is a traditional and important step for performing NLP tasks as it transforms the raw text into more digestible and reliable textual data to make the Deep Learning algorithms perform better. The User Tweets published over social media contain subjective and irrelevant information rather than factual information like hashtags, URLs, emojis, mentions, or symbols which cannot be parsed by basic NLP models. So, we need to clean the tweets to obtain meaningful information from the tweets.

Table 1 Features of the dataset

Selected features	Dropped features
Username	User_ location
Date	User_ description
Text	User_ created
Hashtags	User_ followers
Source	user_ friends
Is retweet	user_ favorites
	user_ verified

I. **The Pre-processing steps taken are stated as follows:**

(i) Almost every user uses hashtags to represent topics on social media platform posts, i.e., #Coronavirus #COVID-19, #Stay Home #Staysafe. Hashtags affect the performance and hence need to be removed) To avoid recognizing the same word as different due to capitalization, we replace them with lowercase. (iii)Word segmentation is performed to replace the concatenated words that are hashtagged such as "coronavirus should be "coronavirus" respectively) Remove stop words to reduce the noise in textual data as they do not affect understanding a sentence's sentiment valence. (v) Replacing Usernames, Emoji, Non-Alphabets, Contractions, consecutive letters, Punctuations, multiple spaces, etc.,

II. **Labeling**

Here, we label the tweets. Based on the nature of the dataset the number of classes depends. This is a Binary-Classification problem. *Text Blob tool* was used to label the emotional sentiment into positive, and negative. We understand that the core problem lies in classifying the polarity of text in a tweet which lies in [-1,1] where -1 defines a negative sentiment and 1 defines a positive sentiment. To overcome this we take into consideration the subjectivity/objectivity activity which is more challenging than the Polarity Sentiment generation. Subjectivity lies in [0,1] which defines the amount of personal opinion and factual information. This means higher subjectivity; the text is influenced by personal opinion rather than factual information.

III. **Final Corpus Generation**

We drop the column data that is irrelevant to the target generation. The final dataset contains only the *Clean Tweets from the original dataset* along with the *Sentiment* generated using the Subjectivity activity using the Text Blob tool.

4 Feature Extraction

For feature extraction, we used word embeddings and vectorization techniques. For vectorization, the term frequency-inverse document frequency (TF-IDF) has been used. Term Frequency (TF) and Document Frequency (DF) are two ideas that are combined in TF-IDF to create practical vectors. The same pre-trained Glove embedding techniques that were trained on Common Crawl and Wikipedia are used for word embeddings. The glove is available in four varieties. Among them, we use 300 Dimension vectors. We created a vocabulary dictionary of size for our Covid_19 training dataset is 96,132. Each word in the Glove file is traversed and compared in a specific dimension with all other words in the dictionary. Whenever there is a match, we copy the equivalent vectors into the embedding Matrix. Finally, Glove captures 400,000-word vectors for our Covid_19 dataset.

We evaluated performance in the sentiment Binary-classification task using ML and DL-based classifiers in order to produce an extensive analysis. ML-based classifiers, such as XGBoost, Random Forest (RF), Logistic Regression (LR), Support Vector Machine (SVM), Naive Bayes (NB), and Decision Tree (DT) are used for our analysis. We implemented two DL-based classifiers, Long-Short Term Memory (LSTM) and Bidirectional Long-Short Term Memory (Bi-LSTM), in addition to traditional machine learning.

TF-IDF combines (TF) and (DF), two ideas. To create feature vectors with TF-IDF values, use the Tfidf Vectorizer class. Less frequently occurring words do not contribute to classification, so the attribute max_features indicates the frequency of the most frequent words to create feature vectors of our choice. Hence, only 2000 features are retained. The word should appear in at least five documents, according to the min_df value of 5. Similar to the previous example, a max_df value of 0.7% indicates that the word must not appear in more than 70% of the documents. Because words that appear in more than 70% of documents are too prevalent and unlikely to have a significant impact on the classification of sentiment, the threshold of 70% was chosen. We call the fit_transform method to convert the Covid_19 dataset into corresponding TF-IDF feature vectors and forward it to our pre-processed dataset on different Machine Learning Classifiers like RF, SVC, LOG, XGBoost, DT, and NB. Random Forest and Logistic Regression classifier has achieved the highest accuracies of 91% followed by Support Vector Classifier and XGBoost classifier with an accuracy of 90% followed by Decision tree and Naive Bayes with an accuracy of 88%. In addition to the traditional ML classifiers, we have applied them to Deep learning classifiers like LSTM and Bi-LSTM. LSTM is an advanced architecture of the RNNs (Recurrent Neural Networks). The first layer of this straightforward binary classification model, known as the Embedded layer, uses 32-length vectors to represent each word. The LSTM layer, which has 100 memory units, is the second layer. Finally, since this is a binary classification problem, we use a dense output layer with a single neuron and a sigmoid activation function to predict whether an observation will be in the positive or negative category. Since it is a binary classification problem, log loss (or binary_crossentropy in Keras) is used as the loss function. The efficient ADAM optimizer algorithm is used. This yields the highest accuracy of 96% when compared to the Bi-direction LSTM. Our Bi-LSTM architecture primarily consists of four components. As previously mentioned, we begin with the embedding layer, which inputs the sequences and produces the word embeddings. The convolution layer then traverses these embeddings, transforming them into tiny feature vectors. The bidirectional LSTM layer comes next. We have a set of Dense (fully connected) layers for binary classification after the LSTM layers. We use a sigmoid activation function before the final output. The model fits for only two epochs as it quickly *overfits* the problem due to class imbalance and other reasons.

5 Implementation Results

The whole number of positive, negative, and neutral tweet counts are shown in Fig. 1.

The maximum number of times repeated positive words is shown in Fig. 2.

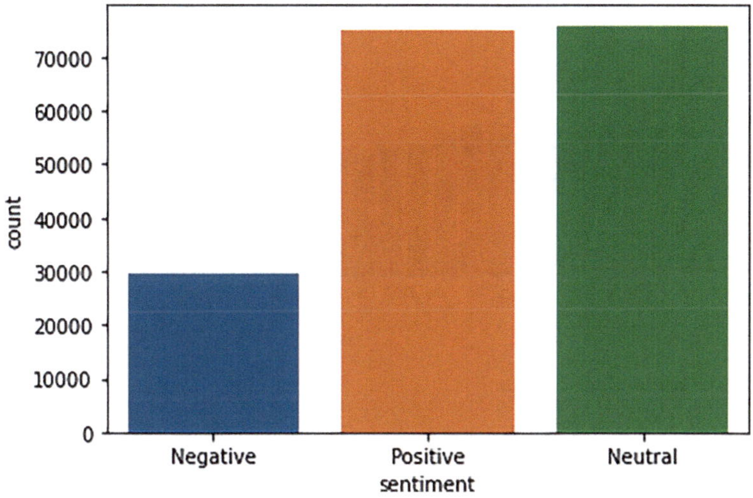

Fig. 1 Tweet counts

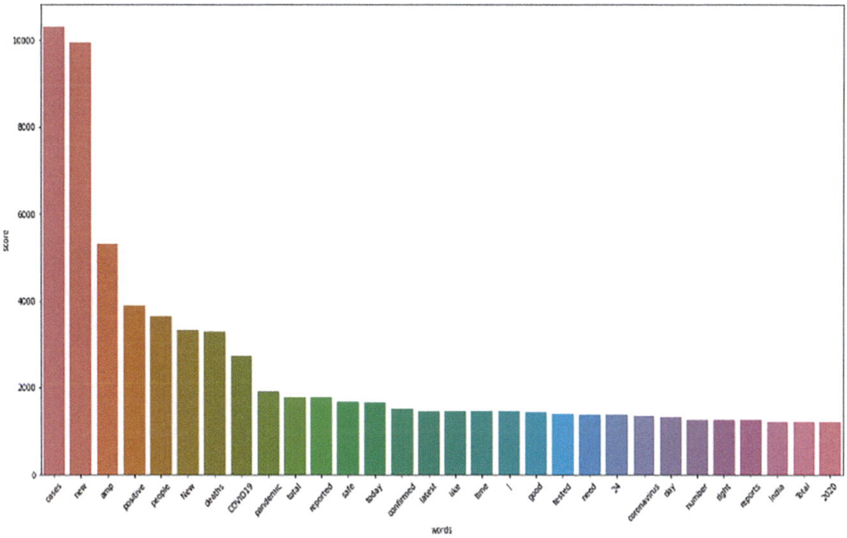

Fig. 2 Max positive word counts

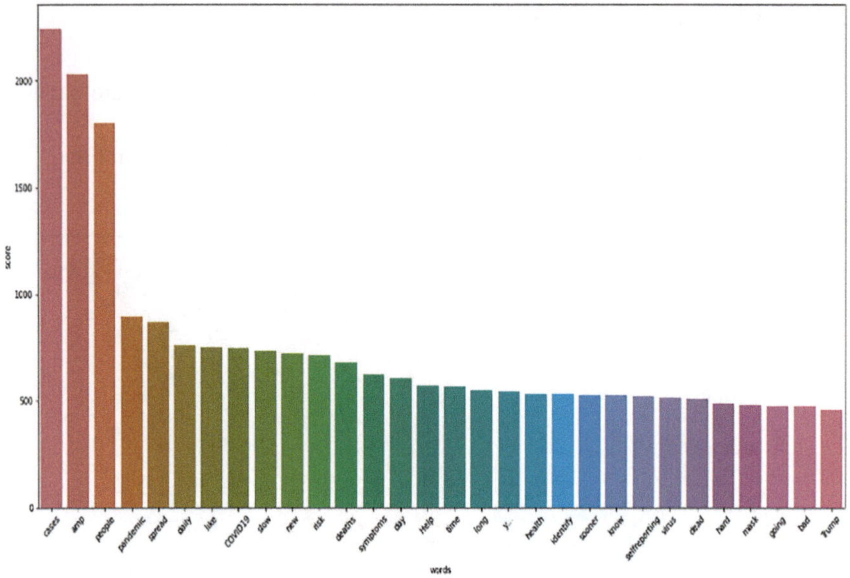

Fig. 3 Max negative word counts

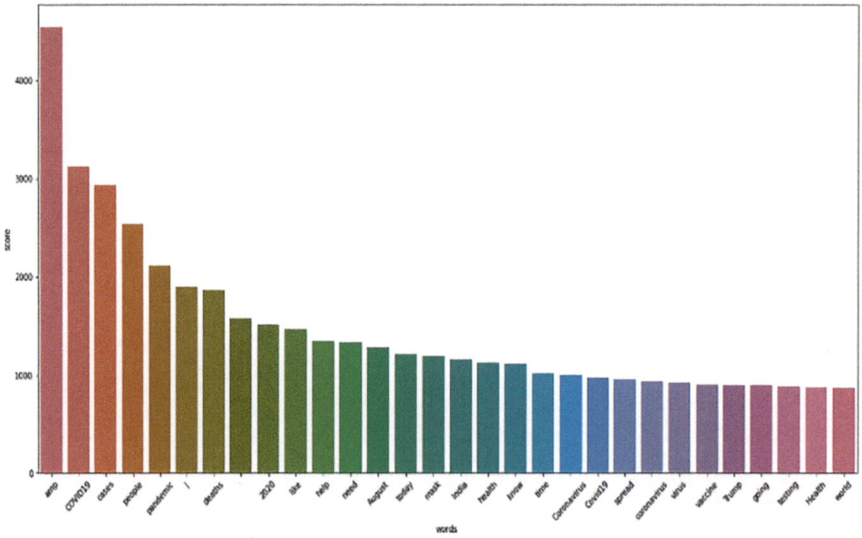

Fig. 4 Max neutral word counts

The maximum number of times repeated negative words is shown in the bar graph in Fig. 3.

The maximum number of times repeated neutral words are shown in the bar graph Fig. 4.

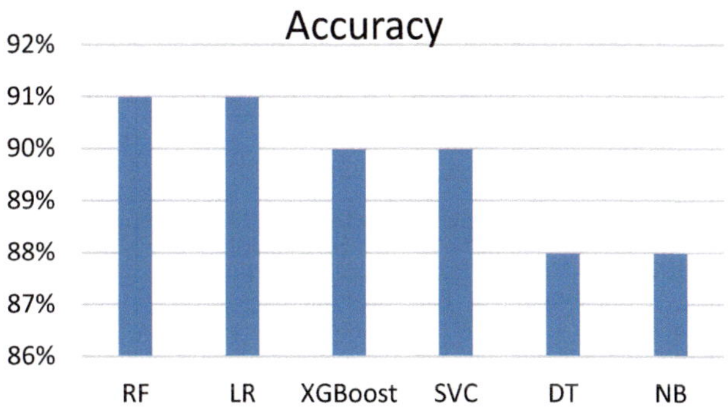

Fig. 5 Comparison of ML classifiers with TF-IDF Vectorizer

Fig. 6 ML classifiers with Glove Word Embeddings

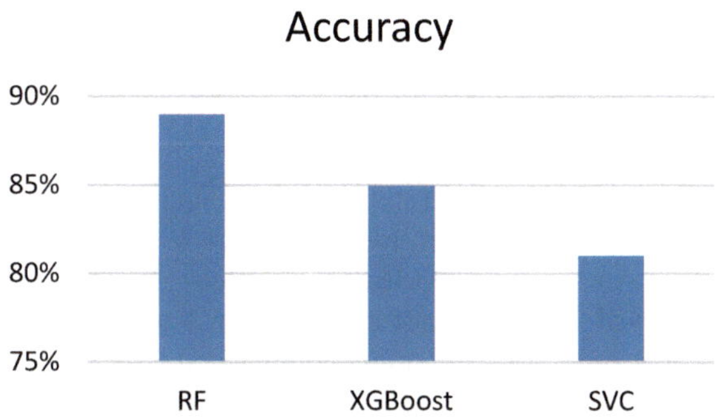

Fig. 7 DL classifiers with Glove Word Embeddings on the Covid_19 dataset

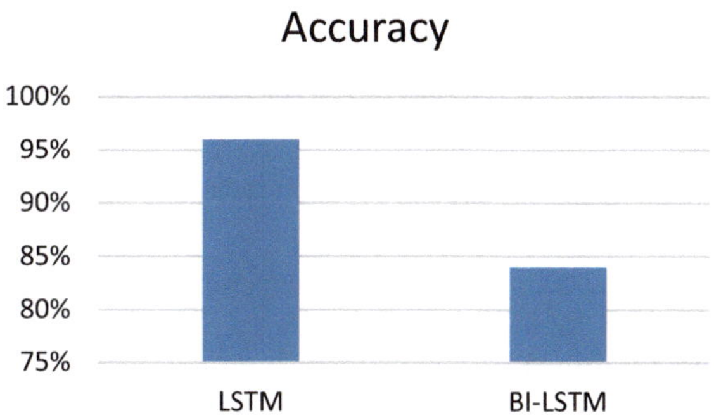

Comparison of ML Classifiers with TF-IDF Vectorizer, ML Classifiers with Glove Word Embeddings, and DL Classifiers with Glove Word Embeddings shown in the following Figs. 5, 6, and 7, respectively, and we found DL Classifiers with Glove Word Embeddings of LSTM performed the highest accuracy of 96% (Tables 2, 3, and 4).

Table 2 ML Classifiers of TF-IDF

ML Classifiers with *TF-IDF Vectorizer* on Covid_ 19 Dataset						
Classifier	RF	LR	XGBoost	SVC	DT	NB
Accuracy	91%	91%	90%	90%	88%	88%

Table 3 ML classifiers with Glove Word Embeddings

ML Classifiers with *Glove Word Embeddings* on Covid_ 19 Dataset			
Classifier	RF	XGBoost	SVC
Accuracy	89%	85%	81%

Table 4 DL classifiers with Glove Word Embeddings

DL Classifiers with *Glove Word Embeddings* on Covid_ 19 Dataset		
Classifier	LSTM	BI-LSTM
Accuracy	96%	84%

6 Conclusion

Our findings from the above results indicate that Machine Learning algorithms are performing better for our Covid_19 dataset due to less complexity in the data. Among the deep learning models—LSTM and BI-LSTM, the LSTM model achieved the highest accuracy with 96% with a good learning rate while the model was overfitted when a more complicated BI-LSTM model was applied to the same Covid_19 dataset. This is due to an unbalanced dataset, class imbalance, no. of samples considered, etc. Bi-LSTM is applicable for more complicated NLP use cases. A balanced dataset with class balance and a considerable input sample size will contribute to achieving better performance results to apply to BI-LSTM and further Transformer Language models.

Conflict of Interest On behalf of all authors, the corresponding author states that there is no conflict of interest.

References

1. Jongeling, R.; Sarkar, P.; Datta, S.; Serebrenik, A. On Negative Results When Using Sentiment Analysis Tools for Software Engineering Research. Empir. Softw. Eng. 2017, 22, 2543–2584. [On negative results when using sentiment analysis tools for software engineering research]
2. Ahmad, M.; Aftab, S.; Ali, I. Sentiment Analysis of Tweets Using SVM. Int. J. Comput. Appl. 2017, 177, 25–29
3. A. Khan and B. Baharudin, "Sentiment classification using The sentence-level semantic orientation of opinion terms from blogs," 2011 Natl. Postgrad. Conf. - Energy Sustain. Explore. Innov. Minds, NPC 2011, 2011, https://doi.org/10.1109/NatPC.2011.6136319.
4. Han, K.-X.; Chien, W.; Chiu, C.-C.; Cheng, Y.-T. Application of Support Vector Machine (SVM) in the Sentiment Analysis of Twitter DataSet. Appl. Sci. 2020, 10, 1125.

5. R. Go, B. Huang. "Twitter Sentiment Classification Using Distant Supervision" Stanford University, Technical Paper. 2009.
6. Rani, S.; Singh, J. Sentiment Analysis of Tweets Using Support Vector Machine. Int. J. Comput. Sci. Mob. Appl. 2017, 5, 83–91.
7. P. D. Turney, "Thumbs up or thumbs down?," Proc. 40th Annu. Meet. Assoc. Comput. Linguist., no. July, pp. 417– 424, 2002, https://doi.org/10.3115/1073083.1073153.
8. D. H. Abd, A. R. Abbas, and A. T. Sadiq, "Analyzing sentiment system to specify polarity by lexicon-based," Bulletin of Electrical Engineering and Informatics (BEEI), vol. 10, no. 1, pp. 283–289, 2021, https://doi.org/10.11591/eei.v10i1.2471.
9. C. L. Liu, W. H. Hsaio, C. H. Lee, G. C. Lu, and E. Jou, "Movie rating and review summarization in a mobile environment," IEEE Trans. Syst. Man Cybern. Part C Appl. Rev., vol. 42, no. 3, pp. 397–407, 2012, https://doi.org/10.1109/TSMCC.2011.2136334.
10. Y. Luo and W. Huang, "Product review information extraction based on adjective opinion words," Proc. - 4th Int. Jt. Conf. Comput. Sci. Optim. CSO 2011, pp. 1309–1313, 2011, https://doi.org/10.1109/CSO.2011.209.
11. R. Liu, R. Xiong, and L. Song, "A sentiment classification method for Chinese document," ICCSE 2010 - 5th Int. Conf. Compute. Sci. Educ. Final Progr. B. Abstr., pp. 918–922, 2010, https://doi.org/10.1109/ICCSE.2010.5593462.
12. L. Ramachandran and E. F. Gehringer, "Automated assessment of review quality using latent semantic analysis," Proc. 2011 11th IEEE Int. Conf. Adv. Learn. Technol. ICALT 2011, pp. 136–138, 2011, https://doi.org/10.1109/ICALT.2011.46.
13. D. Munandar, A. F. Rozie, and A. Arisal, "A multi domains short message sentiment classification using hybrid neural network architecture," Bulletin of Electrical Engineering and Informatics (BEEI), vol. 10, no. 4, pp. 2181– 2191, 2021, https://doi.org/10.11591/EEI.V10I4.2790.
14. D. K. Behera, M. Das, S. Swetanisha, and P. K. Sethy, "Hybrid model for movie recommendation system using content K-nearest neighbors and restricted Boltzmann machine," Indonesian Journal of Electrical Engineering and Computer Science (IJEECS), vol. 23, no. 1, pp. 445–452, 2021, https://doi.org/10.11591/ijeecs.v23.i1.
15. M. A. B. W. Nordin, D. Vedenyapin, M. F. Alghifari, and T. S. Gunawan, "The disruptometer: An artificial intelligence algorithm for market insights," Bulletin of Electrical Engineering and Informatics (BEEI), vol. 8, no. 2, pp. 727–734, 2019, https://doi.org/10.11591/eei.v8i2.1494.
16. A. H. Yassir, A. A. Mohammed, A. A. J. Alkhazraji, M. E. Hameed, M. S. Talib, and M. F. Ali, "Sentimental classification analysis of polarity multi-view textual data using data mining techniques," International Journal of Electrical and Computer Engineering (ICE), vol. 10, no. 5, pp. 5526–5534, 2020, https://doi.org/10.11591/IJECE.V10I5.PP5526-5534.
17. https://ieeexplore.ieee.org/search/searchresult.jsp?newsearch=true&queryText=sentiment%20 analysis
18. https://ieeexplore.ieee.org/document/9458011

Index

A. Patel et al. (eds.), *Advances in Machine Learning and Big Data Analytics II*, Springer Proceedings in Mathematics & Statistics 442,
https://doi.org/10.1007/978-3-031-51342-8

Batch number: 10755329

Printed by Printforce, the Netherlands